D1053220

INORGANIC AND ORGANOMETALLIC POLYMERS

Special Topics in Inorganic Chemistry

Series Editor

R. Bruce King
Department of Chemistry
University of Georgia

Books in the Series

Brian N. Figgis and Michael A. Hitchman
Ligand Field Theory and Its Applications

INORGANIC AND ORGANOMETALLIC POLYMERS

RONALD D. ARCHER
Professor Emeritus
University of Massachusetts, Amherst

A John Wiley & Sons, Inc., Publication

New York • Chichester • Weinheim • Brisbane • Singapore • Toronto

This book is printed on acid-free paper. ♾

Published simultaneously in Canada.

For ordering and customer service, call 1-800-CALL-WILEY.

Library of Congress Cataloging-in-Publication Data:

Archer, Ronald D.
Inorganic and organometallic polymers / Ronald D. Archer.
p. cm — (Special topics in inorganic chemistry)
Includes bibliographical references and index.
ISBN 0-471-24187-3 (cloth : alk. paper)
1. Inorganic polymers. 2. Organometallic polymers. I. Title. II. Series.

QD196 .A73 20001
541.2'254 — dc21 00-043910

Printed in the United States of America.

10 9 8 7 6 5 4 3 2 1

SPECIAL TOPICS IN INORGANIC CHEMISTRY

This text represents the second in a series of one-volume introductions to major areas of inorganic chemistry written by leaders in the field. Inorganic chemistry covers a variety of diverse substances including molecular, coordination, organometallic, and nonmolecular compounds as well as special materials such as metallobiomolecules, semiconductors, ceramics, and minerals. The great structural diversity of inorganic compounds makes them vitally important as industrial feedstocks, fine chemicals, catalysts, and advanced materials. Inorganic compounds such as metalloenzymes also play a key role in life processes. This series will provide valuable, concise graduate texts for use in survey courses covering diverse areas of inorganic chemistry.

R. Bruce King, Series Editor
Department of Chemistry
University of Georgia
Athens, Georgia USA

CONTENTS

PREFACE

If I were to have a special dedication, it would be to the late John C. Bailar, Jr., my Ph.D. mentor. John piqued my interest in the stereochemistry of monomeric coordination compounds initially, and his statement regarding the apparent impossibility of preparing soluble metal coordination polymers of high molecular mass became a challenge that twenty years later put me on the quest for the soluble eight-coordinate polymers. You will find the successful results sprinkled throughout this book.

A number of books and textbooks on inorganic materials chemistry exist. The only recent textbook on inorganic polymers is very heavily weighted toward main group polymers. Recent advances in metal-containing polymers led me to develop a special-topics graduate course on inorganic polymers. The success of this course led Prof. R. Bruce King, the series editor, to suggest that I write "an inorganic polymer book suitable for graduate students." It has been a joy to write the book because so much is happening in the field and I have learned so much more myself.

I thank profusely the research students, postdoctoral associates, visiting scientists, and co-investigators with whom I worked on inorganic polymers and who provided the incentive for producing this text. This includes several short-term undergraduate exchange students from Germany and Britain who made significant research contributions, too. Also, special thanks to the graduate students who took the special-topics graduate course on inorganic polymers and provided valuable input to the manuscript. Thanks also to the University of Massachusetts Polymer Science and Engineering Department and Department of Chemistry colleagues who have aided my knowledge in polymer science and have allowed my group to use their equipment.

Prepublication materials from Leonard Interrante and Charles Carraher are most graciously appreciated. I wish to acknowledge the help received from the extensive reviews by Harry Allcock, (especially his and F. W. Lampe's *Contemporary Polymer Chemistry* textbook published by Prentice-Hall in 1981 and 1990), Charles Carraher, Ian Manners, Charles Pittman, Jan Rehahn, and many others you will find referenced in the text.

The staff at John Wiley have been most helpful, and I especially want to thank Darla Henderson, Danielle Lacourciere, and Amy Romano, all of whom have shown me an extraordinary amount of patience.

Finally, ardent thanks and appreciation to Joyce, my devoted wife since 1954, for all of the sacrifices she has endured to make my career and this book a reality. Without her support, this book could not have been completed.

Ronald D. Archer
Amherst, Massachusetts

CHAPTER 1

INORGANIC POLYMERS AND CLASSIFICATION SCHEMES

1.1 INTRODUCTION

This is an exciting time to be involved in the field of inorganic polymers. The advances being made in the core areas of inorganic polymer chemistry are truly remarkable and outstanding, using any logical definition. Recent synthetic breakthroughs are very impressive. Just a few years ago, no one envisioned the synthesis of polyphosphazenes at room temperature or the ready synthesis of organometallic polymers through ring-opening polymerizations. Both are realities at the present time. These and other examples of both main group and metal-containing polymers are discussed in Chapter 2.

Uses for inorganic polymers abound, with advances being made continually. Polysiloxane and polyphosphazene elastomers, siloxane and metal-containing coupling agents, inorganic dental polymers, inorganic biomedical polymers, high temperature lubricants, and preceramic polymers are examples of major applications for inorganic polymers. Conducting and superconducting inorganic polymers have been investigated as have polymers for solar energy conversion, nonlinear optics, and paramagnets. These uses are detailed in Chapter 4. If we were to include inorganic coordination and organometallic species anchored to organic polymers and zeolites, catalysis would also be a major use.

Inorganic and Organometallic Polymers, by Ronald D. Archer
ISBN 0-471-24187-3

1.1.1 What Is an Inorganic Polymer?

Inorganic by its name implies nonorganic or nonhydrocarbon, and polymer implies many *mers*, monomers or repeating units. Organic polymers are characteristically hydrocarbon chains that by their extreme length provide entangled materials with unique properties. The most obvious definition for an inorganic polymer is a polymer that has inorganic repeating units in the backbone. The intermediate situation in which the backbone alternates between a metallic element and organic linkages is an area where differences in opinion occur. We will include them in our discussions of inorganic polymers, although, as noted below, such polymers are sometimes separated out as inorganic/organic polymers or organometallic polymers or are excluded altogether.

Various scientists have provided widely differing definitions of inorganic polymers. For example, Currell and Frazer (1) define an inorganic polymer as a macromolecule that does not have a backbone of carbon atoms. In fact, several other reviews define inorganic polymers as polymers that have no carbon atoms in the backbone (2–4). Such definitions leave out almost all coordination and organometallic polymers, even though a sizable number of such polymers have backbone metal atoms that are essential to the stability of the polymer chains.

Some edited books (3), annual reviews (5), and the present work include metal-containing polymers in the definition by using titles like inorganic and organometallic polymers. One text includes these polymers but only gives them a few percent of the total polymer coverage (6). Research papers sometimes use the term inorganic/organic polymers, inorganic/organic hybrid polymers, organometallic polymers, or metal-containing polymers for polymers that have both metal ions and organic groups in the backbone. MacCallum (7) restricts inorganic polymers to linear polymers having at least two different elements in the backbone of the repeat unit. This definition includes the coordination and organometallic polymers noted above, but it classifies polyesters and polyamides as inorganic polymers while leaving out polysilanes and elemental sulfur!

Holliday (8) is also very inclusive by including diamond, graphite, silica, other inorganic glasses, and even concrete. Thus it seems that ceramics and ionic salts would also fall under his definition. Anderson (9) apparently uses a similar definition; however, Ray (10) suggests that the term inorganic polymers should be restricted to species that retain their properties after a physical change such as melting or dissolution. Although this would retain silica and other oxide glasses, inorganic salts would definitely be ruled out. Whereas other definitions could undoubtedly be found, the lack of agreement on the definition of inorganic polymers allows for either inclusiveness or selectivity.

This book will explore the classifications of polymers that are included in the more inclusive definitions and will then take a more restrictive point of view in terms of developing the details of inorganic (including metal-containing organometallic) polymer synthesis, characterization, and properties. The synthesis and characterization chapters will emphasize linear polymers that have either at least one metal or one metalloid element as a regular essential part of the backbone and others that have mainly noncarbon main group atoms in the backbone. Inorganic species that retain their polymeric nature on dissolution will be emphasized rather than species that happen to be polymeric in the solid state by lattice energy considerations alone.

For the main group elements, linear chain polymers containing boron, silicon, phosphorus, and the elements below them in the periodic table will be emphasized provided they have sufficient stability to exist on a change of state or dissolution. For transition and inner transition elements, linear polymers in which the metal atom is an essential part of the backbone will be emphasized, with the same restriction noted for the main group elements.

To categorize inorganic polymers further, we must distinguish between oligomers and polymers on the basis of degrees of polymerization. Too often in the literature, a new species is claimed to be polymeric when only three or four repeating units exist per polymer chain on dissolution. For our purposes, we will use an arbitrary cut-off of at least 10 repeating units as a minimum for consideration as a polymer. Anything shorter will be classed as an oligomer.

Note: In step-growth and condensation polymers of the AA + BB type, where the repeating unit is AABB, 10 repeating units, $(AABB)_{10}$, corresponds to a degree of polymerization of 19. That is, $2n - 1$ reaction steps are necessary to assemble the 20 reacting segments that make up the polymer. The reader can verify this relationship with a simple paper-and-pencil exercise. One of the greatest challenges in transition metal polymer chemistry has been to modify synthetic procedures such that polymers rather than oligomers are formed before precipitation (cf. Exercise 1.1).

1.2 CLASSIFICATIONS BY CONNECTIVITIES

N. H. Ray, in his book on inorganic polymers (10), uses *connectivity* as a method of classifying inorganic polymers. Ray defines connectivity as the number of atoms attached to a defined atom that are a part of the polymer chain or matrix. This polymer connectivity can range from 1 for a side group atom or functional group to at least 8 or 10 in some metal-coordination and metal-cyclopentadienyl polymers, respectively. Multihapticity is designated with a superscript following the η for example, the cyclopentadienyl ligand in Figure 1.2b is η^5.

An alternate designation of connectivity of the cyclopentadienyl ring is based on the number of electron pairs donated to the metal ion. Thus a metal species with a bis(cyclopentadienyl) bridge has a connectivity of 6 using this alternate designation. This is more in keeping with its bonding.

Also note that double-ended bridging ligands in linear coordination polymers are classed as bis(monodentate), bis(bidentate), bis(tridentate), bis(tetradentate), etc. and provide connectivities of 2, 4, 6, or 8, respectively.

1.2.1 Connectivities of 1

Anchored metal-containing polymers used for catalysis can have connectivity values as low as 1 with respect to the polymer chain as shown in Figure 1.1.

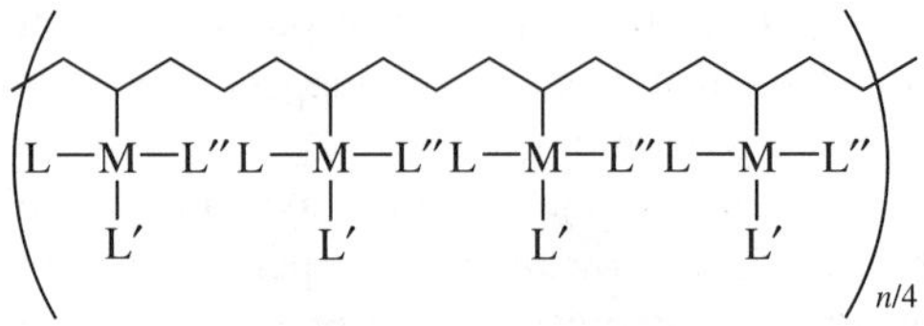

Figure 1.1 Schematic of anchored metal-containing polymer with a connectivity of 1, where M might be palladium or platinum with three other ligands. For catalytic activity, at least one of the three must be easily removed by a substrate.

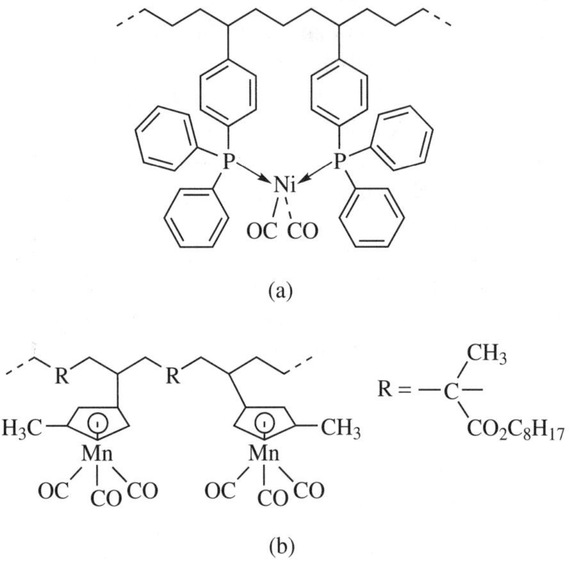

(a)

(b)

Figure 1.2 Higher connectivities for metal-anchored polymers: (a). Schematic representation of an anchored polymer that can convert dienes to cyclohexene aldehydes under the right conditions. (b). Schematic representation of an anchored polymer that can photolytically transport N_2 across membranes. The analogous manganese cyclopentadienyl tricarbonyl monomer decomposes under comparable conditions.

TABLE 1.1 Dentate Number (Denticity) Designation of Metal Chelates.

Donor Atoms on Metal	Designation in This Text[a]	Alternate Designation
one	monodentate [Fig. 1.3d]	unidentate
two	bidentate[b] [Fig. 1.8b,c]	didentate
three	tridentate [Fig. 1.10a]	terdentate
four	tetradentate [Fig. 1.12]	quadridentate
five	pentadentate	quinquidentate
six	hexadentate	sexadentate

[a] 1990 IUPAC nomenclature except when noted otherwise [text examples in brackets]
[b] 1970 IUPAC nomenclature

Note that the metal can have other ligands (groups coordinated to the metal) as well, but inasmuch as they do not affect the polymer connectivity, the metal is defined as having a connectivity of 1. Important connectivities of 1 are fairly rare because the inertness of a single metal connection to a polymer is appreciably less than cases in which multidentate chelation (2 or more ligating atoms from a single ligand are coordinated to the same metal atom; cf. Table 1.1) or multihapticity (2 or more atoms from the same molecule interacting with the same metal atom in an organometallic species; cf. Fig. 1.2) occurs.

1.2.2 Connectivities of 2

Sulfur and selenium in their chain polymer allotropes undoubtedly possess a connectivity of 2. They also have a connectivity of 2 in their ring structures, for example, the crown S_8 structure. Linear polyphosphates, polyphosphazenes, poly(sulfur nitride), polycarboranes, pyroxenes (single-chain silicates), silicones

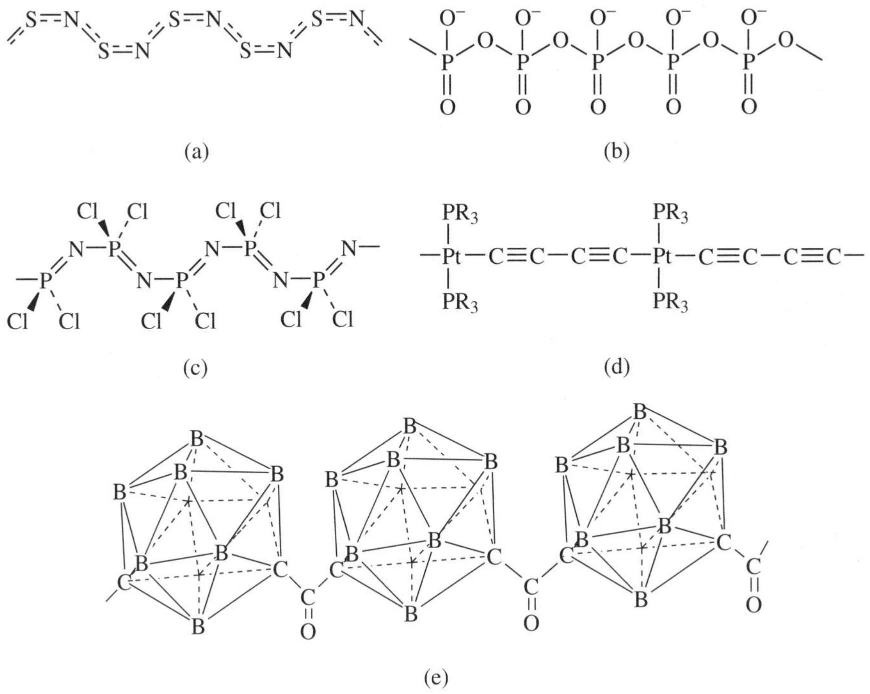

(a)

(b)

(c)

(d)

(e)

Figure 1.3 Examples of inorganic polymeric species with connectivity of 2: (a) poly-(sulfur nitride); (b) linear polyphosphate; (c) poly(dichlorophosphazene); (d) poly[bis-(R_3phosphine)-μ_2-diacetylenato-C_1, C_4(2-)platinum(II)], where R is a large organic group; (e) carborane oligomer with *meta*-$B_{10}H_{10}C_2$ polyhedra linked by CO (although the hydrogens on the boron atoms and the BH groups in the back of the $B_{10}H_{10}C_2$ polyhedra are not shown). Carborane polymers with $-SiR_2(OSiR_2)_n-$ linkages also exist and have been shown to have practical applications (cf. Chapter 4).

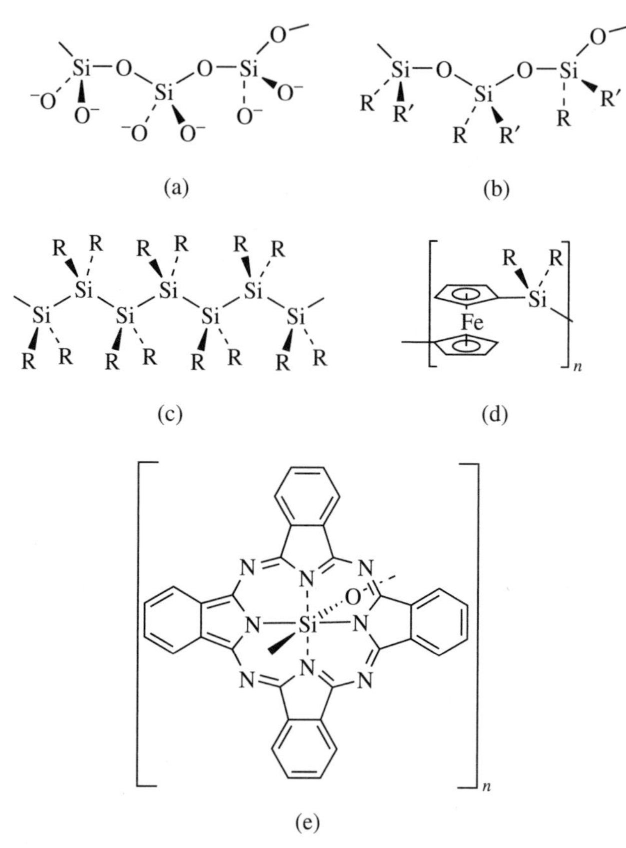

Figure 1.4 Examples of silicon polymers with silicon connectivities of 2: (a) a portion of a pyroxene silicate chain; (b) a portion of a silicone chain where R is typically an alkyl organic group; (c) a portion of a polysilane chain where again R is typically an alkyl organic group; (d) the repeating unit of a high-molecular-weight ferrocene/dialkylsilicon polymer; and (e) the repeating unit of the six-coordinate silicon in poly[oxophthalocyaninatosilicon(IV)] (cf. Figure 1.14c).

(–Si–O– backbones), polysilanes (–Si–Si– backbones), and simple linear coordination and organometallic polymers that are joined by monodentate ligands also have a connectivity of 2. Examples are shown in Figures 1.3 and 1.4. Such polymers will be a primary emphasis of this book.

1.2.3 Connectivities of 3 (Fig. 1.5)

Boron in boric oxide has a connectivity of 3, as do the pnictides (N, P, As, Sb, Bi) in some of their binary chalcogenides (e.g., As has a connectivity of 3 in As_2S_3), silicon in silicates such as mica, talc, and pyrophillite, and carbon in graphite. Such connectivities of 3 provide two-dimensional polymers that

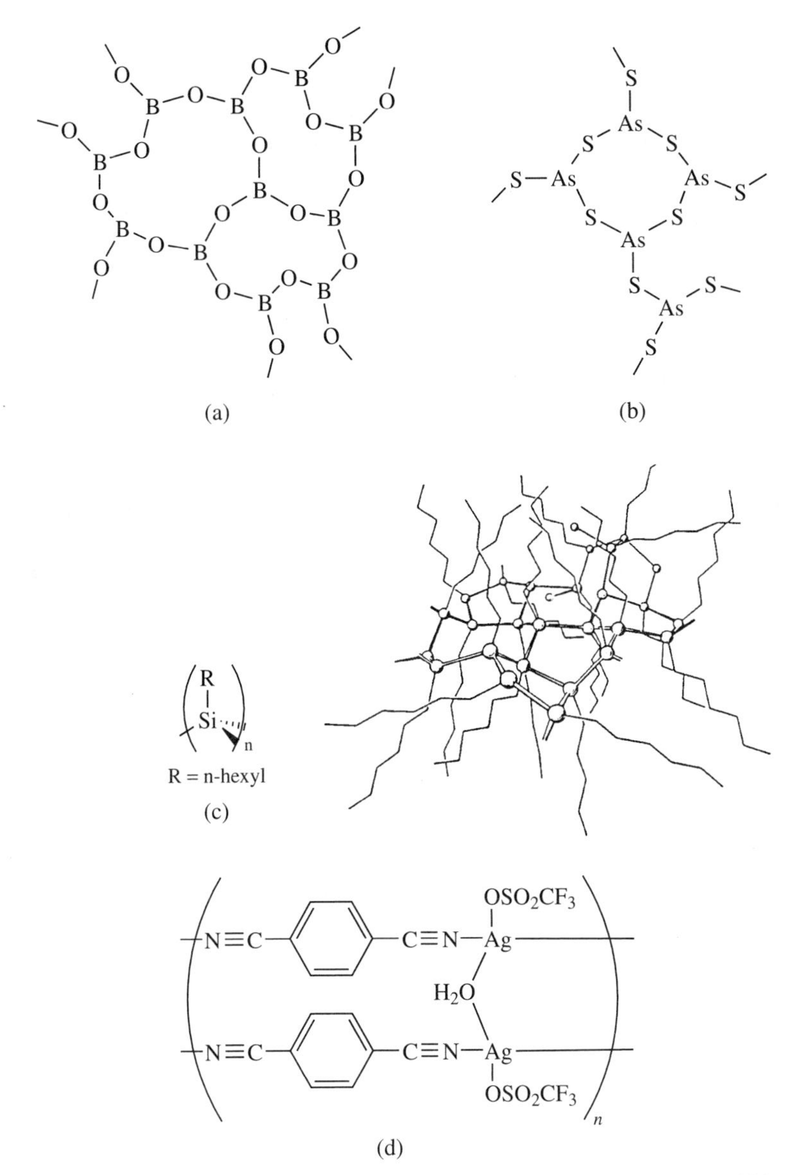

Figure 1.5 Examples of connectivity of 3: (a) boric acid, (b) arsenic(III) sulfide, (c) a synthetic polysilyne (Reprinted with permission from Bianconi et al., *Macromolecules*, 1989, **22**, 1697; © 1989 American Chemical Society); and (d) a synthetic silver polymer (Venkataraman et al., *Acta Cryst*. 1996, **C52**, 2416).

are good lubricants and film- and sheet-forming materials. Polysilynes of the type $[RSi]_n$ and metals [e.g., silver(I)] surrounded with three donors provide synthetic examples of connectivities of 3, although the latter example would not be expected to keep this connectivity in solution.

1.2.4 Mixed Connectivities of 2 and 3

Although both the linear polyphosphoric acids and cyclic metaphosphoric acids have a connectivity of 2 with respect to phosphorus, ultraphosphoric acids exist (Fig. 1.6) that are intermediates in the hydrolysis of P_4O_{10} to simpler phosphoric acids. Note that the connectivity of phosphorus changes from 3 in the oxide through a mixture of 3 and 2 in the ultraphosphoric acids to 2 in the polyphosphoric acids. However, as noted by Ray (10), these are dynamic processes with bond making and bond breaking causing changes in the connectivity of individual phosphorus atoms increasing and decreasing during the hydrolysis process. The phosphate salts possess similar connectivities to the acids.

Amphibole silicates, such as asbestos, have double chains or ladders of silicon and oxygen in which the silicon atoms have connectivities of both 2 and 3. (See Fig. 1.6.) Note that linear polymers with a basic connectivity of 2 typically have mixed connectivities of 2 and 3 when crosslinked because appropriate crosslinking affects only a small portion of the total chain atoms. A number of intractable bis(monodentate) ligand metal coordination species—insoluble, amorphous, uncharacterizable and suspected of being polymers—undoubtedly fall into this class as well.

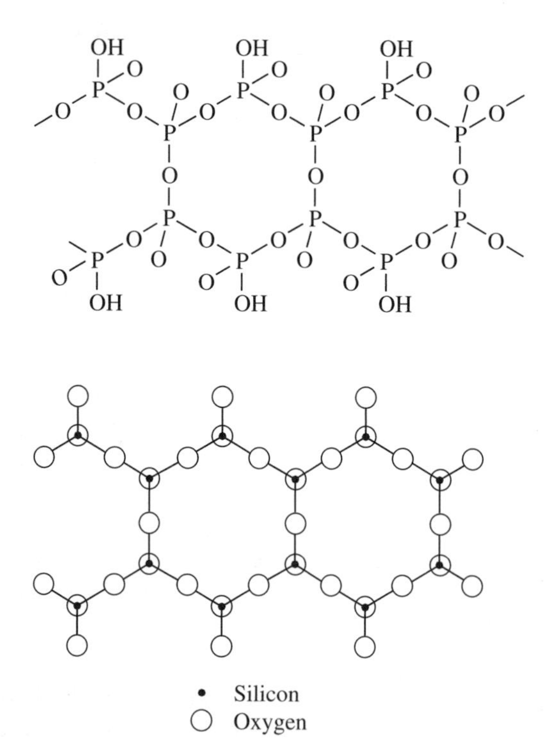

Figure 1.6 Examples of polymeric inorganic species with mixed connectivities of 2 and 3: (a) an ultraphosphoric acid and (b) a portion of an asbestos chain.

1.2.5 Connectivities of 4

Vitreous silica has silicon atoms with a connectivity of 4. Silicate glasses, if counter ions are included (10), also have connectivities of 4. Boron and aluminum phosphates and many other three-dimensional polymers have connectivities of 4 for at least one type of atom in the polymer; cf. Figure 1.7. Another class of inorganic polymers that have connectivities of 4 are metal coordination polymers in which each metal ion in the backbone is coordinated to the polymer chain through two bidentate ligands, where a bidentate ligand is a donor that coordinates to the same metal ion through two donor atoms. Examples are shown in Figure 1.8.

1.2.6 Mixed Connectivities of 3 and 4

A number of polymeric inorganic species have mixed connectivities of 3 and 4, including some borate glasses, where the counter cations provide the counter charges for the four oxide ions connected to at least some of the boron atoms as shown in Figure 1.9. Other examples of mixed connectivity include the silicon atoms in fibrous zeolites and the silicon atoms at the surfaces of silica.

1.2.7 Connectivities of 6

Examples of connectivities of 6 include metal coordination polymers having metal atoms or ions joined with two tridentate ligands. A tridentate ligand is a ligand that has three atoms that are coordinated to the same metal atoms or ion; cf. Figure 1.10.

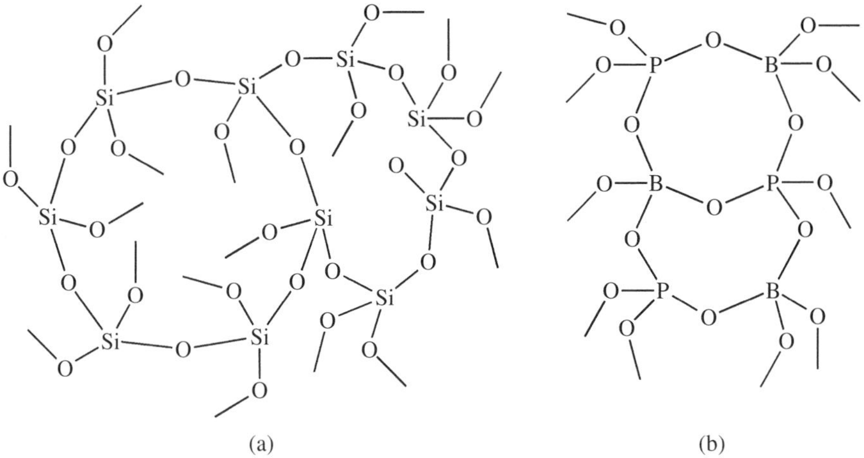

(a)

(b)

Figure 1.7 Examples of polymeric inorganic species with mixed connectivities of 4: (a) silica with silicon atoms of connectivities of 4 and (b) boron phosphate with both phosphorus and boron atoms with connectivities of 4.

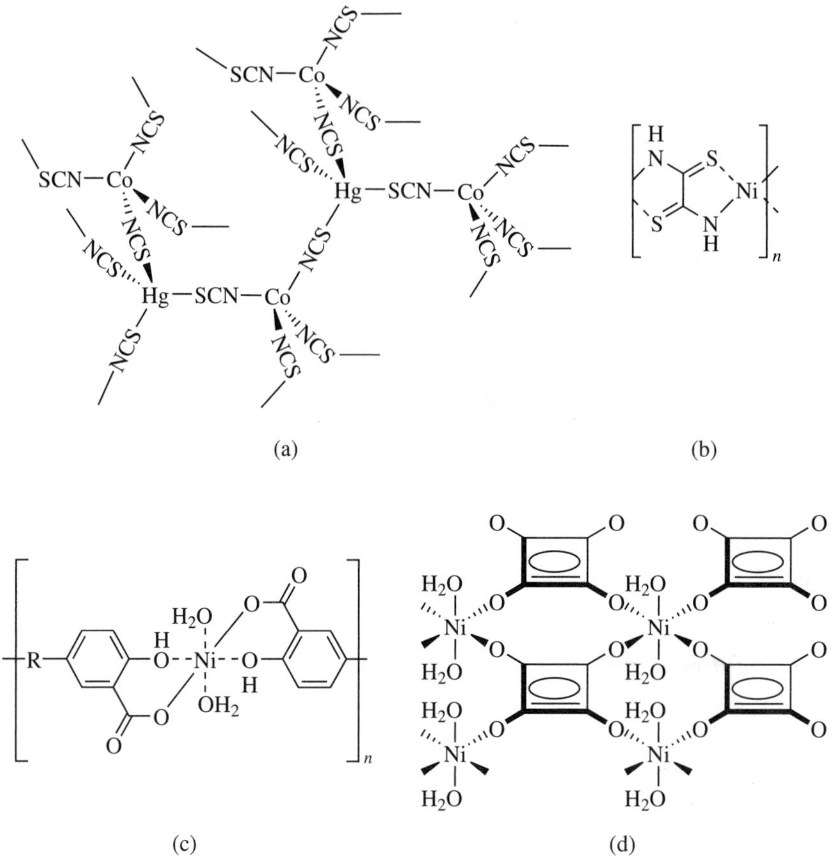

(a)

(b)

(c)

(d)

Figure 1.8 Metal coordination polymers with connectivities of 4 for the metal ions: (a) a small portion of the three-dimensional $[CoHg(SCN)_4]_n$ solid used as a standard for magnetic susceptibility measurements — both Co and Hg are tetrahedrally coordinated; (b) a typical linear polymer for a 4-coordinate metal with a bis-bidentate ligand; (c) a linear polymer for octahedral coordination with two bidentate ligands per metal plus two other ligands not involved in connectivity of the polymer (R can be CH_2, C_3H_8, a large diazo link, etc.); (d) coordination analogous to (c) except that each of the four donors of the ligand are bonded to four different metal ions, which gives a two-dimensional sheet.

Ferrocene polymers (Fig. 1.10) can be considered to have iron atoms with connectivities of 6 if each cyclopentadienyl ring is considered a connectivity of 3 — consistent with bonding considerations. That is, considering the 18-electron rule, iron(II) can accommodate only six pairs of electrons in addition to the six electrons in its $3d^6$ valence electron levels. Thus, although five carbon atoms of each cyclopentadienyl ring are approximately equidistant from the iron, only the three pairs of pi-symmetry electrons are coordinated or bonded to the iron atom from each ring. However, using the number-of-atoms definition, these polymers have a connectivity of 10.

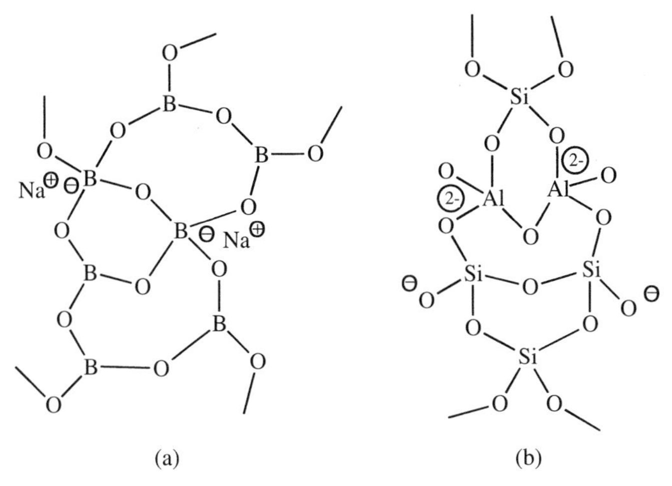

Figure 1.9 An example of mixed connectivities of 3 and 4 for (a) the boron atoms in a typical borate salt and (b) the silicon atoms in a typical fibrous zeolite.

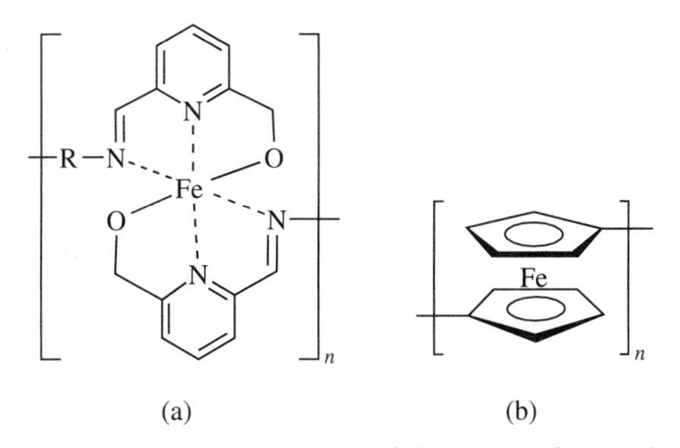

Figure 1.10 Examples of connectivities of 6 (or more) for metal atoms/ions.

Another example of connectivity of 6 can be found in the carborane carbons of the carborane oligomer shown in Figure 1.3e using the atoms-connected definition of connectivity. Naturally, the connectivity would not be more than 4 if the number of electron pairs bonding the carborane carbons to the chain were considered.

1.2.8 Mixed Connectivities of 4 and 6

Orthophosphates and arsenates of titanium, zirconium, tin, cerium, thorium, silicon, and germanium have mixed connectivities of 4 and 6. An example is shown in Figure 1.11.

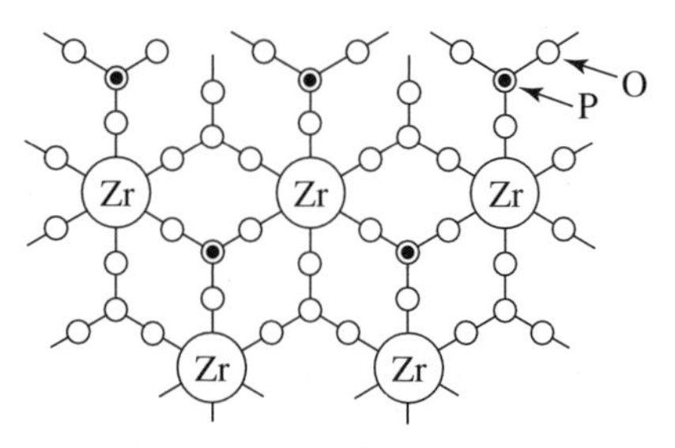

Figure 1.11 An orthophosphate of mixed connectivites of 4 and 6.

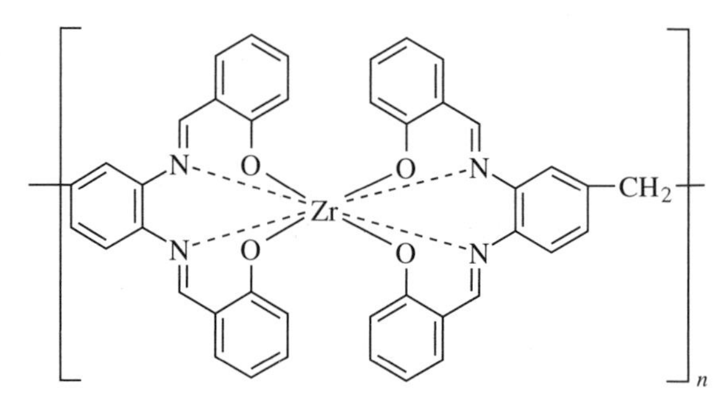

Figure 1.12 A Schiff-base polymer of zirconium with a connectivity of 8.

1.2.9 Connectivities of 8

Metal coordination polymers of zirconium(IV), yttrium(III), and several lanthanide ions [cerium(IV), lanthanum(III), europium(III), gadolinium(III), and lutetium(III)] have been synthesized that possess connectivities of 8 because two tetradentate ligands are coordinated to each metal ion that is part of the polymer chain. An example is shown in Figure 1.12 (cf. Exercises 1.2–1.6).

1.3 CLASSIFICATIONS BY DIMENSIONALITY

Another manner in which polymers can be classed is by dimensionality. Pittman et al. (3) use this classification for polymeric species containing metal atoms in their backbones — one category of metal-containing polymers in the next section. Here we will use the dimensionality for all types of inorganic polymers.

1.3.1 1-D Polymeric Structures

A linear chain polymer is categorized as a one-dimensional (1-D) polymer even though it may have twists and turns in the "linear" chain. Simple polymer chains in which all of the atoms in the chain have a connectivity of 2 are classed as 1-D polymers. However, a linear chain polymer with one or more atoms of each

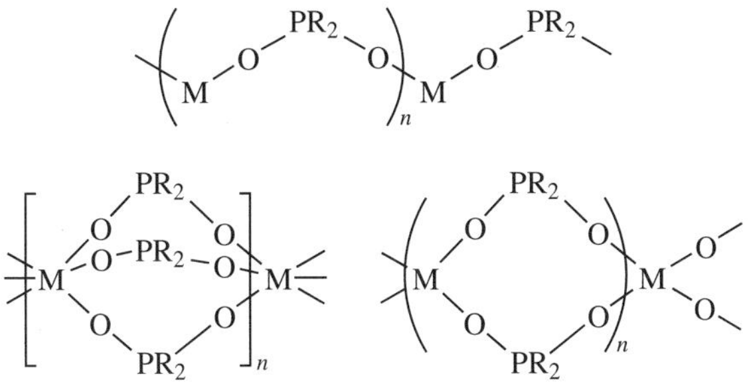

M = Al, Be, Co, Cr, Ni, Ti and Sn

Figure 1.13 Schematic metal phosphonate 1-D polymers with connectivities of 2–6.

repeating unit having a connectivity of more than 2 is also possible. For example, a polymer with benzene rings in the chain will have some carbon atoms with a connectivity of 3. Also, the carborane oligomer in Figure 1.3e has an even higher connectivity for the carbon atoms that are a part of the carborane clusters, as was noted above under the discussion of polymers with connectivities of 6.

A sizable number of other examples of inorganic polymers that fall in the 1-D category have been presented above in this chapter. Figures 1.3 and 1.4 show 1-D inorganic polymers with connectivities of 2, and Figures 1.6b, 1.8, 1.10, and 1.12 illustrate other 1-D inorganic polymers containing atoms with higher connectivities.

The same ligand often can give different connectivities and different dimensionalities with different metal ions. For example, the polymeric metal phosphonates can be singly, doubly, or triply bridged polymers as shown in Figure 1.13.

Rigid-rod (truly linear) metal and metalloid polymers (Fig. 1.14) are also well known. A number of group 10 (Ni, Pd, and Pt) diacetylide derivatives show a polymer-specific parallel or perpendicular chain orientation relative to magnetic fields. Other examples include the analogous octahedral hydrido rhodium(III) derivative, shish kebob phthalocyanine polymers with oxo, pyrazine, and other bridging ligands with both metals and metalloids; the classic 2,5-dioxoquinonates of copper(II), nickel(II), and cadmium(II) and analogous sulfur analogues with several metal ions; and 5-phenyltetrazolates of iron(II) and nickel(II). Whereas the shish kebob polymers have a connectivity of 2 and the dioxoquinonate derivatives have connectivities of 4, the structure with a connectivity of 6 depicted for the phenyltetrazoles requires counter ions (i.e., they are polyelectrolytes) (cf. Exercise 1.7).

1.3.2 2-D Polymeric Structures

Simple inorganic species with a connectivity of 3 often lead to sheet or two-dimensional (2-D) polymers as shown above in Figure 1.5 for boric acid, arsenic

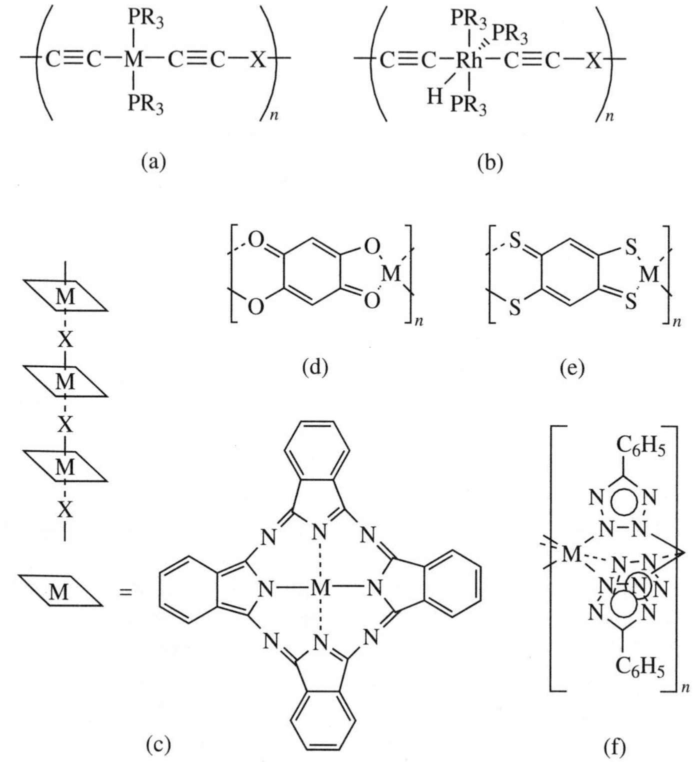

Figure 1.14 Examples of rigid-rod polymers: (a) group 10 diacetylide polymers where $M = Pt^{II}$ or Pd^{II}, X = – or C_6H_4, and R = *n*-butyl or larger for solubility; (b) Rh^{III} diacetylide polymers, where X and R are as in (a); (c) phthalocyanine shish kabob polymers, where the parallelogram represents the phthalocyanine dianion, X = O for $M = Si^{IV}$ or Ge^{IV} or Sn^{IV}, X = F for $M = Al^{III}$ or Ga^{III}, and X = pyrazine (*para*-diazabenzene) for M = divalent metal ion; (d) 2,5-dioxoquinonato polymers where $M = Ni^{2+}$, Cu^{2+}, or Cd^{2+}; (e) suggested sulfur analogue structures for Cu^{2+}, Ni^{2+}, or Fe^{2+} (although the oxygen analogue for iron is a 2-D sheet structure as shown in Fig. 1.15); and (f) 5-phenyltetrazole polymer structures predicted for known Ni^{2+} and Fe^{2+} polymers.

sulfide, and graphite. In fact, at least one type of atom must have a connectivity of 3 or more to obtain a 2-D polymer. For metals coordinated with bidentate ligands, a connectivity of 6 provides the basis for a 2-D polymer.

On the other hand, connectivities do not always determine dimensionality. To illustrate this point, the aqueous iron(II) oxalate polymer has a 1-D linear chain structure, but the analogous 2,5-oxyquinonate complex of iron(II) is a 2-D sheet structure as shown in Figure 1.15.

The 2-D "crossed ladders" structure of copper(II) with dithiooxamides are considered three-dimensional (3-D) polymers by some chemists, whereas the present author prefers to consider them as 2-D in the same sense that ladder polymers are typically considered as 1-D polymers (cf. Exercise 1.8).

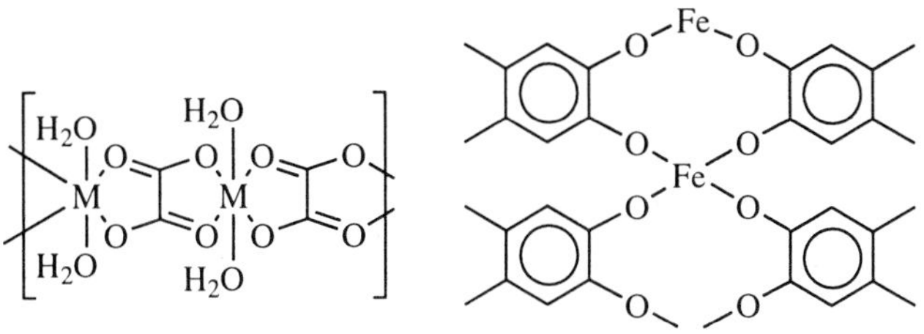

Figure 1.15 The structures of polymeric (a) iron(II) oxalate and (b) 2,5-dioxoquinonate.

1.3.3 3-D Polymeric Structures

Inorganic polymeric networks in which bonding occurs in three dimensions are well known. Starting with quartz (SiO_2) as a prime example (cf. Fig. 1.7a), the most common characteristic of such species is insolubility — unless decomposition occurs during a dissolution process. To have a true 3-D polymer, at least some of the atoms must have a connectivity of 4 or more. Some polymers, such as some of the polysilynes (Fig. 1.5c) are pseudo-3-D as a result of 3-D ring formation to relieve steric strain. Prussian blue is a classic example of a mixed Fe(II) and Fe(III) 3-D polymeric structure, with each iron ion surrounded octahedrally by six cyano ligands; cf. Figure 1.16 (cf. Exercises 1.9 and 1.10).

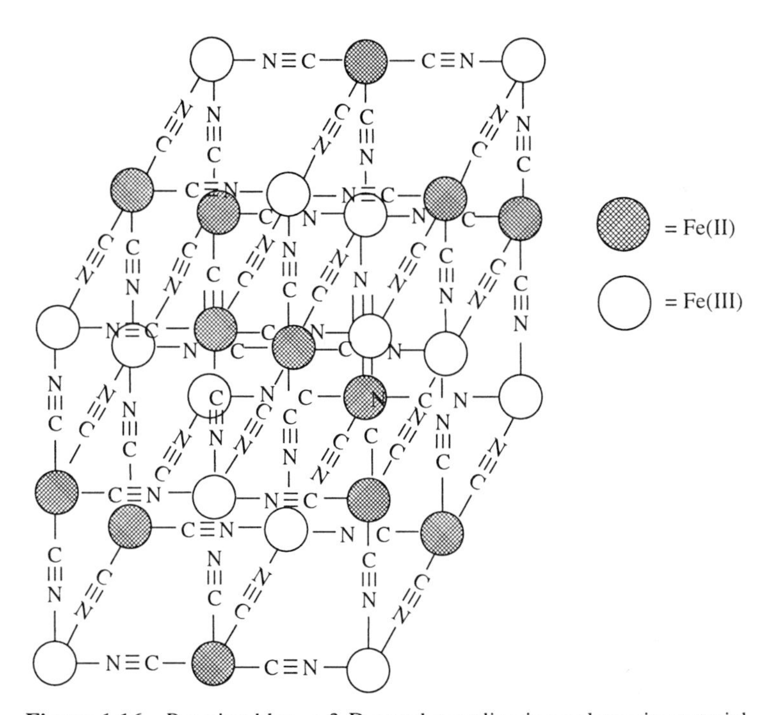

Figure 1.16 Prussian blue, a 3-D metal coordination polymeric material.

Metal oxides and crystalline monomeric inorganic compounds are often 3-D polymeric materials in the solid state. These materials will not be considered in detail in this volume other than to note some of their uses in Chapter 4.

1.4 THE METAL/BACKBONE CLASSIFICATION OF METAL-CONTAINING POLYMERS

Metal-containing polymers can be grouped according to the position or positions of the metal atoms in the polymer structure. At least three such classifications

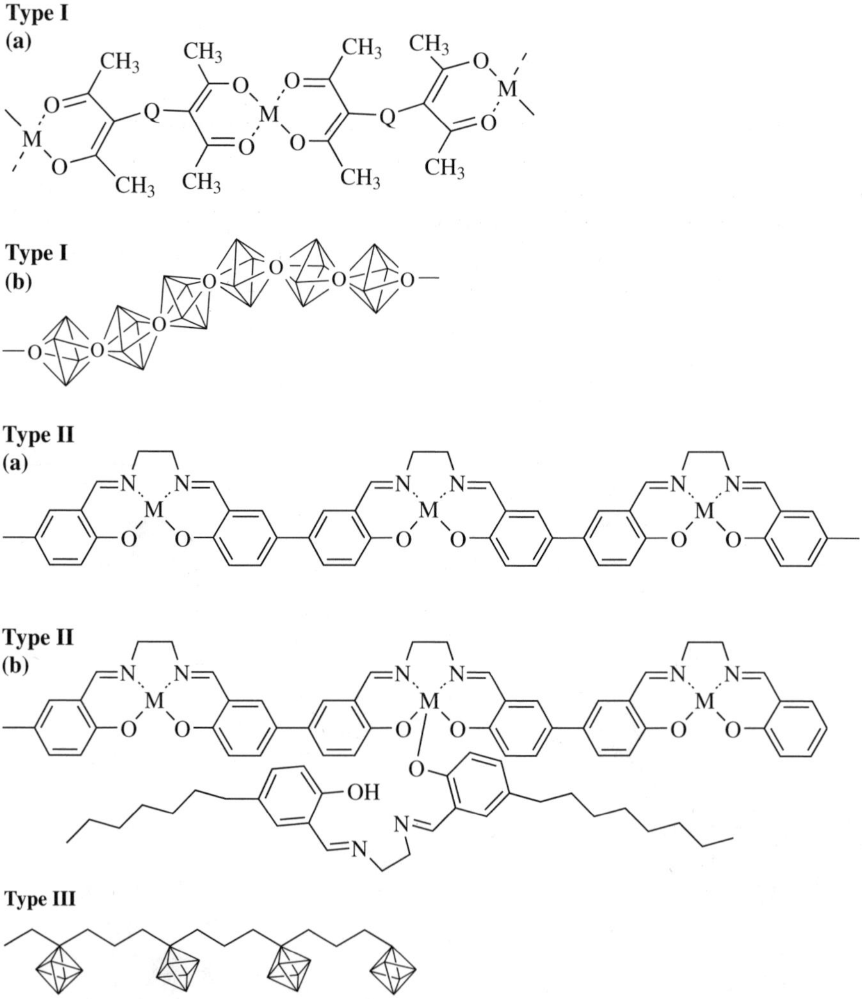

Type I
(a)

Type I
(b)

Type II
(a)

Type II
(b)

Type III

Figure 1.17 Types of metal-containing polymers based on the positions of the metal(s) relative to the polymer backbone — see text for details.

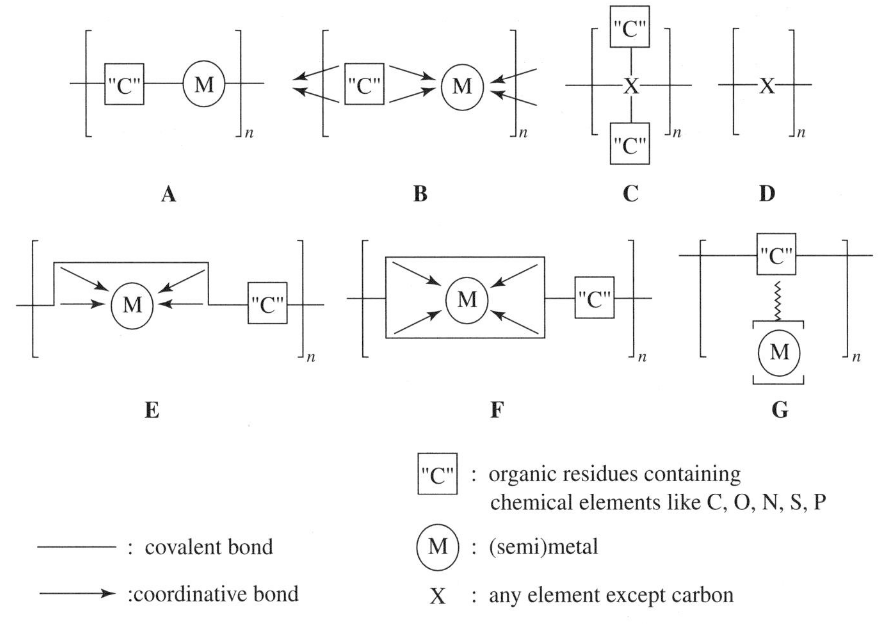

Figure 1.18 Rehahn's (1998) classification of organic/inorganic hybrid polymers. Note that A, B, C, and D are subcategories of Type I, E and F of Type II, and G of Type III.

have been made (11–13). All include at least the three types of metal-containing polymeric species in which the metal atoms or ions are (**I**) in the backbone and essential to maintenance of the backbone, (**II**) modifiers of the backbone but not essential to the backbone, and (**III**) pendent to the backbone. A schematic representation of this type of metal-containing polymer is found in Figure 1.17. Rehahn (13) further subdivides these into a total of seven categories (Fig. 1.18).

1.4.1 Type I Metal-Backbone Polymers

Type I metal-backbone polymers have metal atoms or ions that are an essential part of the polymer backbone such that the polymeric nature of the species would be destroyed if the metal atoms or ions were removed. This type of polymer can be subdivided into (**a**) polymers with organic bridging groups and (**b**) polymers with inorganic bridging groups. These two groups are sometimes referenced as inorganic/organic polymers and inorganic backbone polymers, respectively. Most of the metal-containing polymers that have metal atoms or ions in the backbones are **Type I(a)**; however, several metal shish kabob polymers with oxo or fluoro bridges and other polymers with oxo bridges from dihydroxo condensations are **Type I(b)** polymers. Many examples of these types of polymers will be given throughout this volume.

1.4.2 Type II Metal-Enmeshed Polymers

Type II metal-enmeshed polymers involve metal ions that are enmeshed into the polymeric organic macromolecule and thus may modify the properties of the organic polymer without being essential to the maintenance of the polymer

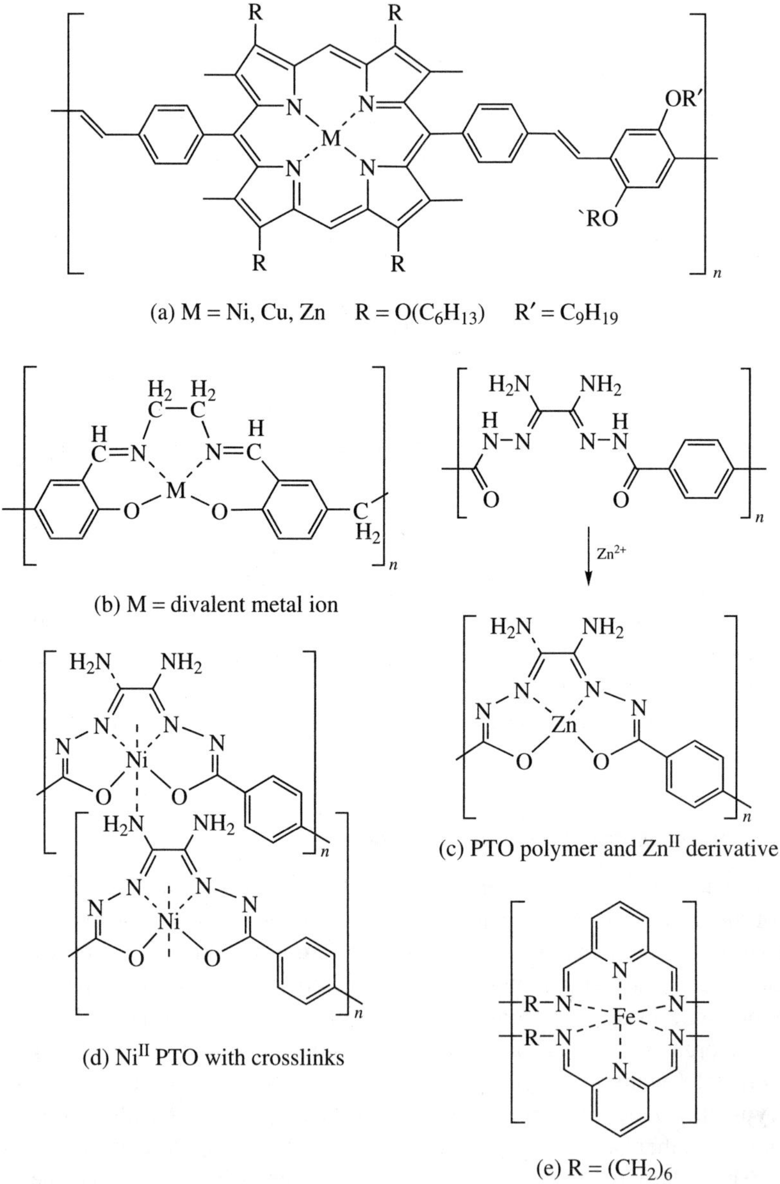

(a) M = Ni, Cu, Zn R = $O(C_6H_{13})$ R′ = C_9H_{19}

(b) M = divalent metal ion

(c) PTO polymer and Zn^{II} derivative

(d) Ni^{II} PTO with crosslinks

(e) R = $(CH_2)_6$

Figure 1.19 Examples of Type II metal coordination polymers.

backbone; that is, the polymer would still be a polymer if the metal were removed. This type of polymer can be subdivided into (**a**) polymers that have metal species that are an intimate part of the polymer without being involved in polymer crosslinking and (**b**) analogous polymers in which the metal atoms or ions are involved in crosslinking. **Type II(a)** would include what Foxman and Gersten have called *parquet polymers*. Examples include poly(phthalocyanato) or poly(porphyrinato) metal-containing polymers linked via a ligand functionality; difunctional salicylaldehyde/diamine Schiff base polymers that can encase metal ions; and the thermally stable zinc(II) derivative of the poly(terephthaloyl oxalic-bis-amidrazone) (PTO) polymer (Fig. 1.19). The conjugation of the PTO backbone is modified through chelation of the zinc ions, but the fibrous polymer chains are neither lengthened nor crosslinked by the zinc ions. Given its thermal stability and fire-resistant nature, it would be an excellent fiber were it not for the amphoteric nature of zinc, which allows the zinc to be slowly extracted during washing cycles. However the analogous nickel(II) derivative, which should be more inert, is very highly crosslinked (thus a Type II(b) polymer) and so brittle that it is useless as a fiber. Other examples have been given elsewhere (12).

1.4.3 Type III Anchored Metal Polymers

Type III anchored metal polymers, in which the metal ions are pendent to the polymer backbone, have been extensively studied. In fact, a book (14) has been

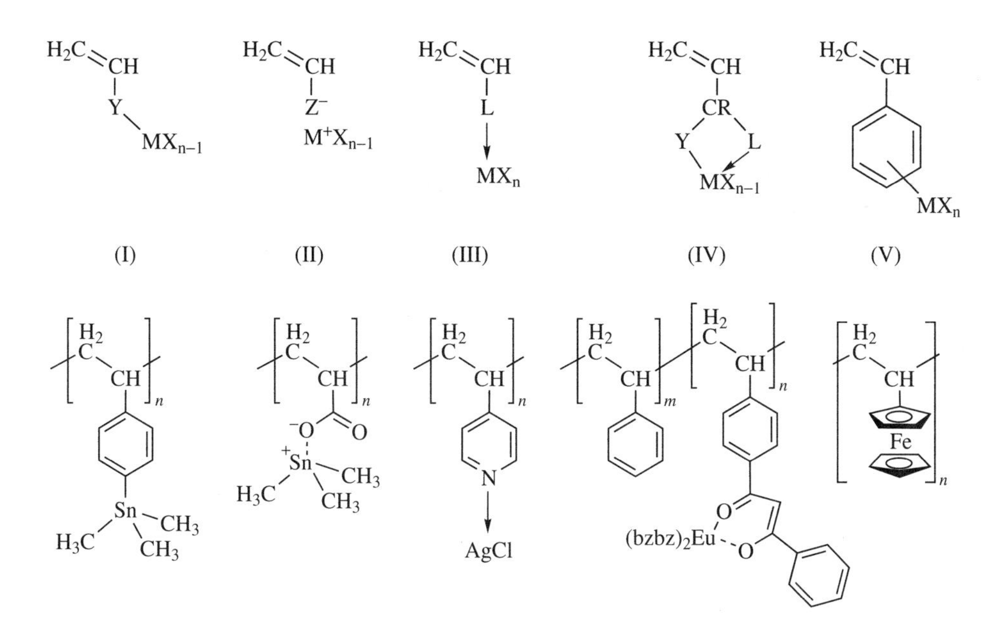

Figure 1.20 Metal-containing monomer types for vinyl-type polymerizations (Pomogailo, 1994) (M = metal; n = metal valence; X, Y, and Z are functional groups; and L = electron-pair donor ligand) and an example of each.

published on the synthesis and polymerization of metal-containing monomers that have a vinyl or similar polymerizable unsaturation. The metal-containing monomers are divided into five classes: (I) covalent, (II) ionic, (III) donor-acceptor, (IV) chelate, and (V) π types as shown in Figure 1.20. The book mentioned above only emphasizes the first four types. It should be evident that polymerization of the any of these will provide polymers with the metal entities on side chains.

Anchored metal catalysts have proven useful in immobilizing homogeneous catalysts. The anchoring of a catalyst allows for the easier recovery of the catalyst at the end of the reaction.

Until recently, polymerization of vinyl derivatives of metal cyclopentadienyls (one major class of type V monomers) was the only effective way of making high-molecular-weight metal cyclopentadienyl polymers. Other, earlier methods of synthesizing metal cyclopentadienyl polymers normally provided only moderate-molecular-weight materials. [However, the recently developed ring opening polymerization reactions (Section 2.3) have changed that.]

1.5 LINEAR INORGANIC POLYMERS—THE THRUST OF THIS BOOK

Linear (1-D) tractable (or soluble) inorganic polymers of high molecular weight with metals as part of the backbone (Type I above) have proven to be very difficult to synthesize and will be a primary thrust of the synthetic chapter (Chapter 2). Such metal-containing polymers can be broken down into coordination polymers and organometallic polymers, even though the dividing line is sometimes very blurry. Categorization of the linear inorganic polymers containing only main group elements provides a variety of polymers including the commercially significant polysiloxanes (or silicones), polyphosphazenes, and polysilanes. A number of less-known varieties also exist. Incidentally, "linear" is a misnomer, as noted below in the discussion of the siloxane polymers.

1.5.1 Metal-Containing Polymers

Metal-containing polymers can be subdivided into (1) metal coordination or Werner coordination polymers and (2) organometallic polymers with M–C bonds in the backbones. The dividing line is not always straightforward, as the σ-type diacetylide bonding between the platinum metal ions in the rigid rod polymers already noted [Fig. 1.14a,b] seems more like Werner type species than organometallic ones even though the chain is constructed only with metal-carbon bonds.

Metal coordination polymers have already received a reasonable amount of coverage in the other classification schemes. From the preceding classification sections, it should be evident that linear 1-D metal coordination polymers have a connectivity twice that of the dentate number of the individual ligand-to-metal bonding interaction. That is, a bis-bidentate bridging ligand occupies four coordination sites around each metal ion, a bis-tridentate ligand occupies six coordination sites, and so forth. However, if the coordination number of the metal ion is greater than double the dentate number of the bridging ligand to each metal, the remaining sites must be occupied by ligands that will not foster extensive crosslinking and insolubility. Strategies for preventing extensive crosslinking and insolubility will be explored in Chapter 2.

As noted by Foxman and Gersten (12), "A general problem in the characterization of coordination polymers arises because of the extreme intractability of many of the materials and the consequent lack of available structural information." The large number of polymers synthesized with planar aromatic bis-bidentate ligands with divalent d^8 ion metals that form flat square-planar coordination linkages show such intractability (insolubility) and precipitate before the attainment of true polymeric molecules. The stacking that arises is depicted in Figure 1.21.

The inertness of the metal ligand bonds is another factor of importance in defining the usefulness of coordination polymers. If the polymers fall apart on dissolution in polar solvents, they are not "good" inorganic polymers for most purposes. Low-spin four-coordinate d^8 platinum(II) and, to a lesser extent, palladium(II) provide inertness when the intractability is overcome with large nonstacking side groups (Fig. 1.14a). For octahedral metal centers, d^3 [e.g., chromium(III) (Fig. 1.13c)] and low-spin d^6 [e.g., rhodium(III) (Fig. 1.14b)] provide substitution inertness. Low-spin eight-coordinate d^2 species of tungsten(IV) also provide substitution inertness, but they are susceptible to oxidation to eight-coordinate tungsten(V) and octahedral tungsten(VI) at elevated temperatures — both of the oxidized states are much less inert. A tabulation of substitution-inert centers is given in Table 1.2.

Another method for obtaining such inertness is to use multidentate ligands — the higher the denticity, the greater the inertness — other things being equal. For example, bis-tetradentate Schiff-base bridging ligands provide polymers of zirconium(IV) (Fig. 1.12) and cerium(IV) that are quite stable. A similar set of polyelectrolytes of several labile trivalent transition and lanthanide metal ions have also shown solution stability, and their sodium salts have been characterized by nuclear magnetic resonance (NMR) spectroscopy, gel

TABLE 1.2 Substitution-Inert Metal Ions for Common Coordination Centers That Might Be Used in Metal-Containing Polymers.

Configuration	Coordination Number (Symmetry)	Relative Inertness
d^2	low-spin 8-coordinate (D_{2d} or D_{4d})	$W^{IV} > Mo^{IV}$
d^3	6-coordinate (O_h)	$Cr^{III} > Mo^{III} > V^{II}$
d^4	low-spin 6-coordinate (O_h)	Mn^{III} ($< Cr^{III}$ or Fe^{III})
d^5	low-spin 6-coordinate (O_h)	$Ru^{III} > Fe^{III}$
d^6	low-spin 6-coordinate (O_h)	$Pt^{IV} > Rh^{III} > Co^{III} > Ru^{II} > Fe^{II}$
d^8	low-spin 4-coordinate (D_{4h})	$Pt^{II} > Pd^{II} > Ni^{II}$; $Ir^{I} > Rh^{I}$

[a]D_{2d} = trigonal-faced dodecahedral; D_{4d} = square (or Archimedes) antiprismatic; O_h = octahedral; D_{4h} = square or tetragonal-planar

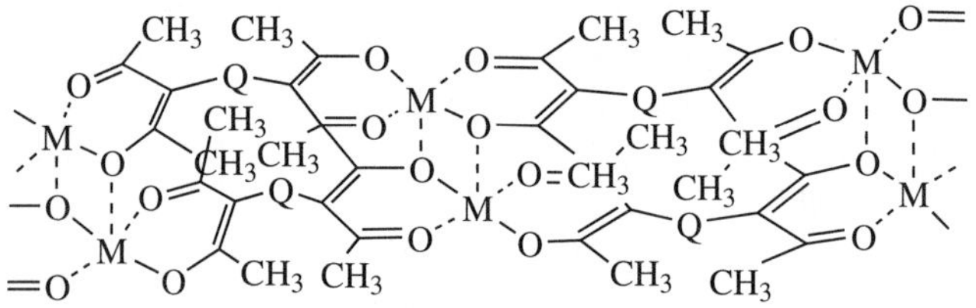

Figure 1.21 Stacking of planar coordination polymer segments that cause intractability (insolubility).

permeation chromatography (GPC), viscosity measurements, and absorption and emission spectroscopies in dimethyl sulfoxide (DMSO) or *N*-methylpyrrolidone (NMP).

Organometallic polymers have been synthesized for a wide variety of metals, but as in the case of the coordination polymers, all too often, oligomers have resulted rather than high-molecular-weight materials. Also, as in the case of the coordination polymers, it is not possible from the characterizations provided to determine the size of the organometallic polymers. The reader is referred to a very thorough review provided by Pittman, Carraher, and Reynolds (15), with more recent reviews (3, 5, 16) being less complete or annual in nature but obviously more up to date, especially with regard to ring-opening organometallic polymerizations that were unknown earlier. Even more recently, Manners and coworkers (17) provided an excellent review on organometallic polymers with transition metals in the main chain.

Metallocene polymers dominate the organometallic polymer field with ferrocene polymers in great abundance. Figure 1.22 provides a summary of a variety of metallocene polymers of other elements, and Figure 1.23 provides a sampling of ferrocene polymers. Figure 1.22a–d provides four examples of group 4 metallocene polymers obtained by condensation polymerization of metallocene dichlorides with diols, dicarboxylates, dithiols, and diamines. Note that these bridging groups are coordinated directly to the metal itself, as opposed to the ruthenocene and cobalticenium derivatives (Fig. 1.22e–h), in which the bridge is attached to the cyclopentadienyl moiety. In the former case, the liability of the group 4 metal ion provides minimal stability as opposed to the covalent bonds of the latter type. The ruthenocene analogue with a direct alkyl bridge (Fig. 1.22e) is just one of many ferrocene polymer analogues of ruthenium and osmium that can be obtained.

Among the three cobalticenium polymers shown, one is a mixed titanocene-cobalticenium polymer. Alternating metal polymers often provide both advantages and disadvantages relative to the individual polymers. In this case, the different electronic properties of the two metals might be an advantage for certain applications, but the liability of the titanium carboxylate bond could be a problem in some solvents and situations where other ligands might be present.

Ferrocene polymers predominate the metallocene polymer field for a variety of reasons including (1) the stability of the ferrocene center itself, (2) the extensive

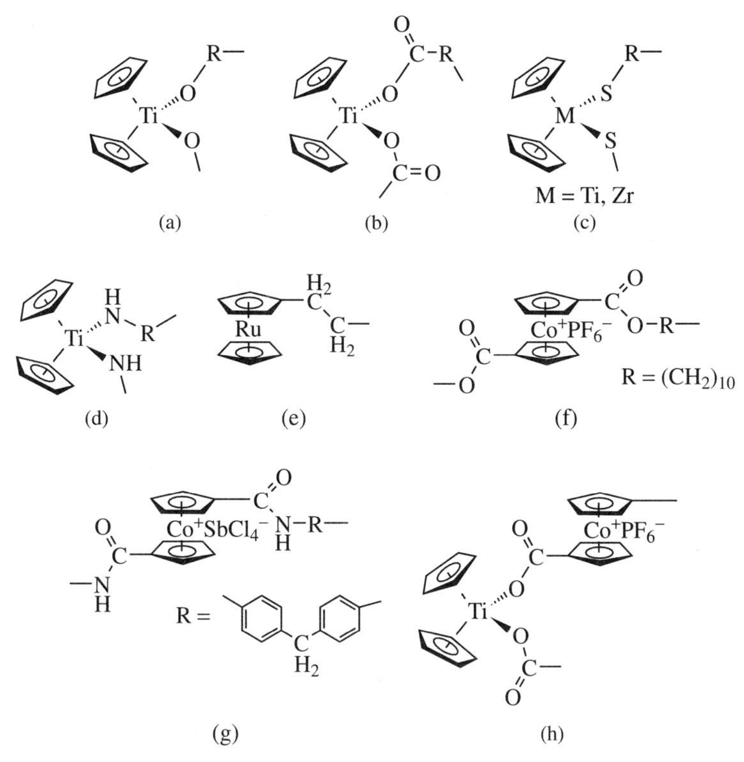

Figure 1.22 Repeating units for a variety of known Type I metallocene polymers.

monomer chemistry of ferrocene derivatives, (3) the low cost of iron chemicals relative to most of the other transition and inner transition metal derivatives, and (4) the virtual nontoxicity of iron. Although it is impossible to include all of the ferrocene polymer derivatives that have been prepared, a wide variety of them can be found in Figure 1.22. The simplest poly(ferrocene) (Fig. 1.23a) is difficult to prepare with high molecular weights, and the maximum is usually under 5000, unlike the poly(dimethylsilaferrocene) shown above (Fig. 1.4d). for which high ($>10^6$)-molecular-weight materials are readily obtained through ring-opening polymerization (Section 2.3). The condensation polymerizations used for most of the derivatives in Figure 1.23 provide only modest molecular weights. The wide variety of polymers that have been synthesized should be evident from the species shown. Note that either the polyphosphineoxide or polyphosphinesulfide derivative (Fig. 1.23b,c) can be prepared, but the reaction is not as clean as given.

Some of the ferrocene is cleaved, and cycloalkyl-bridged units are also a part of the polymer. Note that carboxylate bridges can be with the carbonyl either on the cyclopentadiene ring (Fig. 1.23d) or on the bridging group (Fig. 1.23e). Polyamide linkages on the cyclopentadienyl ring or in a β-position of an alkyl group on the ring are stable (Fig. 1.23f,g), whereas β-substituents are unstable. Note the R bridges can be either alkyl or aryl (Fig. 1.23g), and polyureas and polycarboxylates can also be obtained with β-substituents(Fig. 1.23h,i). A boron derivative (Fig. 1.23j) has been synthesized.

The recent ring-opening polymerization of 1,1-chloromethylsilaferrocene at room temperature with a $PtCl_2$ catalyst, now thought to be colloidal platinum (21) (Section 2.3.2.1), provides a pathway for many other derivatives by substitution of the chloro group (Fig. 1.23k). Block organometallic polymers, such as that shown in Figure 1.23l, can be prepared by multistep syntheses that include ring-opening polymerization. Finally, a ferrocene siloxane polymer (Fig. 1.23m) prepared by melt polymerization at 100 °C provides a 50,000-molecular-weight polymer that is hydrolytically stable in THF – H_2O and thermally stable to over 400 °C.

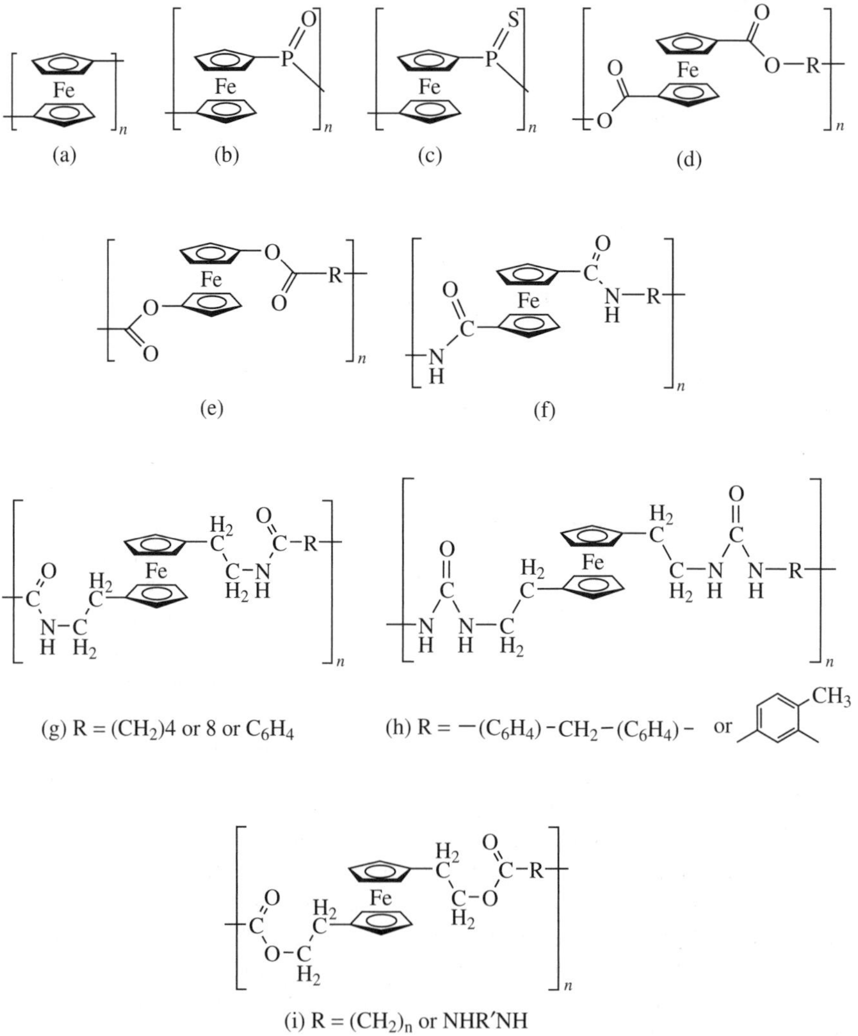

(a) (b) (c) (d)

(e) (f)

(g) R = $(CH_2)4$ or 8 or C_6H_4 (h) R = $-(C_6H_4)-CH_2-(C_6H_4)-$ or

(i) R = $(CH_2)_n$ or NHR′NH

Figure 1.23 A variety of known Type I ferrocene polymers. Continues.

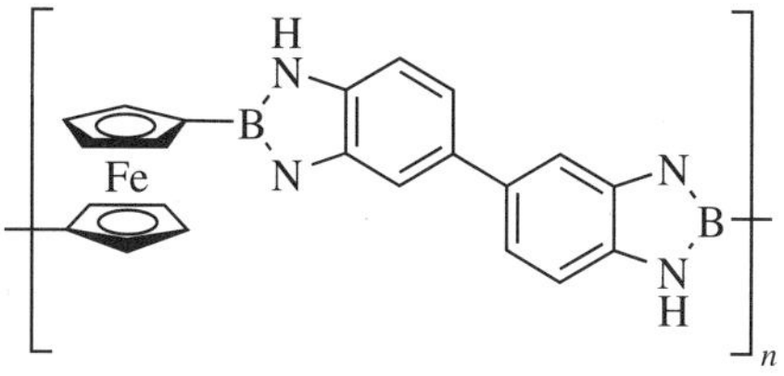

H N B N Fe N B N H n

(j)

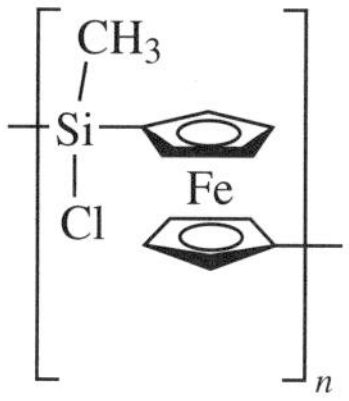

CH_3 Si Cl Fe n

(k)

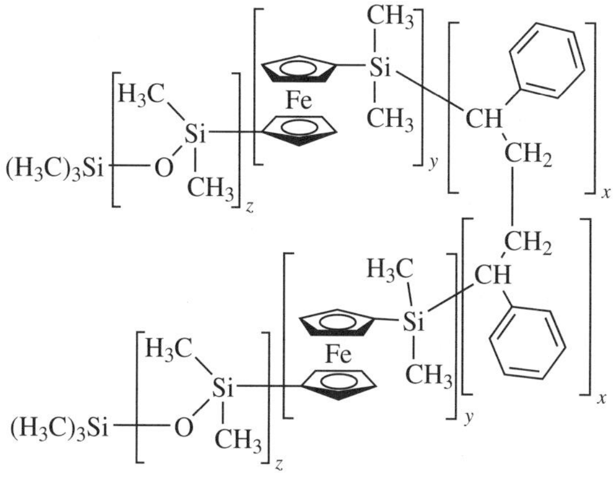

CH_3 Si CH_3 Fe H_3C Si $(H_3C)_3Si$ O CH_3 z y CH CH_2 x

CH_2 CH H_3C Si Fe CH_3 H_3C Si $(H_3C)_3Si$ O CH_3 z y x

(l)

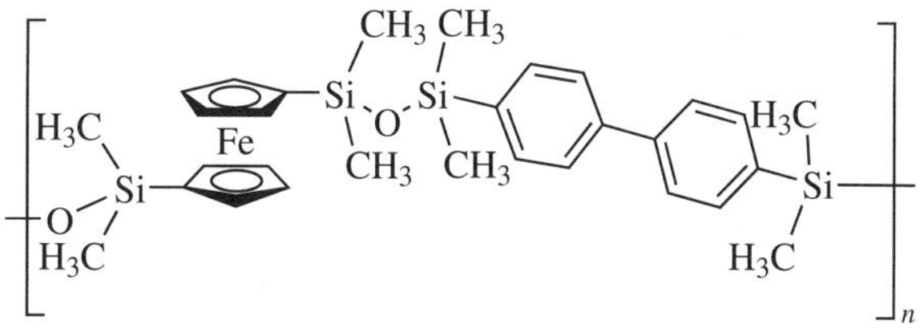

CH_3 CH_3 Si O Si CH_3 CH_3 Fe H_3C Si O H_3C H_3C Si H_3C n

(m)

Figure 1.23 *(Continued).*

1.5.2 Main Group Inorganic Polymers

Main group inorganic polymer chemistry has been the primary thrust of an earlier volume on inorganic polymers (6). Polysiloxane (silicone) (see, e.g., Fig. 1.4b) and polyphosphazene (phosphonitrile) (see, e.g., Fig. 1.3c) chemistry are well-developed areas of inorganic polymers. Polysiloxane applications are widespread, and the polyphosphazenes are finding interesting technical niches as well. The polysilanes (see, e.g., Fig. 1.4c) have a fairly extensive chemistry and are also being developed commercially. The polyoxothiazenes $[-N{=}S(O)(R)-]_n$, in which both an oxygen atom and an alkyl or aryl group are attached to the sulfur, are a fairly new class of inorganic polymers. Earlier sulfur-containing inorganic polymers include elemental sulfur and sulfur nitride (Fig. 1.3a). The carborane polymers (see, e.g., Fig. 1.3e), although initially synthesized much earlier, have not been extensively exploited to date.

Polysiloxanes, polysilanes, and related polymers are well known polymers. Linear polysiloxanes are an important part of modern civilization. They are used as high-performance fluids (commonly called silicones, even though they do not have the Si=O unit that gave them their name relative to organic ketones), soft contact lenses, elastomers (silicon rubber and caulking materials), surface coatings and modifiers, and body implants. More uses will be noted in Chapter 4. In the linear polysiloxanes, the R groups can range from methyl groups in poly(dimethylsiloxane) to other alkyl or aryl groups, or the two groups on a given silicon can be nonequivalent as in poly(methylphenylsiloxane) (Fig. 1.24a). Closely related polymers include the sesquisiloxanes (Fig. 1.24b), the poly(silalkylenes) (Fig. 1.24c), the poly(siloxane-silarylenes) (Fig. 1.24d), polysilanes or more properly poly(silylenes) such as poly(dimethylsilylene-co-methylphenylsilylene) or "polysilastyrene" (Fig. 1.24e), poly(silazanes) (Fig. 1.24f). A number of germanium and tin polymers are also known.

The *trans, trans* approximately planar zigzag noted in Figure 1.24a and 1.24e for siloxanes and silanes is quite typical for these species, although rotation occurs about the bonds. The siloxanes are typically atactic (random side groups as depicted in Fig. 1.24a) with Si–O–Si angles of about 143° and O–Si–O angles of about 110° that together cause a cyclical projection that comes back to itself after about 11 repeat units and thus becomes helical for long "chains." The large angle at oxygen is usually credited to O(pπ) $\rightarrow$ Si(dπ) bonding, that is, bonding of π-symmetry electrons on an oxygen atom to one of the empty 3d orbitals of a silicon atom. Because all of the silicon d orbitals are empty, there is no restriction to rotation by the π-bonding. Thus rearrangements on surfaces such that all of the oxygen atoms are rotated enough to stay in contact and bond to hydrophilic (polar surfaces) leaving hydrophobic organic groups projecting out from the surface provide the waterproofing that silicones are so well known to impart to fabrics.

The angles in the polysilanes and poly(silmethylenes) are all approximately tetrahedral because no d orbital participation is important in these systems. Note that rotations about the bonds are free. Note also that neither the polysilanes nor the poly(silmethylenes) have natural hydrophilic sites and their structures are more sterically dependent on the pendent groups. All atoms in the backbones have two pendent atoms or groups of atoms on them, just like organic hydrocarbon polymers.

To date, over one hundred polysilanes have been synthesized to provide a wide range of materials. Some are amorphous glasses or elastomers, some are partially crystalline flexible solids, and yet others are highly crystalline intractable materials. Although their ultraviolet sensitivity has limited their use as "normal" polymers, they do find some use as computer chip photoresists, photoconductors, photoinitiators, and nonlinear optical materials. Details will be found in Chapter 4.

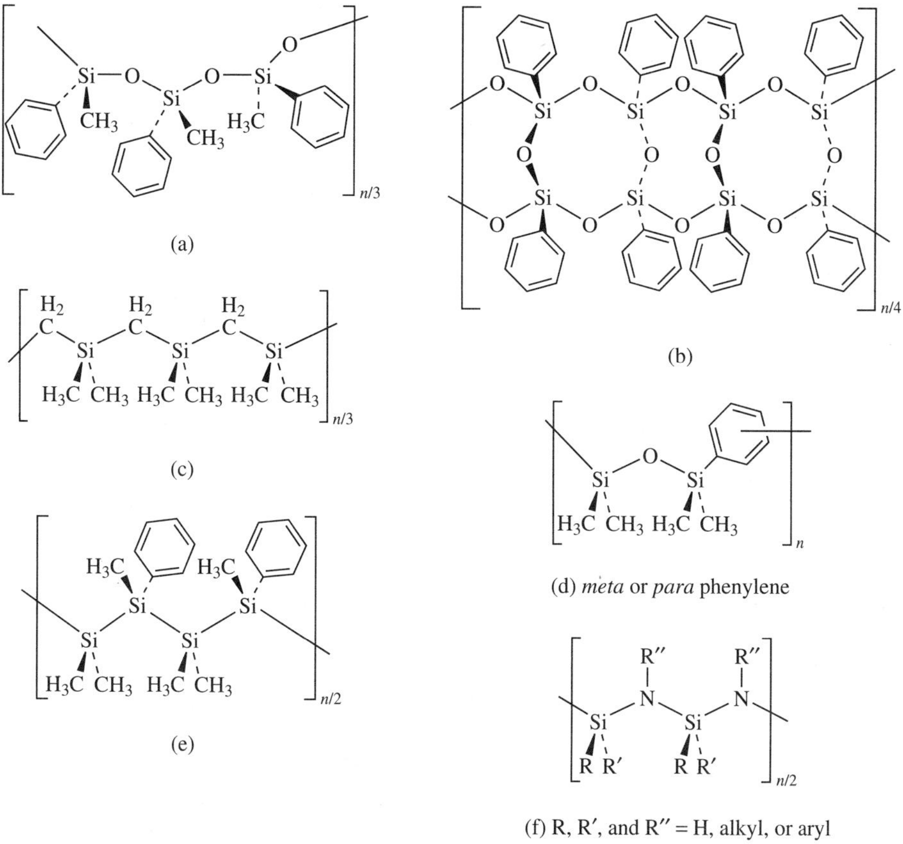

(a)

(b)

(c)

(d) *meta* or *para* phenylene

(e)

(f) R, R′, and R″ = H, alkyl, or aryl

Figure 1.24 Silicon/main group polymers: (a) poly(methylphenylsiloxane); (b) a poly-(sesquisiloxane) $[Si_2(C_6H_5)_2O_3]_n$; (c) poly(dimethylsilmethylene); (d) poly(dimethylsiloxane-dimethylsilphenylene); (e) poly(dimethysilylene-co-methylphenylsilylene) or polysilastyrene; and (f) a polysilazene.

Polymeric phosphazenes and related polymers (18–20) include the parent poly(dichlorophosphazene), $(PNCl_2)_n$ {or better formulated as $[NPCl_2]_n$ to indicate that the chloro substituents are on the phosphorus; cf. Fig. 1.3c)}. This polymer has been known for more than a century as polyphosphonitrilic chloride or inorganic rubber. But because $[NPCl_2]_n$ hydrolyzes in water, the polymer was simply a laboratory curiosity until 1965, when Allcock and his coworkers prepared a noncrosslinked $[NPCl_2]_n$ polymer that had solubility in benzene, toluene, and tetrahydrofuran (THF). They also took advantage of the reactivity of the phosphorus-chlorine bond to prepare the first of over 300 derivatives synthesized by the 1990s. The classical $[NPCl_2]_n$ polymer that was prepared along with the lower-molecular-weight $[NPCl_2]_{3-7}$ cyclic units, or prepared by heating any of the cyclic species to about 250 °C, was insoluble in all normal solvents because of extensive crosslinking. Some properties of a number of the regular polyphosphazenes that have been synthesized are given in Table 1.3. The trends in the glass transition temperature (T_g, above which the polymer has flexibility to reorient its chains) show some interesting variations related to intra- and interchain interactions. Wide-ranging properties (e.g., crystallinity vs. amorphous behavior and hydrophilicity vs. hydrophobicity) can be introduced into the polyphosphazenes through substituent variations. This leads to a wide variety of potential uses. The cost of the polymers has limited their practicality, but the innocuous nature of their hydrolysis products in biological systems provides impetus for biomedical uses of these polymers where cost is less important than effectiveness and lack of toxicity. Early attempts to commercialize the polyphosphazenes were economically unsuccessful, but new ventures are underway as this book is being written — not too surprising given their potential as elastomers, solid polymer electrolytes, hydrogels, microencapsulators, bioerodibles, adhesives, liquid crystalline polymers, nonlinear optical materials, and fire-resistant materials. More details on the potential usefulness of these polymers will be found in Chapter 4.

The structures of typical polyphosphazenes are approximately *cis, trans* planar structures as shown in Figure 1.3c for the chloride derivative. This structural

TABLE 1.3 Properties of Selected Symmetrical Polyphosphazenes[a].

R of NPR_2	Properties	T_g(°C)	T_m(°C)
Br	Leathery material, unstable in water-	15	—
Cl	Elastomer, unstable in water	−66	−7(+39)[b]
F	Elastomer, unstable in water	−96	−68, −40
	Aryl Groups		
OC_6H_4-p-C_6H_5	Microcryst. thermoplastic, high refractive index	+93	>350
NHC_6H_5	Glass	+91	—
$OC_6H_4COOC_2H_5$	Microcryst. thermoplastic (films)	+7.5	127
OC_6H_4COOH	Glass, soluble in aqueous base	−4.5	390
OC_6H_5	Microcryst. thermoplastic (films, fibers)	−8	—
	Alkyl Groups		
NHC_2H_5	Glass, soluble in aqueous acid	+30	—
$NHCH_3$	Glass, water soluble	+14	—
OCH_2CF_3	Microcryst. thermoplastic (films, fibers)	−66	242
OCH_3	Elastomer	−76	—
OC_2H_5	Elastomer	−84	—
$O(C_2H_4O)_2CH_3$	Elastomer, water soluble	−84	—
OC_3H_7	Elastomer	−100	—
O-*n*-C_4H_9	Elastomer	−105	—

[a]Based on Allcock's earlier compilations.[6,19] Microcryst. = Microcrystalline; [b]For the stretched polymer.

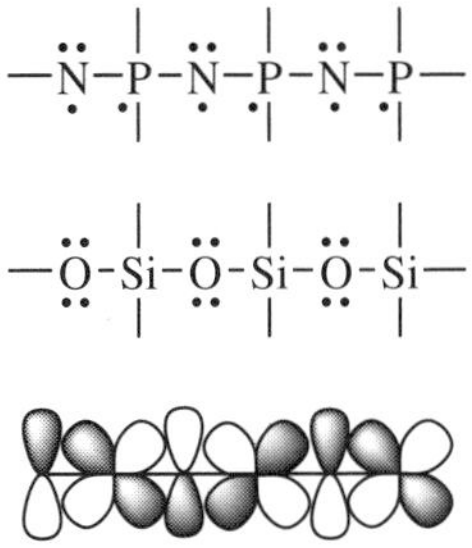

—N–P–N–P–N–P—

—O–Si–O–Si–O–Si—

Figure 1.25 Lewis structures for the backbone sigma bonding in phosphazenes and siloxanes (with the other valence electrons shown as dots on their respective atoms) and a simplified representation of the $p\pi$-$d\pi$ bonding that is postulated to occur is shown down a linear portion of the chains (perpendicular to the views of Fig. 1.3c and Figure 1.23a).

feature provides a lower interaction between the side groups than the *trans, trans* structure observed for most of the siloxanes.

Whereas Si–O and P–N pairs are isoelectronic (each supply 10 valence electrons), traditionally the phosphazenes are considered as unsaturated compounds with one pair of the electrons per repeating unit relegated to $P(d\pi)$–$N(p\pi)$ bonding and the other non-sigma bonding pair as a lone pair on the nitrogen. Although this provides alternating single and double bonds, the bond lengths are equal in the ring systems that are analogous to the polymer chains, so delocalization seems logical as shown in Figure 1.25. However, no matter how large the phosphazene ring or chain, no shifts in π–π^* electronic transistions are observed. This leads to the speculation that the d-orbital nodes stop the delocalization at each phosphorus atom. On the other hand, siloxanes are considered saturated compounds with the opening of the Si–O–Si bond angle by $Si(d\pi)$–$O(p\pi)$ bonding, with the electrons being from the so-called oxygen lone-pairs. If all four valence electrons not involved in sigma bonding in the phosphazenes were placed as lone pairs on the nitrogen atoms in an analogous fashion, each nitrogen would have an electron excess (a negative charge) and each phosphorus would have an electron shortage (a positive charge).

Substituents on polyphosphazenes are not limited to organic groups. Derivatives with iron and ruthenium bonded directly to the phosphorus atoms are known (Fig. 1.26a) and cyclopentadienyl groups on the phosphorus atoms have been used to prepare anchored ferrocene derivatives (Fig. 1.26b,c), too. Note that both cyclopentadienyl (Cp) entities are on the phosphorus atoms in the latter case. The PNPCpFeCp ring strain in the trimer that is used to prepare the polymer allows low-temperature ring-opening polymerization as noted in Chapter 2.

Attempts to bind metal ions on the lone pairs of the nitrogen atoms have not been overly successful. However, a derivative related to cisplatin, *cis*-diamminedichloroplatinum(II) (Fig. 1.26d), has been prepared with a maximum loading of about 1 platinum center per 17 phosphazene units.

Polyheterophosphazenes (16) are closely related to the polyphosphazenes. They include the polycarbophosphazenes, the polythiophosphazenes, and the

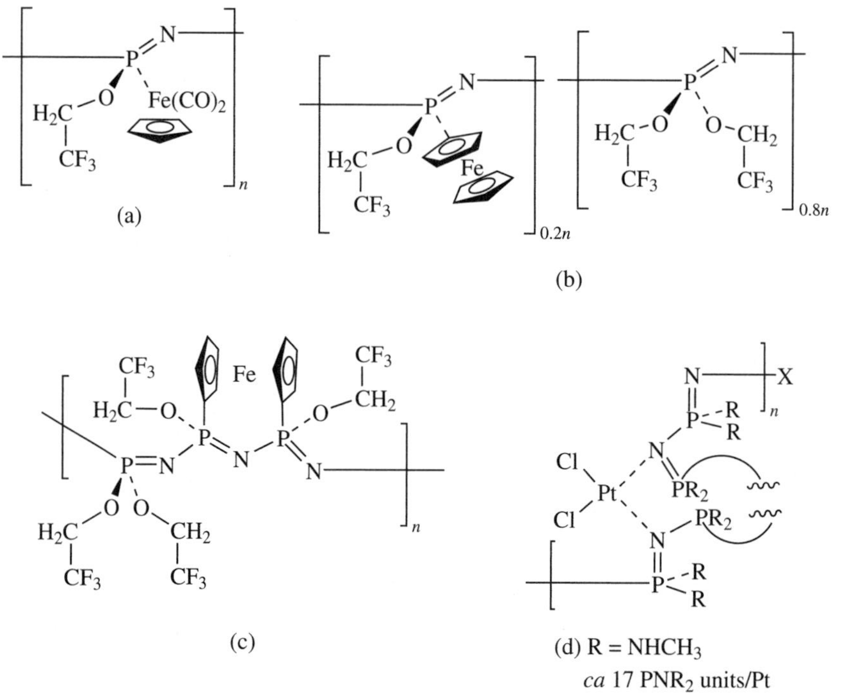

(a)

(b)

(c)

(d) R = $NHCH_3$
ca 17 PNR_2 units/Pt

Figure 1.26 Selected polyphosphazenes containing metal atoms.

metallophosphazenes. The precursor chloro polymers are prepared from trimers in which one of the PCl_2 groups has been replaced by a CCl, an S(O)Cl, or an MCl_3 group, respectively, where M = Mo or W. Substitution of organic groups for the chloro groups occurs on both the phosphorus and the carbon atoms simultaneously. The chloro groups on the sulfur analog are less readily substituted; therefore, different substituents can be placed on the phosphorus and the sulfur. Sample polyheterophosphazenes are shown in Figure 1.27. Such syntheses are described in Chapter 2. Analogous SCl derivatives [sulfur(IV) rather than sulfur(VI) of the S(O)Cl type noted above] are also known, but they are hydrolytically unstable.

Polyoxothiazenes are a new class of inorganic polymers (4). They can be looked at as a total substitution of S(O)R groups for all of the PR_2 groups of a polyphosphazene rather than just one-third of them in the polythiophosphazenes. Both poly(alkyloxothiazenes) and poly(aryloxothazenes) have been synthesized with alkyl and aryl groups on the sulfur atoms, respectively. Some of the polymers are only soluble in solvents of high polarity, such as DMSO or dimethyl formamide (DMF). The glass transition temperatures (T_g's) decrease as alkyl substituents increase in size; although considered unusual (4), the trend is analogous to the alkoxy phosphazenes shown in Table 1.3.

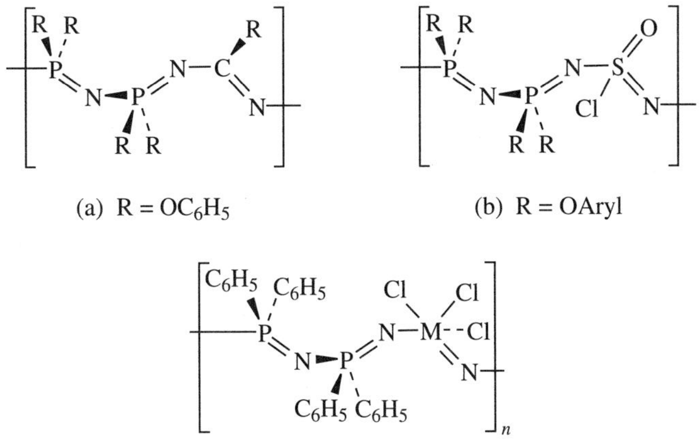

(a) R = OC_6H_5

(b) R = OAryl

(c) M = Mo or W

Figure 1.27 Polyheterophosphazenes: (a) a polycarbophosphazene; (b) a polythiophosphazene; and (c) a metallophosphazene.

Polycarboranes with a variety of bridging groups for both *meta*- and *para*-carboranes have been known for many years (3). In addition to the CO bridges shown in Figure 1.3e, siloxane $-[Si(R)_2O]_n-Si(R)_2-$ chain bridges with the dicarbodecaborane $-CB_{10}H_{10}C-$ provide copolymers with very high thermal stabilities (>400 °C) and softening temperatures but with relatively low T_g's (−42 to −88 °C for $n = 2$ to 6 and R = CH_3). These polymers have been used as O-rings, gaskets, wire coatings, and gas chromatography stationary phases. Other bridges including PCl bridges, which can be modified to PR bridges, are also known.

REFERENCES

1. Currell, B. R., Frazer, M. J. *Roy. Inst. Chem. Rev.* 1969, **2**, 13.
2. Carraher, C. E., Jr., Pittman, C. U., Jr. in *Ullmann's Encyclopedia of Industrial Chemistry*; VCH Press: Weinheim, 1989; Vol. A14, pp 241–262.
3. Pittman, C. U., Jr., Carraher, C. E., Jr., Sheats, J. E., Timken, M. D., Zeldin, M. in *Inorganic and Metal-Containing Polymeric Materials*; Sheats, J. E., Carraher, C. E., Jr., Pittman, C. U., Jr., Zeldin, M. and Currell, B., Ed., Plenum Press: NY, 1990, pp 1–27.
4. Roy, A. K. in *Kirk-Othmer Encyclopedia of Chemical Technology*; Kroschwitz, J. I. and Howe-Grant, M., Ed., John Wiley & Sons: New York, 1995; Vol. 14, pp 504–23.
5. Manners, I. *Ann. Rep. Prog. Chem., Sect. A, Inorg. Chem.* 1991–1997, **88–94**, 77, 93, 103, 131, 127, 129, 603, respectively.
6. Mark, J. E., Allcock, H. R., West, R. *Inorganic Polymers*; Prentice Hall: Englewood Cliffs, NJ, 1992.

7. MacCallum, J. R. in *Kinetics & Mechanisms of Polymerizations*; Solomon, D. H., Ed., Dekker: NY, 1972; Vol. 3, pp 333–69.
8. Holliday, L. *Inorg. Macromol. Rev.* 1970, **1**, 3.
9. Anderson, J. S. *Introductory symposium lecture*; Anderson, J. S., Ed.: Nottingham, 1961.
10. Ray, N. H. *Inorganic Polymers*; Academic Press: NY, 1978.
11. Archer, R. D. *Coord. Chem. Rev.* 1993, **128**, 49.
12. Foxman, B. M., Gersten, S. W. in *Encyclopedia of Polymer Science and Engineering*; Kroschwitz, J. I., Mark, H. F., Bikales, N. M., Overberger, C. G. and Menges, G., Ed., John Wiley & Sons: NY, 1986; Vol. 4, pp 175–191.
13. Rehahn, M. *Acta Polymer.* 1998, **49**, 201.
14. Pomogailo, A. D., Savost'yanov, V. S. *Synthesis and Polymerization of Metal-Containing Monomers*; CRC Press: Boca Raton, 1994.
15. Pittman, C. U., Jr., Carraher, C. E., Jr., Reynolds, J. R. in Encyclopedia of Polymer Science and Engineering; Kroschwitz, J. I., Mark, H. F., Bikales, N. M., Overberger, C. G. and Menges, G., Ed., John Wiley and Sons: NY, 1987; Vol. 10, pp 541–594.
16. Sheats, J. E., Carraher, C. E., Jr., Pittman, C. U., Jr., Zeldin, M., Culbertson, B. M. in *Metal-Containing Polymeric Materials*; Pittman, C. U., Jr., Carraher, C. E., Jr., Zeldin, M., Sheats, J. E. and Culbertson, B. M., Ed., Plenum Press: NY, 1996, pp 3–37.
17. Nguyen, P., Gómez-Elipe, P., Manners, I. *Chem. Rev.* 1999, **99**, 1515.
18. Mark, J. E. A., H. R., West, R. in *Inorganic Polymers*; Mark, J. E., Allcock, H. R. and West, R., Ed., Prentice Hall: Englewood Cliffs, NJ, 1992, pp 61–140.
19. Allcock, H. R. in *Encyclopedia of Polymer Science and Engineering*; Kroschwitz, J. I., Mark, H. F., Bikales, N. M., Overberger, C. G. and Menges, G., Ed., John Wiley & Sons: NY, 1987; Vol. 13, pp 31ff.
20. Wisian-Neilson, P. in *Encyclopedia of Inorganic Chemistry*; King, R. B., Ed., John Wiley & Sons: Chichester, 1994; Vol. 6, pp 3371–89.
21. Temple, K., Jakle, F., Lough, A. J., Sheridan, J. B., Manners, I. *Polymer Preprints* 2000, **41** (1), 429.

EXERCISES

1.1. Verify the statement "In step-growth polymers of the AA + BB type, where the repeating unit is AABB, ten repeating units, $(AABB)_{10}$, corresponds to a degree of polymerization of 19. That is, $2n - 1$ reaction steps are necessary to assemble the 20 reacting segments that make up the polymer."

1.2. What is the connectivity of the nickel atom in Figure 1.2a?

1.3. What is the connectivity of the manganese atom in Figure 1.2b **if connectivity is defined as the number of atoms linking manganese with the polymer chain?**

1.4. What is the connectivity of the manganese atom in Figure 1.2b **if connectivity is defined as the number of electron pair bonds linking manganese with the polymer chain?**

1.5. Find the connectivity of the Si atoms in Figure 1.23b.

1.6. Note that the cobalt ions of Figure 1.8a are surrounded by four nitrogen donors and the mercury ions by four sulfur donors. Use the hard-soft acid-base concept to explain this.

1.7. Using either Pittman review (3, 15) or the more recent Rehahn article (13), find at least two main group and two transition metal-containing polymers that can be classed as one-dimensional that are not given in this chapter.

1.8. Do the same as in Exercise 1.7 for two-dimensional inorganic polymers.

1.9. See Figures 1.14 and 1.15 for examples of structures for 2,5-dioxoquinonate metal polymers. Foxman and Gersten (12) have the Cu^{2+} 2,5-dioxoquinonate structure as a planar sheet, quoting Wroblesky and Brown [*Inorg. Chem.*, 1979, **18**, 498 and 2738], whereas Pittman et al. (3) have it as a rigid rod structure and quote S. Kanda and Y. Saito [*Bull. Chem. Soc., Japan*, 1957, **30**, 192] or Bailes and Calvin [*J. Am. Chem. Soc.*, 1947, **69**, 1886] for these types of species. Who is right? Justify your conclusion.

1.10. What structures would you predict for the trivalent-metal oxalates? Check the chemical literature for verification of your prediction.

CHAPTER 2

INORGANIC POLYMER SYNTHESES

A wide variety of synthetic methods have been used for synthesizing inorganic main group and organometallic polymers. Syntheses can be classified according to the type of synthesis (step-growth, chain-growth, or ring-opening) as is done in this chapter, by the polymerization process (bulk, solution, suspension, emulsion, or phase transfer), or by the type of polymer (coordination, organometallic, siloxane, silane, phosphazene, etc.) as is often done, especially in multiauthored books. We will limit our discussion of syntheses primarily to linear polymers that are potentially soluble and therefore quite easily and definitively characterized. The synthesis classification scheme is shown in Figure 2.1.

2.1 STEP-GROWTH SYNTHESES

A sizable number of linear inorganic polymers (including most that have metal ions as part of the polymer backbone) are synthesized by step-growth syntheses. These polymerizations are primarily condensation reactions, although step-growth addition polymerizations are also known (1). [The terms step-growth polymerization, step-growth condensation, and step-growth addition are redundant because polymerization, condensation, and addition all imply growth; therefore, these terms can be shortened to step polymerization, step condensation and step addition (2).]

Inorganic and Organometallic Polymers, by Ronald D. Archer
ISBN 0-471-24187-3

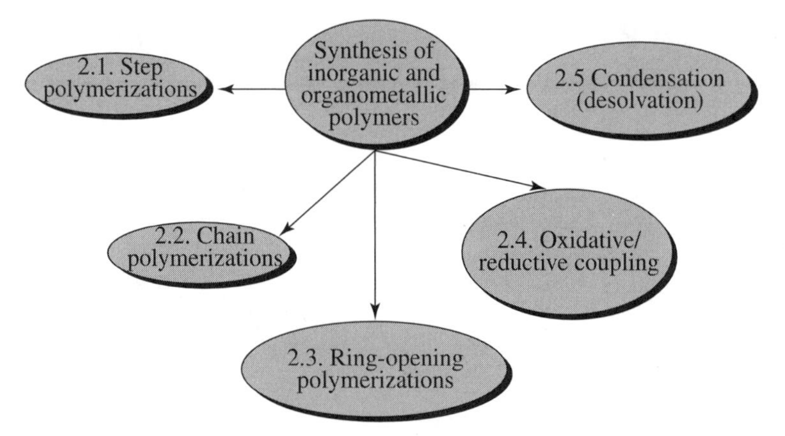

Figure 2.1 A classification of polymerization reactions that are important for inorganic and organometallic polymers.

2.1.1 Step Condensation Synthesis Generalities

In a step condensation polymerization, water or another small molecule is displaced from the growing polymer chain in each polymerization step. The small molecule production provides an entropic driving force in opposition to the negative entropy that results from the growing polymer chain. On the other hand, no byproduct is formed in a step addition polymerization.

Metal-ion electronic configurations, coordination geometries, and chelation must be considered in designing metal-containing polymers and in developing synthetic procedures for the polymers. Details on this aspect of coordination and organometallic compounds can be found in any good inorganic textbook, for example, Cotton and Wilkinson (3) (cf. Exercise 2.1). Although inert species provide the most stable polymers, inert species do not react well with bridging ligands unless special circumstances exist. Oxidation-reduction reactions can overcome this sluggishness. Alternatively, labile (noninert) metal centers can be used in such syntheses, but increased chelation is normally required to provide stability for such polymers. These aspects are important in providing polymers that are more than short oligomers on dissolution in a suitable solvent.

Several methods are possible for removing the water that is formed in step condensation polymerizations-such as the esterification that is shown schematically in Eq. 2.1.

$$n\text{HOOC–M–COOH} + n\text{HO–R–OH} \longrightarrow \text{H[–OOC–M–COO–R–]}_n\text{OH} + 2n - 1\text{H}_2\text{O} \qquad (2.1)$$

where M is a metal ion and its coordinated ligands (including 2 with carboxylic acid functionality) or an analogous organometallic species and HO–R–OH is a diol or a second metallic species with dihydroxy functionality.

Water removal methods for such condensation reactions include:

1. the removal of water azeotropically with a solvent such as toluene (that forms an 80%/20% toluene/water azeotrope);
2. the removal of water with a solid desiccant or molecular sieve that absorbs the water;
3. the deactivation of the water through the use of a solvent that has a high affinity for water, such as dimethyl sulfoxide (DMSO); and
4. the removal of water by volatilization using heat and/or vacuum; e.g., the synthesis of polymeric siloxophthalocyanine (Fig. 1.4e) from dihydroxo(phthalocyanine)silicon *in vacuo* at 400 °C (4).

Water removal is usually necessary to obtain a high extent of reaction in such condensation reactions.

An example of a step condensation reaction where the small molecule is HCl is shown in Eq. 2.2

$$n\text{HO–M–OH} + n\text{Cl–R–Cl} \longrightarrow \text{H[–O–M–O–R–]}_n\text{Cl} + 2n - 1\text{HCl} \qquad (2.2)$$

where M is a metal ion and its coordinated ligands including diol or dihydroxo functionality and Cl–R–Cl is an activated dichloro species such as $ClC_6H_4SO_2C_6H_4Cl$ or $ClC(O)(C_6H_4)C(O)Cl$. Alternatively, R could be another coordination or organometallic species including two chloro ligands or ligands with activated chloro groups.

To obtain a high extent of reaction, a base (either in homogeneous solution or as a heterogeneous solid) is often required to remove the HCl formed in condensation reactions of the type shown in Eq. 2.2. Thermal or vacuum removal is possible in some cases.

A large number of step condensation reactions have been used to synthesize organic polymers (Table 2.1). Most, if not all, of these reactions have been adapted to inorganic and organometallic polymers.

Step condensation polymerizations of the type shown in Eqs. 2.1 and 2.2 require very careful reactant ratio control as well as a high extent of reaction to obtain polymers rather than just short oligomers. The effects of both extent of reaction and reactant ratio on average chain length are given in Table 2.2 for the general step condensation reaction:

$$n\text{AMA} + n\text{BRB} \longrightarrow \text{A(MR)}_n\text{B} + 2n - 1\text{AB} \qquad (2.3)$$

where AB is the condensation byproduct, M is as in Eqs. 2.1 and 2.2 with functional group A, and R is the second component with the complementary functional group B and may be either an organic or an inorganic species. However, as shown in Figure 2.2, the average degree of polymerization (DP) is almost twice as large as the average number of repeating units (n), because every MR bond that is formed is concurrent with an increase in the size of the growing polymer unit.

TABLE 2.1 Typical Step-Growth Condensation Polymerizations from a Reference Volume[a].

Reactants	Reaction By-Product	Characteristic Linkage Formed	Type of Polymer
Dicarboxylic acids + diols	H_2O	$-CO_2-$	Polyesters
Dialkyl esters of dicarboxylic acids + diols	ROH	$-CO_2-$	Polyesters
Diacyl chlorides + diols	HCl	$-CO_2-$	Polyesters
Diacyl chlorides + diamines	HCl	–CONH–	Polyamides
Dicarboxylic acids + diamines	H_2O	–CONH–	Polyamides
Bis(chloroformate)s + diamines	HCl	$-O_2CNH-$	Polyurethanes
Bis(chloroformate)s + diols	HCl	$-OCO_2-$	Polycarbonates
Diols (or bisphenols) + phosgene	HCl	$-OCO_2-$	Polycarbonates
Bisphenols + diphenyl carbonate	PhOH	$-OCO_2-$	Poly(arylene carbonate)s
Dichlorosilanes + H_2O	HCl	–SiO–	Polysiloxanes
Organic dichlorides + Na_2S_x	NaCl	$-S_x-$	Polysulfides
Phenols + O_2	H_2O	–ArO–	Poly(arylene ether)s
Urea (or melamine) + formaldehyde	H_2O	$-NHCH_2-$	'Amino-resins'
Phenols + formaldehyde	H_2O	$-ArCH_2-$	'Phenolic resins'
Diols + aldehydes	H_2O	–OCH(R)O–	Polyacetals
Diketones + diamines	H_2O	$>C{=}N-$	Poly(azomethine)s
Aromatic sulfonyl chlorides	HCl	$-ArSO_2-$	Poly(arylene sulfone)s
Aromatic acyl chlorides	HCl	–ArCO–	Poly(arylene ketone)s
Benzyl chlorides	HCl	$-C_6H_4CH_2-$	Polybenzyls
Diisocyanates	CO_2	–N=C=N–	Poly(carbodiimide)s
Tetracarboxylic acid + diamines	H_2O	(–CO)₂N– (imide: two –CO– groups bonded to N–)	Polyimides
Dinitriles + hydrazine	H_2	4-amino-1,2,4-triazole ring (N–N, with N–NH_2)	Poly(aminotriazole)s
Dicarboxylic acid diphenyl esters + dihydrazines	PhOH	–CONHNH–	Polyhydrazides

[a]Reprinted from Eastomond, G. C., Ledwith, A., Russo, S., Sigwalt, P. in *Comprehensive Polymer Science*; Allen, G. and Bevington, J. C., Ed., Pergamon Press: Oxford, 1989; Vol. 5, p 7 with permission from Elsevier Science.

TABLE 2.2 Average Degree of Polymerization as a Function of Extent of Reaction and Reactant Ratios in Step-Growth Reactions[a].

	Reactant Ratios			
Extent of Reaction	Exactly 1.00:1.00	1.01:1.00	1.02:1.00	1.05:1.00
99.9 %	1000	168	92	39
99.	100			
98.	50	40	34	23
97.	33			
96.	25			
95.	20			14
90.	10	<10	9	8

[a]These values are based on the formula $DP = (N_M + N_R)/(N_M + N_R - 2r)$ where N_M and N_R are the moles of the two components M and R, r is the extent of reaction, and DP is the average degree of polymerization.

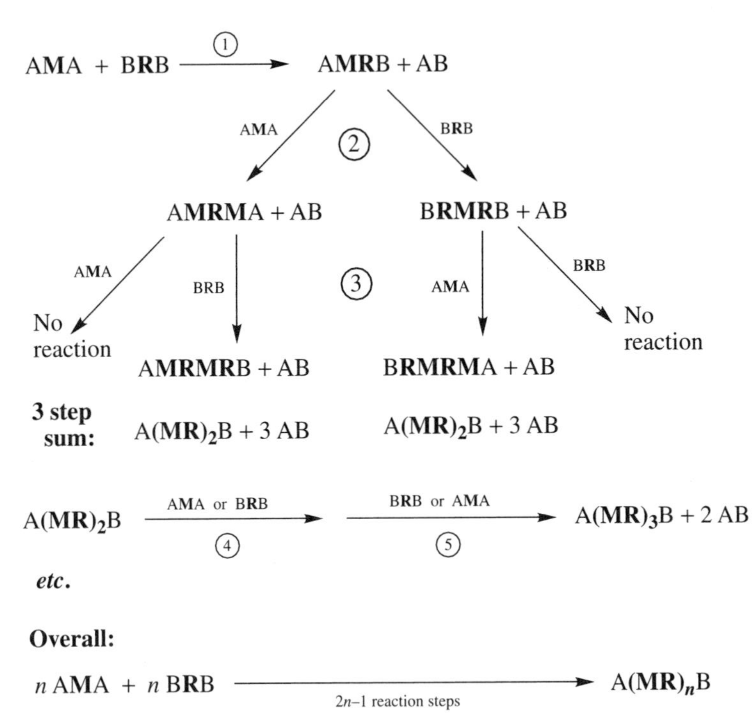

Figure 2.2 A schematic step condensation polymerization of two monomers AMA and BRB that react to form **M–R** bonds plus the AB condensation product. The number of reaction steps is designated as the degree of polymerization (DP). $DP = 2n - 1$, where n is the number of **MR** repeating units.

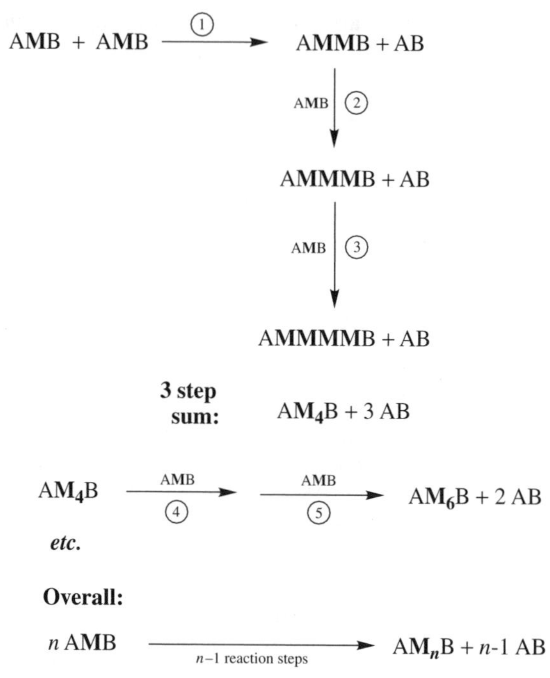

Figure 2.3 The step condensation polymerization of a difunctional monomer AMB: AMB reacts with other identical monomers to produce a $\mathbf{M}_n$ polymer and the AB condensation product. Again, the number of reaction steps is the degree of polymerization (DP), but in this case DP $= n - 1$, where n is the number of monomer (**M**) repeating units in the chain.

Note that the two components must react in a definite alternating sequence that produces a regular . . .M–R–M–R–M–R. . . chain. Details with regard to molecular distributions and end-groups can be found in Chapter 3.

Figure 2.3 shows the analogous situation for a monomer that reacts with itself yielding an . . .M–M–M–M–M. . . polymer. In this case, the average degree of polymerization and the average number of repeating units are almost identical (DP $= n - 1$).

2.1.2 Step Condensation Syntheses of Metal-Containing Polymers

Solubility is another problem that affects the degree of polymerization of polymers with metal ions in the polymer backbones. In fact, solubility often limits the length of the polymer chain even more than either the extent of reaction or the reagent ratios. The problem is particularly acute with planar metal ions, such as the divalent d^8 species (Ni^{II}, Pd^{II}, and Pt^{II}) with bis(bidentate-monoanionic) conjugated bridging ligands that tend to stack like polyaromatics; however, ligands with multiple electron pairs on the donor atoms have interchain metal-ligand interactions as well (cf. Fig. 1.21). Together these two effects cause insolubility to be the norm for such species.

A number of methods can be used (5) to overcome the insolubility problem.

1. Bulky ligands minimize the stacking interactions and provide soluble (tractable) planar divalent d^8 polymers, as noted earlier for the diacetylide polymers. (see the caption for Fig. 1.14a.) A similar approach has been used to prepare soluble phthalocyanine polymers (6) (see Fig. 2.4). Bulky ligands have also been used to provide solubility for polymers made by other methods as well; cf. Figure 2.37 below.
2. Eight-coordinate centers that tend to be nonrigid have been used for synthesizing soluble metal-containing polymers. One example was noted in Chapter 1 (Fig. 1.12) and more are noted in this chapter.
3. Octahedral coordination centers consisting of a metal ion surrounded by three bidentate ligands can also be used to produce soluble polymers (see Fig. 2.5).
4. Strong solvent interactions with metal coordination centers also aid in the solubility of metal-containing polymers. For example, $[UO_2(O_2CCH{=}CHCO_2)(DMSO)_2]_n$ with a number-average molecular mass ($\overline{M}_N$) of 10,000 is soluble in nucleophilic organic solvents such as DMSO and *N*-methylpyrrolidone (NMP) (7). See Figure 2.6 for the diaqua solid-state equivalent species.
5. Small tetrahedral centers (beryllium, boron, carbon, etc.) provide ideal centers for soluble polymers. See Figure 2.7 for a beryllium coordination polymer. Although beryllium would appear to be an ideal center for metal-containing polymers, the toxicity of beryllium compounds has precluded the extensive development and use of beryllium polymers.

2.1.2.1 Condensation of Functionalized Metal-Containing Species

One method of synthesizing metal coordination polymers involves the coupling of a monomeric coordination or organometallic species possessing two active sites with a bridging group that can react with these active sites. In the examples shown in Figure 2.8, the 3-CHO and the 3-H groups of the β-diketone ligands exchange roles in the two reactions.

> **Caution:** Synthesis using preformed coordination species normally requires kinetically inert metal complexes. Otherwise, scrambling of the metal complexes during the polymerization will preclude well-ordered polymeric species. Most of the following are "18-electron" systems, either d^6-octahedral or the analogous organometallic species, where the anionic cyclopentadienyl ring is considered as contributing 6 electrons. The 15-electron d^3-octahedral chromium(III) complexes can also be used in a similar fashion. Moving away from octahedral coordination, low-spin d^8-tetragonal planar and low-spin d^2-dodecahedral species also provide the necessary inertness. See Table 1.2 for a summary of the inert centers and the relative inertness within an electronic configuration.

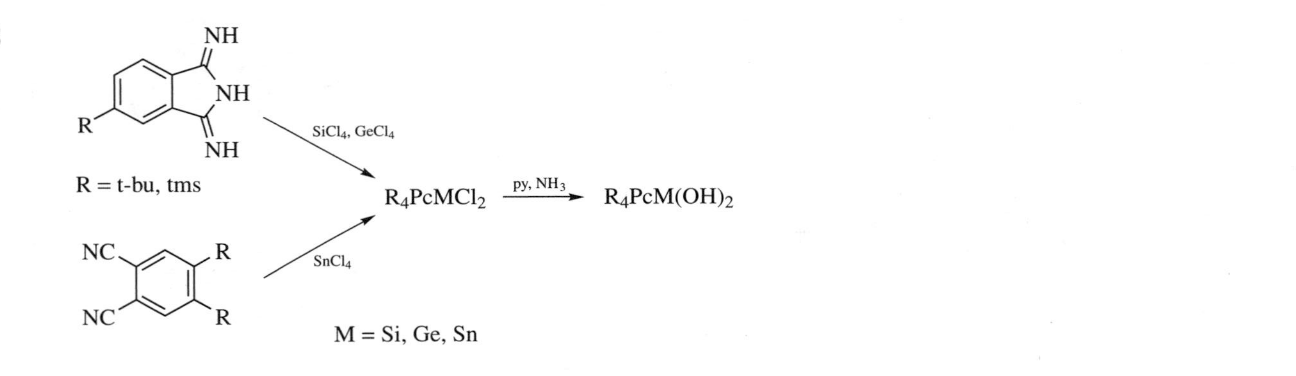

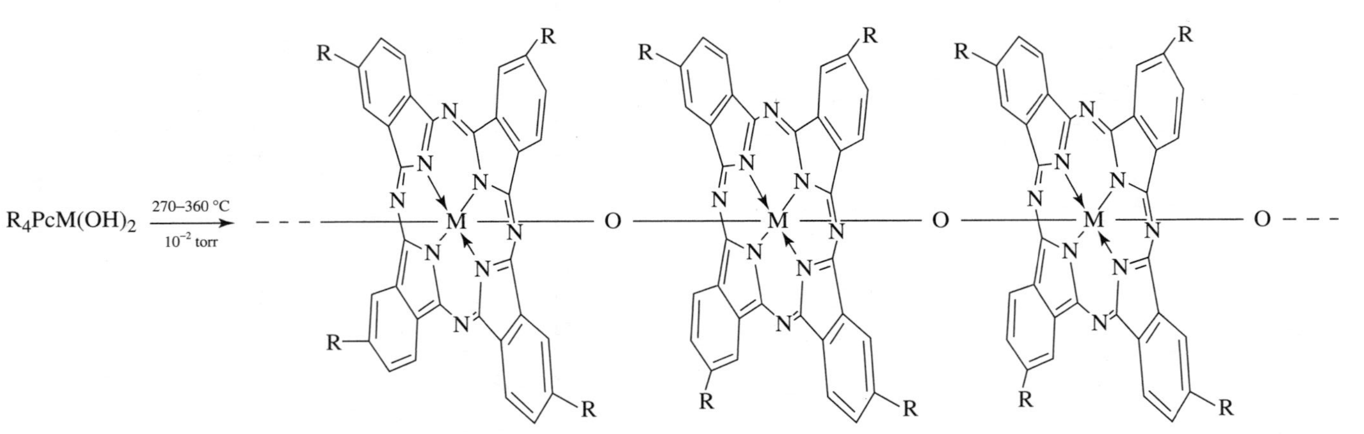

Figure 2.4 The synthesis of solubilized $[MPcO]_n$ where M = Si, Ge, Sn; Pc = tetra-*t*-butylphthalocyanato (6); and n = ~25 [J. W. Buchler, *Metal Complexes with Tetrapyrrole Ligands II*, Springer-Verlag: Berlin, 1991, pp 41-146].

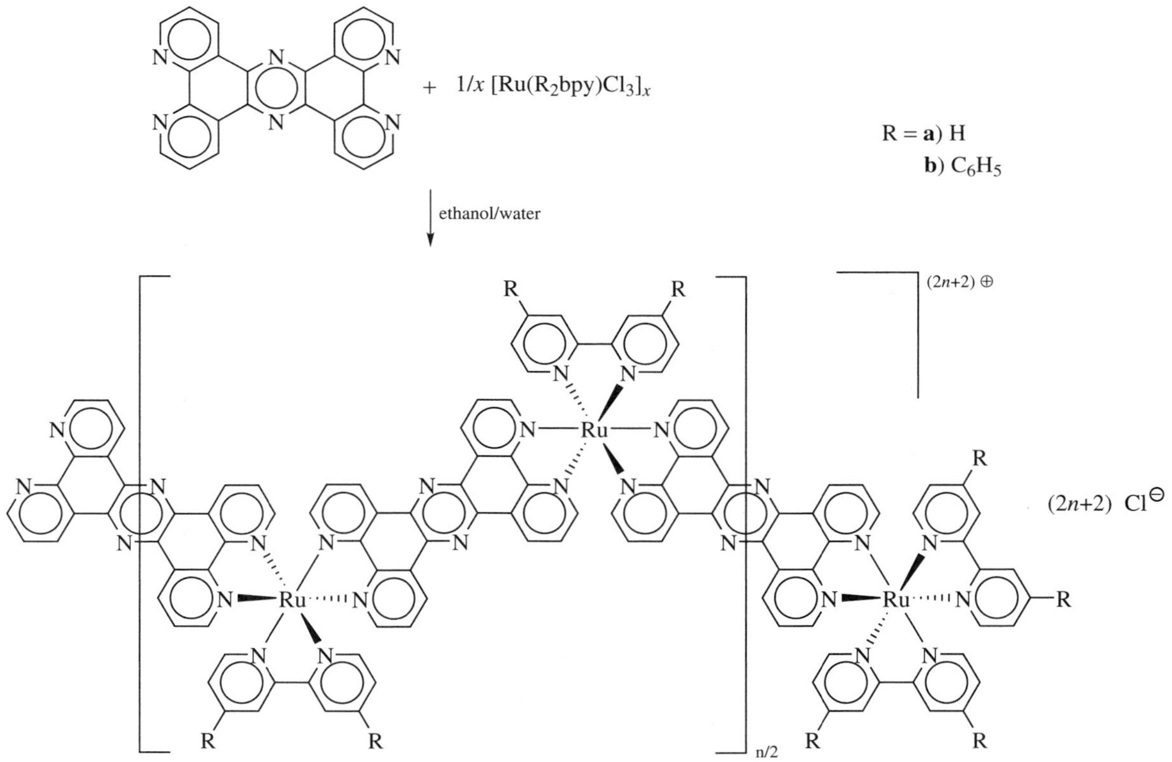

$+ \ 1/x\ [Ru(R_2bpy)Cl_3]_x$

R = **a**) H
b) C_6H_5

ethanol/water

$(2n+2)\oplus$

$n/2$

$(2n+2)\ Cl^{\ominus}$

Figure 2.5 The synthesis of a soluble octahedral metal coordination polymer with three-bidentate ligands per metal (reprinted with permission from Ref. 19; © 1997 American Chemical Society).

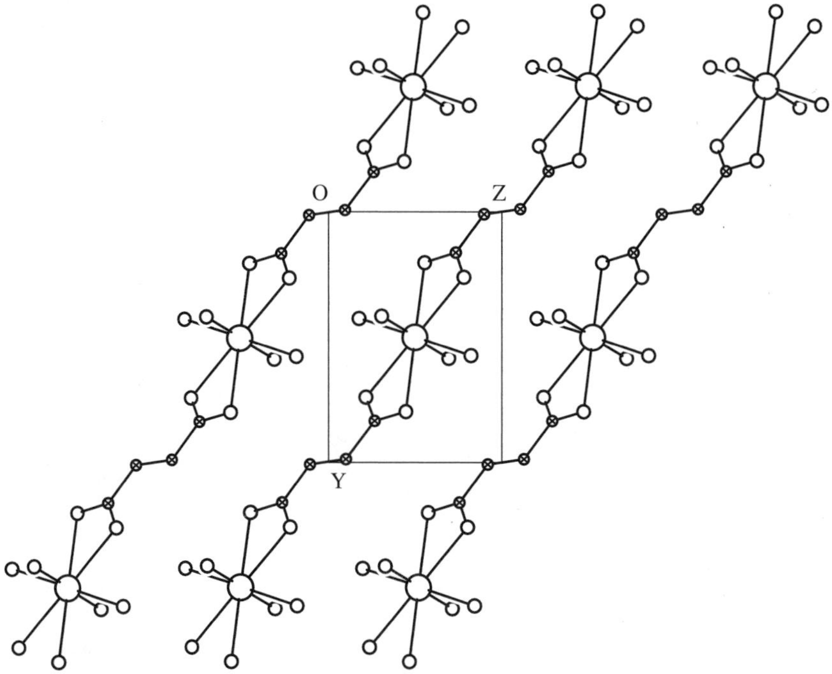

Figure 2.6 Uranyl fumarate, $[UO_2(O_2CCH{=}CHCO_2)(H_2O)_2]_n$, (Leciejewicz, J., Alcock, N. W., Kemp, T. J. *Structure and Bonding*, 1995, **82**, 43; © Springer-Verlag), a polymer that is soluble in nucleophilic organic solvents such as dimethyl sulfoxide forming $[UO_2(O_2CCH{=}CHCO_2)(DMSO)_2]_n$ (7).

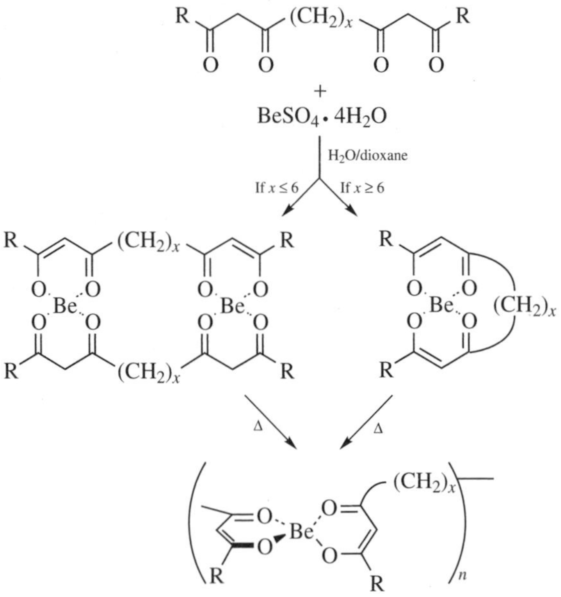

Figure 2.7 The ring-opening polymerization of beryllium β-diketones (54).

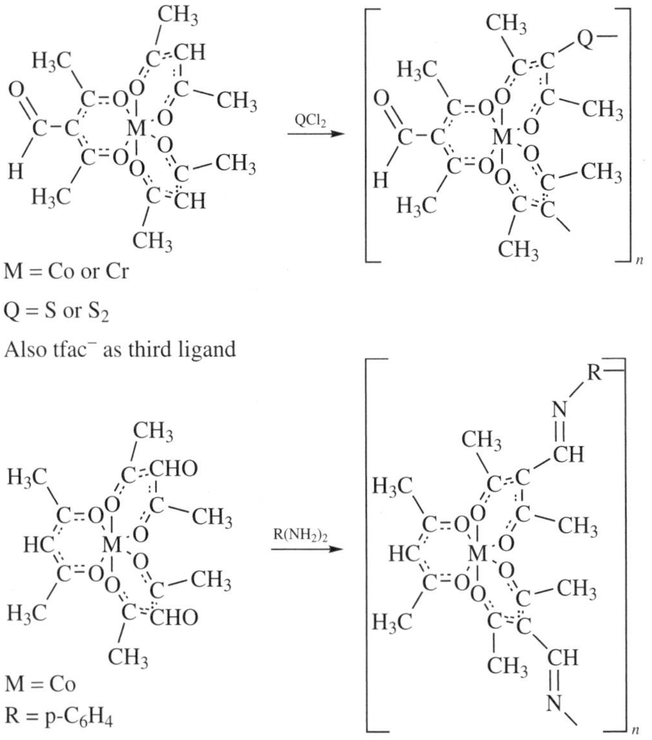

M = Co or Cr

Q = S or S_2

Also $tfac^-$ as third ligand

M = Co

R = p-C_6H_4

Figure 2.8 Step polymerization reactions of inorganic and organic reactants with preformed metal coordination species.

In the first reaction (Fig. 2.8), the condensation reactions between sulfur dichloride and sulfur monochloride (S_2Cl_2) with the ring hydrogen atoms of two metal β-diketone ligands provide inorganic thio and dithio bridges between the β-diketone ligands of the metal coordination monomers. Either a homogeneous base (e.g., triethylamine) or a heterogeneous solid base (e.g., sodium carbonate) can be used to remove (neutralize) the HCl that is formed. The 3-CHO group of the third ligand is unreactive with the sulfur ligands. Note that the third ligand must be unreactive for linear polymers (i.e., noncrosslinked) polymers to be synthesized. Alternatively, a β-diketone ligand with a CF_3 group can be used to deactivate the ring hydrogen (3 position in a 2,4-diketone) and prevent condensation with the sulfur halide as well. The CF_3 group withdraws electron density from the β-diketone ring and virtually stops the electrophilic reaction between the sulfur atoms and the ring.

On the other hand, a metal β-diketone with two aldehyde groups can undergo a Schiff-base condensation reaction with an aryl diamine (also shown in Fig. 2.8). This latter reaction has the 3-CHO-containing β-diketones as active ligands and the 3-H-containing β-diketone as unreactive. Therefore, linear polymers result with the bis(3-aldehydic)-β-diketone. The water can be removed azeotropically or tied up with either a molecular sieve or a solvent such as DMSO. In fact, DMSO is a good solvent for many metal coordination polymerizations. Its polarity and its partial organic character seem to be just the right mix to keep the growing polymers in solution in many cases. The mechanism of Schiff-base condensation involves nucleophilic attack by the lone pair electrons of the nitrogen base on the carbonyl carbon. Protonation of the carbonyl oxygen enhances the reaction, but full protonation of the amine slows the reaction. Therefore, the reaction rate goes through a maximum when less than a stoichiometric amount of acid is available. One counterintuitive method of increasing the Schiff-base reaction rate is to provide part of the amine as the analogous onium (RNH_3^+) salt. Even though protonated amines slow the Schiff-base reaction, proton transfer is so fast that the Schiff-base reaction can go virtually to completion faster than the neutral amine itself.

Organometallic examples are shown in Figures 2.9 and 2.10. The functional groups on the cobalticenium ion (Fig. 2.9) react with a diol to provide a polymeric organometallic species of 2500–4000 average molecular mass (8) and with aromatic diamines to form more soluble polyamides (9, 10). A polyferrocene condensation reaction is shown in Figure 2.10. The $\overline{M}_N$ is only about 4000, but the authors fractionated the product and obtained fractions at 1000-mass unit intervals up to and including 10,000 (11).

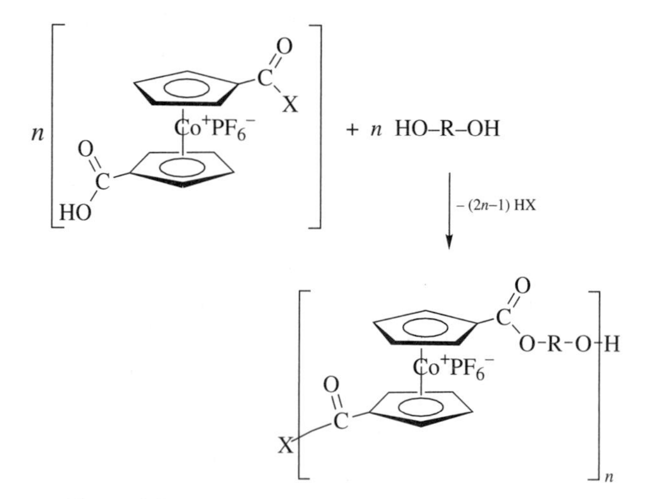

Figure 2.9 Cobalticenium polyelectrolyte syntheses.

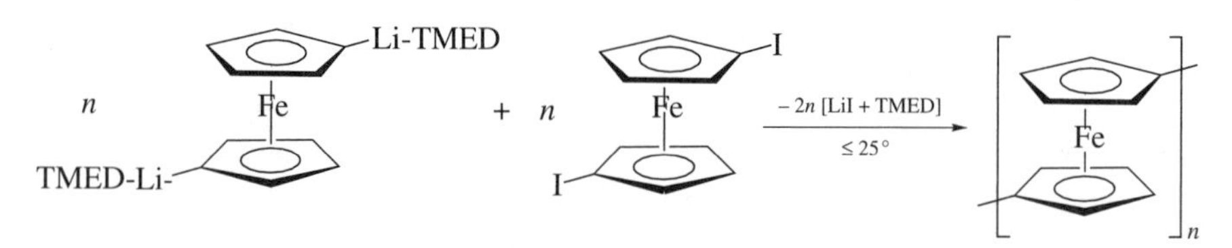

Figure 2.10 A step-growth polyferrocene synthesis.

2.1.2.2 Condensation Through Bridging Ligand Coordination

Completing the coordination sphere around a metal ion with two bridging ligands is a second method of synthesizing step-condensation metal coordination or organometallic polymers. The bridge can be a simple bridging ligand, such as the dianion of a diol or hydroxybenzoic acid (Fig. 2.11) or a multidentate ligand such that the two bridging ligands encompass the entire coordination sphere as shown in Figure 2.12 (12). The synthesis of the dianion (Fig. 2.11) from a virtually

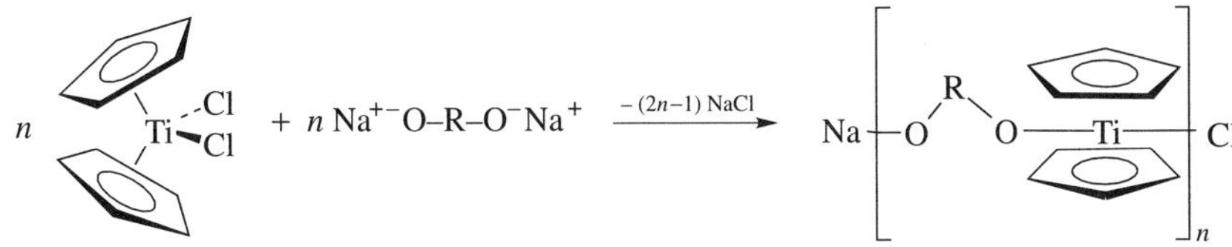

$$n\,Cp_2TiCl_2 + n\,Na^{+}\,{}^{-}O\text{–}R\text{–}O^{-}\,Na^{+} \xrightarrow{-(2n-1)\,NaCl} Na\text{–}[O\text{–}R\text{–}O\text{–}TiCp_2]_n\text{–}Cl$$

Figure 2.11 A schematic step-growth titanocene polymer synthesis, where R = alkyl, aryl, or benzoyl group.

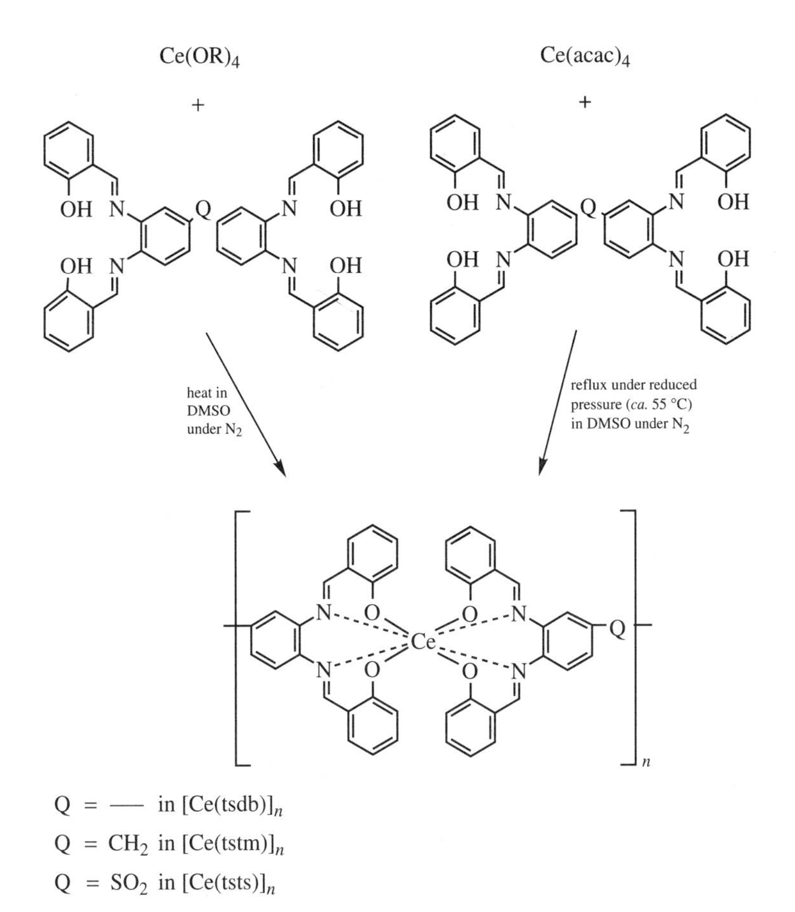

Figure 2.12 The use of bis(tetradentate) bridging ligands to synthesize cerium polymers.

nonacidic diol is shown in Eq. 2.4.

$$\text{HOROH} + 2\text{Na} \longrightarrow {}^{-}\text{ORO}^{-} + \text{H}_2 + 2\text{Na}^{+} \tag{2.4}$$

The organometallic reactions depicted in Figure 2.11 produce NaCl, and the coordination reactions in Figure 2.12 produce either an alcohol or a β-diketone in addition to the polymer; thus, they are all condensation reactions. The hot DMSO solvent used in the reactions shown in Figure 2.12 solubilizes the bridging ligand, the metal monomer, and the growing polymer chain. Note that d^0 and f^0 electron configurations are considered labile metal ions. The large charge and small size of titanium(IV) provide more inertness than might be anticipated for d^0 species, but the tetradentate chelation in the cerium(IV) polymer provides even more inertness. A large number of dihaloorganometallic species in addition to the titanium species shown in Figure 2.11 have also been polymerized with diamines, dithiols, diols, hydrazines, dithioamides, dioximes, dithiohydrazides, hydrazides, and urea (13). However, virtually no organometallic polymers that are soluble with high molecular mass have been prepared via this route.

Reactions analogous to those in Figure 2.12 with trivalent lanthanide species and sodium salts of bridging Schiff-base ligands form lanthanide polyelectrolytes as shown in Figure 2.13 (14, 15). The sodium polyelectrolytes are more soluble than either the lithium or the potassium polyelectrolytes (cf. Exercise 2.2). Again, DMSO is used to solubilize the bridging ligand, the metal monomer, and the growing polymer chains. Once again, f^n ions are expected to be quite labile, but the tetradentate chelation provides more inertness.

A reaction that mixes a functionalized metal-containing species with a second metal-containing species that can accept two bridging donors from the first species to complete its coordination sphere is shown in Figure 2.14. The cobalticenium ion, formally cobalt(III), is an inert d^6 18-electron ion, and the dihalobis(cyclopentadienyl)titanium(IV) is a more labile 16-electron species that readily gives up its chloro ligands to the protons of the carboxylates on the cobalticenium ion. The carboxylate anions coordinate to the titanium to provide a high-molecular-mass polymer. Viscosity measurements suggest a high $\overline{M}_N$ ($\sim$80, 000), although the solubility of this polymer is quite limited (16).

The inert polymers synthesized with low-spin tetragonal-planar d^8 $Pt(PR_3)_2Cl_2$ ($R = n$-butyl) (or the analogous Pd compound) and diacetylide (HC≡C–C≡CH) yields a polymer of high molecular weight along with HCl (see Fig. 1.3d). The butyl groups minimize the stacking forces that often limit the solubility of polymers synthesized with such d^8-planar complexes. An analogous polymer made from the low-spin (or spin-paired) d^6-octahedral $RhH(PR_3)_3Cl_2$ and diacetylide is also known (Fig. 1.14b). Note that both the d^6-octahedral and the d^8-planar coordination centers are inert metal centers (17,18). Such metal centers are typically very inert to substitution and provide quite stable species, whether monomeric, oligomeric, or polymeric. Fortunately, the chloro ligands are good leaving groups and allow the substitutions to occur.

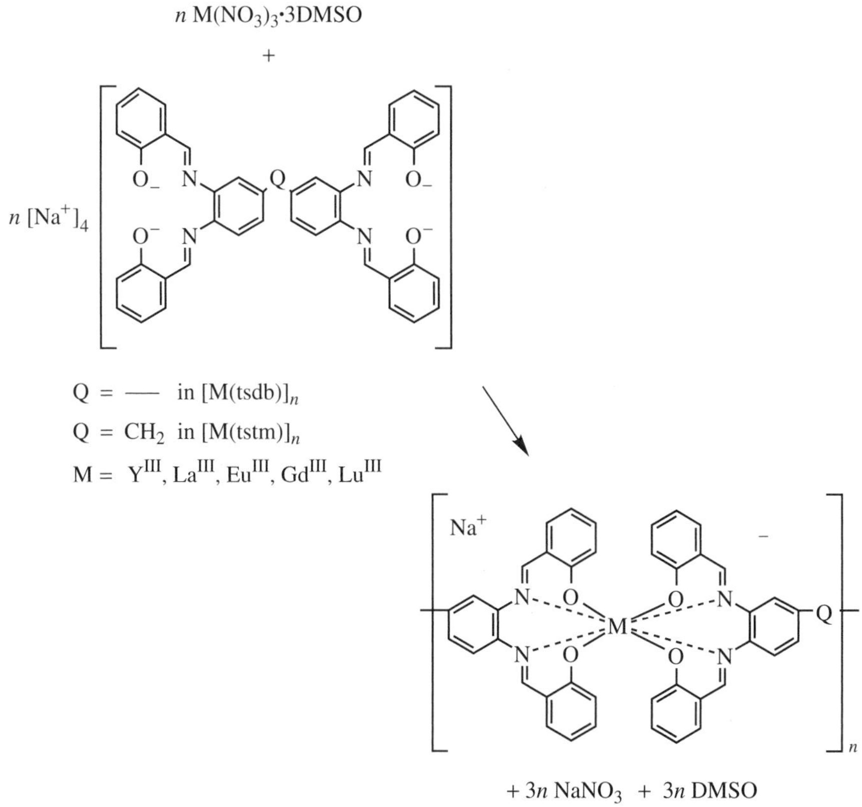

$n\ M(NO_3)_3 \cdot 3DMSO$

+

$n\ [Na^+]_4$

Q = — in $[M(tsdb)]_n$

Q = CH_2 in $[M(tstm)]_n$

M = Y^{III}, La^{III}, Eu^{III}, Gd^{III}, Lu^{III}

$+ 3n\ NaNO_3 + 3n\ DMSO$

Figure 2.13 The synthesis of lanthanide polyelectrolytes with bis(tetradentate) Schiff-base bridging ligands.

A more complicated example is the reaction used to prepare the soluble ruthenium(II) polymers shown in Figure 2.5. Ruthenium(II) is an ideal octahedral coordination center for heterocyclic nitrogen donors because of its low-spin (or spin-paired) d^6 electronic configuration. However, as noted above, the monomers are also inert. To synthesize a polymer of ruthenium(II) through ligand substitution in a reasonable time period, an oxidation-reduction reaction is the preferred route. The polymer shown in Figure 2.5 is synthesized from $Ru(bpy)Cl_3$, where bpy is 2,2′-bipyridine and the metal is ruthenium(III) (19). The low-spin d^5 ruthenium(III) is inert enough to keep the bpy coordinated during the reaction with the bis(bidentate) bridging ligand (cf. Exercise 2.3).

The purity of the starting ruthenium(III) complex is quite critical — in agreement with the discussion earlier in this chapter. In the first report of this reaction (20), only an oligomer with degrees of polymerization of 10 to 15 could be prepared because of two bpy ligands on some of the ruthenium starting material.

Substitution inertness is also known for eight-coordinate metal complexes with d^2 electronic configurations (18). For example, octacyanomolybdate(IV)

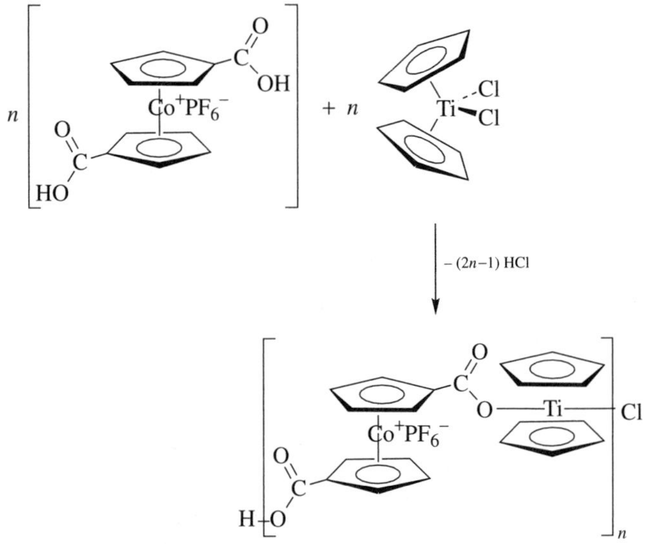

Figure 2.14 A mixed cobalticenium/titanicene polymer synthesis.

exchanges none of its cyanide ligands in 30 days (at room temperature in the absence of light). The inertness is also true for tungsten(IV) chelates (21, 22), thus an oxidation-reduction reaction provides the best answer for the synthesis of a polymer with the eight-coordinate d^2 electronic configuration. A seven-coordinate tungsten(II) complex reacts with quinoxalinedione and produces a tungsten(IV) polymer with quinoxalinediolato ligands. The viscosity implies an average molecular mass of about 25,000 for the DMSO-soluble fraction (Fig. 2.15) (23).

Other high-coordination-number metal-ion polymers include the uranyl dicarboxylate polymers, such as the one shown in Figure 2.6. These species are synthesized by a simple condensation reaction between the diacid and uranyl

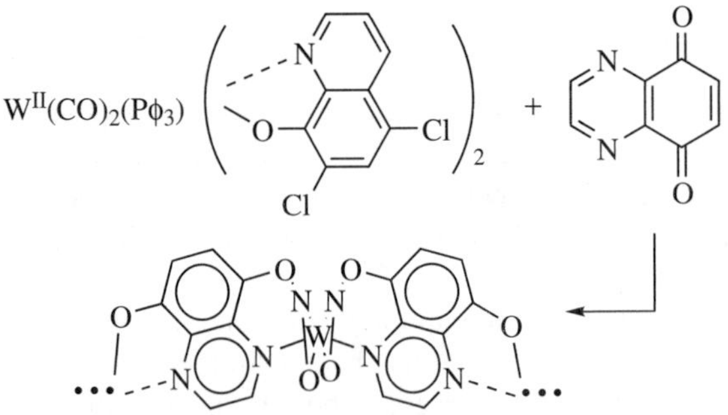

Figure 2.15 The synthesis of a tungsten(IV) polymer from tungsten(II) monomer, where ϕ = phenyl in the reactant and N–O$^-$ = 5,7-dichloro-8-quinolinolato in the product. Note that 5,8-quinoxalinedione is reduced to 5,8-quinoxalinediolate(2−) during the reaction.

acetate (7):

$$n[\mathrm{UO_2(O_2CCH_3)_2}] + n\mathrm{HO_2CCH{=}CHCO_2H} \longrightarrow$$
$$[\mathrm{UO_2(O_2CCH{=}CHCO_2)(DMSO)_2}]_n + 2n\mathrm{CH_3CO_2H} \qquad (2.5)$$

2.1.2.3 Bridging Ligand Formation During Condensation

When bridging ligands have low solubility, as is the case for some of the bis(tetradentate Schiff-base) ligands, it is often difficult to get solutions that are concentrated enough to provide polymerization reactions that are rapid enough and complete enough to make true polymers. An alternate approach is to use more soluble ligand precursors and form the bridging ligands during the polymerization as shown in Figure 2.16 (24). DMSO is an ideal solvent for this reaction, too,

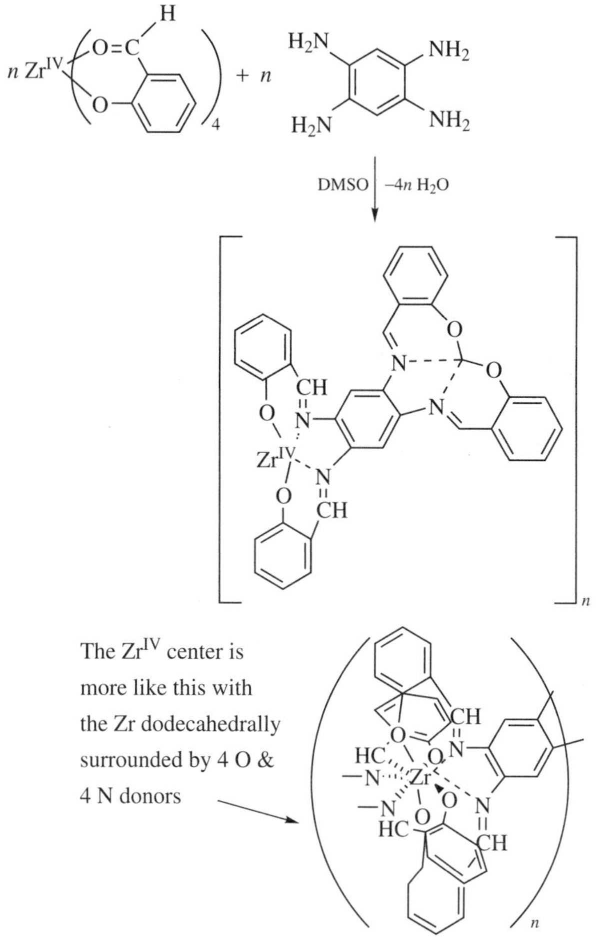

The $\mathrm{Zr^{IV}}$ center is more like this with the Zr dodecahedrally surrounded by 4 O & 4 N donors

Figure 2.16 The synthesis of zirconium(IV) Schiff-base polymers.

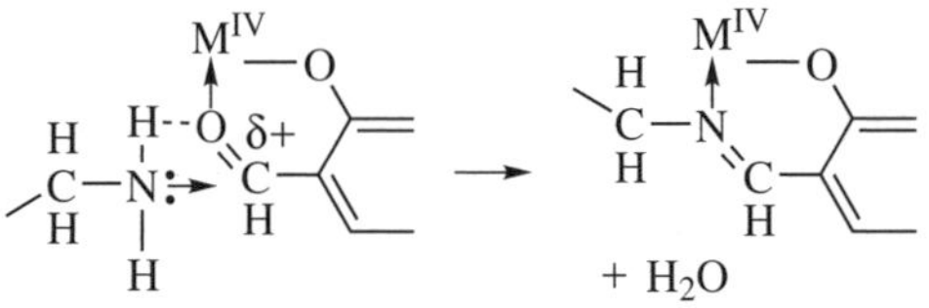

Figure 2.17 The schematic representation of the enhancement of a Schiff-base reaction between an aldehyde and an amine by a highly charged metal center.

because it ties up the water that is formed and it solubilizes the growing polymer chain at the same time. Even though the Schiff-base ligand is virtually insoluble in all common organic and inorganic solvents, the zirconium polymer is orders of magnitude more soluble in either DMSO or NMP.

Other zirconium and cerium Schiff-base polymers have been prepared in DMSO at elevated temperatures in an analogous manner (12):

> These syntheses are repeated at stoichiometric ratios from 0.98:1.00 to 1.02:1.00 in 0.01 increments and at 5 °C temperature intervals to find the best conditions. (Silanized glassware is necessary for the zirconium species because the zirconium polymers condense with the hydroxyl groups on silica in DMSO and become firmly attached to the glassware. Silanization is accomplished by coating the glassware with a chloromethyl silane.) The best temperature for the polymerization reactions with these metal ions appears to be somewhere between 55 and 70 °C and at exactly 1.00:1.00 stoichiometry.
>
> Once reaction rates were ascertained an alternate approach was used. Slightly less than a stoichiometric amount (~97%) of the amine was slowly added to the metal species, and the reaction was allowed to go to completion. A little more amine (~1%) was then added several times but allowing time for completion of the reaction between each addition. This approach proved less satisfactory for these reactions — apparently because of the admission of moisture to the reaction vessel or to the amine.

The best stoichiometric ratio can vary if the purities of the metal complex and the amine are different. The efficiency of these Schiff-base reactions between coordinated salicylaldehyde and aromatic amines is apparently due to the mechanism noted above in the discussion of Schiff-base condensation reactions. However, in the present situation, the interaction between the multicharged ion and the aldehydic oxygen provides an analogous rate enhancement similar to that provided by the proton in the Schiff-base synthesis as shown schematically in Figure 2.17 (cf. Exercise 2.4).

2.1.3 Main Group Step Condensation Polymer Syntheses

A sizable number of main group polymers are best (or only) synthesized via condensation reactions. For example, although scores of polyphosphazenes have been synthesized through ring-opening polymerizations discussed in Section 2.3, polyphosphazenes with direct carbon-phosphorus bonds, whether poly(alkylphosphazenes) or poly(alkyl/arylphosphazenes), have normally been synthesized by condensation reactions, most of which appear to be chain-growth reactions and are discussed in Section 2.2. Novel polymers combining cluster carboranes with a variety of other constituents are also synthesized by condensation reactions as are polyphthalocyanines and the relatively new poly(oxothiazenes). The reductive coupling polysilane polymerizations are also condensation reactions although probably not step-growth polymerizations; these are discussed separately in Section 2.4. Other condensation reactions, such as the condensation of orthophosphate to polyphosphates, should be noted (cf. Section 2.5).

2.1.3.1 Polycarboranes and Polycarbosilanes

The *meta*-carborane oligomer given as an example of connectivity of 2 in Chapter 1 (Fig. 1.3e) is formed by a condensation reaction between the $Li_2(meta\text{-}C_2B_{10}H_{10})$ and phosgene $COCl_2$ (cf. Exercise 2.5). Whereas the $(\text{–}C(O)CB_{10}H_{10}C\text{–})_n$ oligomer consists of $n \leq 5$, longer polymeric species are obtained in analogous reactions between $Li_2(meta\text{-}C_2B_{10}H_{10})$ and $(CH_3)_2SnCl_2$ or $(CH_3)_2GeCl_2$.

Carborane polymers with siloxane bridges have been obtained through a multi-step process involving condensation and using two moles of $(CH_3)_2SiCl_2$ per mole of the same carborane as shown in Eqs. 2.6–2.8 below. The *meta* designation has been dropped for simplicity, and the carbons have been separated to show the connectivities better.

$$\begin{aligned}&LiCB_{10}H_{10}CLi + 2(CH_3)_2SiCl_2 \longrightarrow\\&\qquad ClSi(CH_3)_2\text{–}CB_{10}H_{10}C\text{–}Si(CH_3)_2Cl + 2LiCl\end{aligned}\tag{2.6}$$

$$\begin{aligned}&ClSi(CH_3)_2\text{–}CB_{10}H_{10}C\text{–}Si(CH_3)_2Cl + 2CH_3OH \longrightarrow\\&\qquad CH_3OSi(CH_3)_2\text{–}CB_{10}H_{10}C\text{–}Si(CH_3)_2OCH_3 + 2HCl\end{aligned}\tag{2.7}$$

$$\begin{aligned}&CH_3O\text{–}Si(CH_3)_2\text{–}CB_{10}H_{10}C\text{–}Si(CH_3)_2\text{–}OCH_3\\&\qquad + ClSi(CH_3)_2\text{–}CB_{10}H_{10}C\text{–}Si(CH_3)_2Cl \longrightarrow\\&[\text{–}O\text{–}Si(CH_3)_2\text{–}CB_{10}H_{10}C\text{–}Si(CH_3)_2\text{–}O\text{–}Si(CH_3)_2\\&\qquad \text{–}CB_{10}H_{10}C\text{–}Si(CH_3)_2\text{–}]_n + 2CH_3Cl\end{aligned}\tag{2.8}$$

A partially crystalline elastomer can be prepared from a condensation reaction between dichlorodimethylsilane and the dimethoxy derivative synthesized in Eq. 2.7. This provides an $\text{–}Si(CH_3)_2\text{–}O\text{–}Si(CH_3)_2\text{–}O\text{–}Si(CH_3)_2\text{–}$link between each carborane. *Dexsil* is the generic name given to these carborane/siloxane copolymers. Copolymers with longer siloxane links are also known.

Condensation of Cl_3SiCH_2Cl with a Grignard-type reaction (Mg, diethyl ether) followed by reduction with lithium aluminum hydride produces the silaethylene carbosilane "$[SiH_2CH_2]_n$," which is not a simple linear polymer but instead is a hyperbranched polymeric suitable for pyrolysis to silicon carbide in high yield (25, 26) (cf. Exercise 2.6). In fact, the lack of a simple linear carbosilane structure improves the pyrolysis yield. Although RMgBr Grignard reagents were used in attempts to modify the synthesis of alkyl derivatives, lithium reagents (LiR) were found necessary for the substitution of larger substituents (26).

2.1.3.2 Polyphthalocyanines

The condensation reaction that converts dihydroxo(phthalocyanine)silicon to the polymeric siloxane (Fig. 1.14c) is also an example of a one-component condensation polymerization that can grow to longer chain lengths without a stoichiometry problem. The only limitations are the extent of reaction and the mobility of the molecules to tumble to an appropriate position for reaction to occur. An average chain length of greater than 100 repeating units is known for this siloxane polymer. This means that the extent of reaction is greater than 99% under the conditions of the polymerization. Germanium and tin analogs have been synthesized with an average of 70 and 100 units, respectively (27).

2.1.3.3 Polysiloxanes

Classically the polysiloxanes (commonly called silicones) were also prepared by condensation reactions. In fact, Friedel and Crafts (28) reported the condensation of $Si(C_2H_5)_2(OR)_2$, where R = alkyl, with water producing $[(C_2H_5)_2SiO]_n$ and the alcohol in 1866:

$$n\mathrm{Si(C_2H_5)_2(OR)_2} + n\mathrm{H_2O} \longrightarrow [(\mathrm{C_2H_5})_2\mathrm{SiO}]_n + 2n\mathrm{ROH} \qquad (2.9)$$

The condensation of dichlorodiethylsilane with water to produce the same polymer was reported by Ladenburg (29) 6 years later. At present, polysiloxane polymers are usually prepared by ring-opening polymerizations (Section 2.3) of small cyclic oligomers that have been prepared by hydrolysis (a condensation reaction) of the appropriate dihalodialkylsilane. But even today some special siloxanes use condensation reactions for the polymerization step.

For example, silphenylene-siloxane polymers can be prepared by a low-temperature step condensation polymerization of a bis(ureido)silane with 1,4-bis(hydroxydimethylsilyl)benzene at −20 °C as shown in Figure 2.18. Eight different polymers were prepared by the following procedure that maximizes the molecular mass of the polymer product:

> The bis(hydroxydimethylsilyl)benzene component was dissolved in dry chlorobenzene and cooled to −20 °C. The bis(ureido)silane, similarly dissolved and cooled, was added dropwise until about 96% of the stoichiometric amount had been added. Small amounts (1%) were subsequently added until the molecular mass reached a maximum [as monitored by gel permeation chromatography (GPC)*].

* See Section 3.2.1 for information on this technique.

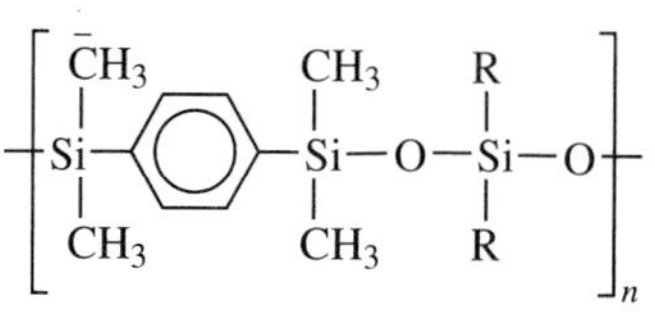

I

	R	R′
Ia	CH_3	$CH_2CH_2CH_3$
Ib	CH_3	$CH_2CH{=}CH_2$
Ic	CH_3	CH_2CH_2CN
Id	CH_3	$CH_2CH_2CH_2CN$
Ie	CH_3	$CH_2CH_2CF_3$
If	C_6H_5	$CH{=}CH_2$
Ig	C_6H_5	$CH_2CH{=}CH_2$
Ih	$CH_2CH_2CH_2CN$	$CH_2CH_2CH_2CN$

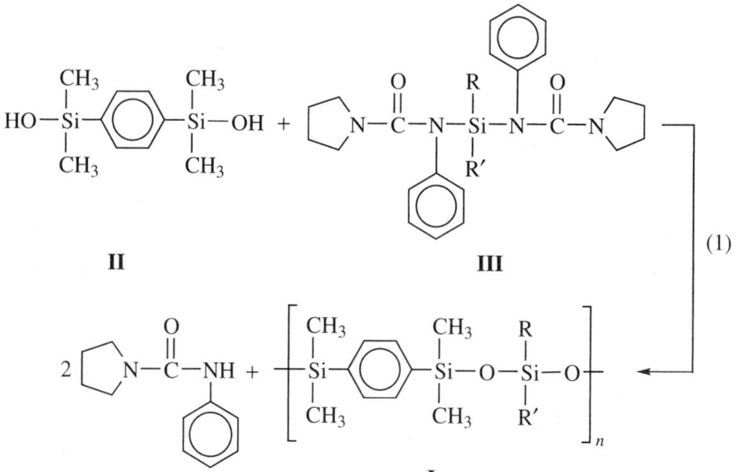

II + III → (1) 2 (pyrrolidinyl-C(=O)-NH-phenyl) + I

Figure 2.18 The synthesis of silphenylene-siloxane polymers (reprinted with permission from Ref. 30; © 1989 American Chemical Society).

The bis(ureido)silanes were prepared from bis(pyrrolidinyl)silanes that had been prepared from pyrrolidine and the appropriate $RR'SiCl_2$ species (30).

Also, poly(dimethylsiloxane)-polyamide multiblock copolymers are synthesized by condensation reactions (31), and the condensation of trichlorophenylsilane

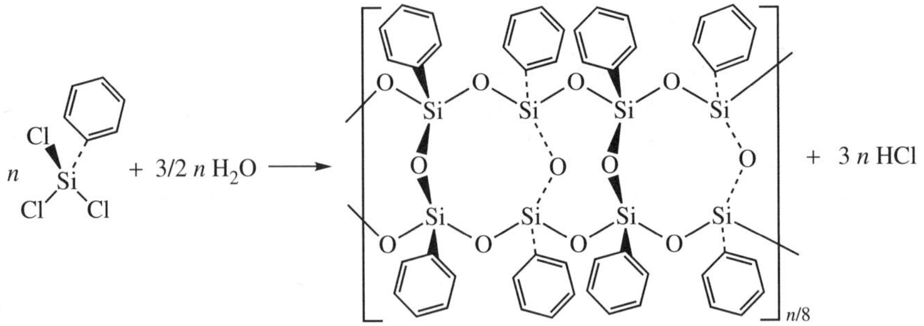

Figure 2.19 Reaction for producing a sesquisiloxane polymer.

with water is thought to produce the ladder-type sesquisiloxane polymer shown in Figure 2.19 (32). Hybrid organic-inorganic composites based on polysiloxane condensation with silica fillers provide elastomer-filler composite structures of industrial interest (33).

A simple laboratory experiment making the poly(dimethylsiloxane) or "silly putty" involves the hydrolysis of dichlorodimethylsilane at elevated temperatures followed by a thermal borate crosslinking at 180 °C (34). Both steps are condensation reactions. The resulting polymer can be characterized using methods noted in Chapter 3 as well as by the qualitative characterizations noted in the laboratory manual.

2.1.3.4 Other Group 14 Step Polymerizations

A number of germanium(IV) and tin(IV) polymers with $-O-MR_2-O-R'-$ backbones, where $M = Ge$ or Sn and R and R' are a variety of organic groups, have been synthesized by step-growth synthetic methods analogous to the titanocene(IV) synthesis shown in Figure 2.11. Carraher and Scherubel (35) synthesized diphenyltin derivatives with 2-butenediolato and 1,3-propanediolato bridges with DP values of 78 and 1400, respectively, with viscosity measurements that should be reasonably accurate based on a calibration method using freezing point depressions for lower-molecular-mass species ($\overline{M} < 10^4$) and extrapolated to the higher polymers. However, other polymerizations resulted in solubilities too low to obtain reliable mass estimates. Similar inorganic-organic hybrid polymers have been prepared from germanium(II) and tin(II) precursors with oxidative addition procedures noted below (Section 2.4.2).

2.1.3.5 Polyoxothiazenes

Although polymers with an alternating $-N{=}S^{VI}(O)R-$backbone, where R is F or NH_2 or phenyl, were described in the patent literature in the 1960s, only recently have the syntheses of these polymers been systematically investigated (36). Analogous to the polyphosphazene synthesis from $Me_3SiN{=}PR_2X$ (noted in Section 2.2), polyoxothiazene polymers could be synthesized through condensation of $Me_3SiN{=}S(O)ROC_6H_5$ at 120–170 °C for several days. Although adding a Lewis acid or Lewis base catalyst lowers the temperature and time somewhat,

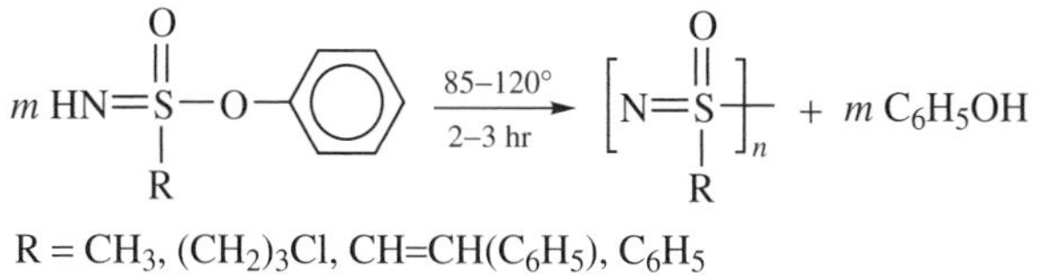

$$m\,\mathrm{HN{=}S(O)(R){-}O{-}C_6H_5} \xrightarrow[2\text{–}3\ \mathrm{hr}]{85\text{–}120^\circ} \mathrm{[N{=}S(O)(R)]_n} + m\,\mathrm{C_6H_5OH}$$

$R = CH_3$, $(CH_2)_3Cl$, $CH{=}CH(C_6H_5)$, C_6H_5

Figure 2.20 The synthesis of poly(alkyl/aryloxothiazenes).

switching to a simpler sulfonimidates, HN=S(O)RX derivatives (where X = OCH_2CF_3 or OC_6H_5), provides polymers at much low temperatures and for much shorter times as shown schematically in Figure 2.20. The reactions are initiated under nitrogen, but reduced pressure is used to remove the alcohol or phenol product and drive the reaction to near-quantitative condensation based on number-average degrees of polymerization of well over 100.

2.1.3.6 Polyborazines

Thermal dehydrogenation of liquid borazine, $B_3N_3H_6$, *in vacuo* at 70 °C provides a soluble polyborazylene estimated to have an average degree of polymerization of over 40. However, the elemental analyses are approximately $B_3N_3H_3$, indicative of some branching or crosslinking because the linear polymer should be $(B_3N_3H_4)_n$. Either is a suitable precursor for boron nitride ceramics. Organosubstituted borazines have also been synthesized in a similar manner.

2.1.4 Step Addition Syntheses

Urethane and urea-type linkages can be formed between metal coordination compounds with diol and diamine functionality and a diisocyanate as shown schematically in Eqs. 2.10 and 2.11.

$$n\mathrm{HOMOH} + n\mathrm{OCNRNCO} \longrightarrow (\mathrm{OMOCONHRNHCO})_n \qquad (2.10)$$

$$n\mathrm{H_2NMNH_2} + n\mathrm{OCNRNCO} \longrightarrow (\mathrm{NHMNHCONHRNHCO})_n \qquad (2.11)$$

where M is a difunctional coordination or organometallic species and R = aryl or a metal-containing species with two isocyanate groups. A listing of organic polymer step addition reactions is provided in Table 2.3.

Intractable beryllium polymers have been synthesized from diisocyanates and beryllium β-diketone derivatives (Fig. 2.21) (37). However, because these are step syntheses, stoichiometric control is essential for obtaining more than oligomers. The intractability may have been the result of hydrolysis at the small beryllium centers. This type of reaction could be exploited more widely than it has been in the past. The limit appears to be the availability of appropriate monomer materials. Step addition reactions are less favorable from entropy considerations than step condensation reactions; thus an enthalpic driving force is necessary for the step addition reactions.

TABLE 2.3 Typical Two Component Step-Growth Addition Polymerizations from a Reference Volume[a].

Reactants	Characteristic Linkage Formed	Type of Polymer
Diisocyanates + diols	$-O_2CNH-$	Polyurethanes
Diisocyanates + diamines	$-NHCONH-$	Polyureas
Diisothiocyanates + diamines	$-NHCSNH-$	Poly(thiourea)s
Diepoxides + diamines	$-CH_2CH(OH)N<$	
Diepoxides + diisocyannates	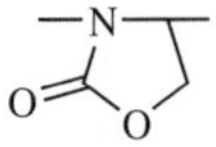 2-oxazolidone ring (–N–, C=O, –O–)	Poly(2-oxazolidone)s
Dithiols + unconjugated dienes	$-CH_2CH_2S-$	Polysulfides
Conjugated bis(diene)s + bis(dienophile)s	Various	
Divinyl sulfones + diols	$-OCH_2CH_2SO_2-$	
Organic dihalides + diamines	$-N^+-X^-$	Poly(ammonium halide)s
Dinitriles + diols	$-CONH-$	Polyamides

[a]Reprinted from Eastmond, G. C., Ledwith, A., Russo, S., Sigwalt, P. in *Comprehensive Polymer Science*; Allen, G. and Bevington, J. C., Ed., Pergamon Press: Oxford, 1989; Vol. 5, p 7 with permission from Elsevier Science.

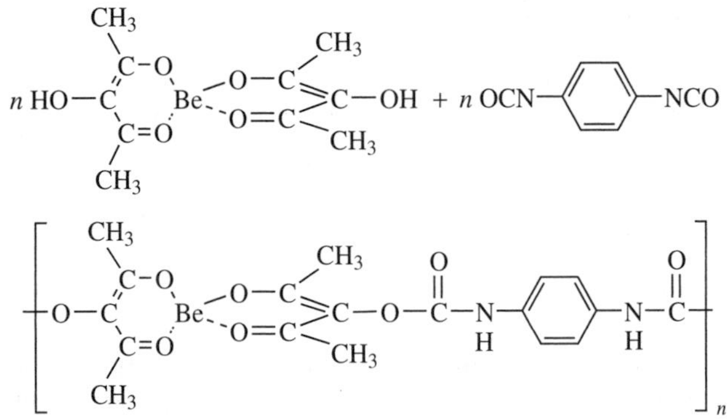

Figure 2.21 The synthesis of beryllium step addition polymers.

2.2 CHAIN POLYMERIZATIONS

Chain polymerizations are often called addition polymerizations, but as noted above, some addition polymerizations are step polymerizations; therefore, the term chain polymerization is preferred. The chain is extended through either a radical or an ionic initiator. The monomers that produce chain polymers must have unsaturated functionality to form the bonds that make up the backbones of the polymer chains. Chain polymerizations have the potential for providing much higher molecular masses than do the step-growth polymerizations because with a single monomer, stoichiometry does not have to be matched. However, chain polymerizations fight entropy, because a large number of molecules are reacting to form one chain. [Step polymerizations have an entropic advantage over chain polymerizations because a small molecule is formed each time either monomer in a step condensation reacts with the growing polymer (Fig. 2.2).

The chain polymerization process is normally considered to consist of four reaction types: chain initiation, chain propagation, chain transfer, and chain termination. Chain initiation can be the result of a free radical initiator or an ionic initiator. The simplest example of a free radical initiator is the thermal decomposition of a peroxide such as benzoyl peroxide (Eq. 2.12):

$$(C_6H_5)C(O)O\text{–}OC(O)(C_6H_5) \longrightarrow 2(C_6H_5)C(O)O\bullet \qquad (2.12)$$

The radical initiator that is formed in Eq. 2.12 combines with the monomer (M) to form a chain initiator as shown in Eq. 2.13:

$$(C_6H_5)C(O)O\bullet + M \longrightarrow (C_6H_5)C(O)O\text{–}M\bullet \qquad (2.13)$$

A cationic initiator can be a simple protonic acid (HX), such as sulfuric, perchloric, or hydrochloric, where the proton reacts with the monomer (M) (Eqs. 2.14 and 2.15):

$$HX \longrightarrow H^+ + X^- \qquad (2.14)$$

$$H^+ + M \longrightarrow HM^+ \qquad (2.15)$$

An anionic initiator can be as simple as *n*-butyllithium. The butyl anion then reacts with the monomer (M) (Eqs. 2.16 and 2.17):

$$n\text{-}C_4H_9Li \longrightarrow n\text{-}C_4H_9^- + Li^+ \qquad (2.16)$$

$$n\text{-}C_4H_9^- + M \longrightarrow n\text{-}C_4H_9M^- \qquad (2.17)$$

These chain initiation steps are followed by hundreds or thousands of chain propagation steps per initiator until the supply of monomer is exhausted or the initiator is terminated or transferred from the chain. These reactions can be simplified as shown in Eqs. 2.18–2.21, where I = initiator (regardless of whether the initiator is a radical, an anion, or a cation) and M = monomer.

$$I + M \longrightarrow IM \qquad (2.18)$$

$$IM + M \longrightarrow IM_2 \qquad (2.19)$$

$$IM_2 + M \longrightarrow IM_3 \qquad (2.20)$$

$$\bullet\bullet\bullet$$

$$IM_{n-1} + M \longrightarrow IM_n \qquad (2.21)$$

The n-mer (polymer with n repeating units) eventually either transfers its active initiator group to another monomer (as shown in Eq. 2.22), or, in the case of radical initiators, it can react with another radical chain to either combine (as in Eq. 2.23) or simply be deactivated (Eq. 2.24).

$$IM_n + M \longrightarrow M_n + IM \quad \text{(chain transfer)} \tag{2.22}$$

$$IM_n + IM_m \longrightarrow M_{n+m} \quad \text{(coupled termination)} \tag{2.23}$$

$$IM_n + IM_m \longrightarrow M_n + M_m \quad \text{(deactivated termination)} \tag{2.24}$$

The fate of the initiator is not indicated in the termination step equations shown, although the initiators may remain on the polymer in an inactive form.

Thousands of metal-containing polymers have been synthesized by chain polymerizations. Virtually all of these polymers are anchored metal polymers (Type III) that have vinyl precursors (cf. Fig. 1.19).

2.2.1 Radical Polymerizations

Radical initiation is most often started by homolytic cleavage of a bond in a suitable molecule, that is, one with a weak sigma bond such as a peroxide (Eq. 2.25) or a nitrogen precursor such as an azo compound (Eq. 2.26).

$$\text{RO–OR} \longrightarrow 2\text{RO}\bullet \tag{2.25}$$

$$\text{R–N=N–R} \longrightarrow 2\text{R} + N_2 \tag{2.26}$$

Table 2.4 provides a listing of suitable radical initiators along with useable temperatures for those obtained by thermal cleavage of a weak bond. Peroxides

TABLE 2.4 Free Radical Initiators.

1.	Organic peroxides or hydroperoxides		
	Dibenzoyl peroxide	PhC(O)OOC(O)Ph	useful from 330 to 350 K
	Di-t-butyl peroxide	t-C_4H_9OO-t-C_4H_9	useful from 377 to 395 K
	Diethyl peroxide	C_2H_5OO-t-C_2H_5	useful from 382 to 402 K
2.	Azo compounds, R–N=N–R		
	2,2′-azobisisobutronitrile (AIBN) R = $(CH_3)_2C(CN)$		useful from 313 to 333 K
	R = Ph_2CH		useful from 293 to 308 K
	R = $PhCH(CH_3)$		useful from 378 to 398 K
	R = $(CH_3)_2CH$		useful from 453 to 473 K
3.	Redox agents (e.g., persulfates plus reducing agents)		
	$S_2O_8^{2-} + HSO_3^- \longrightarrow SO_4^{2-} + SO_4^-\bullet + HSO_3\bullet$		
	$S_2O_8^{2-} + S_2O_3^{2-} \longrightarrow SO_4^{2-} + SO_4^-\bullet S_2O_3^-\bullet$		
4.	Organometallic reagents (Ag alkyls; e.g., ethyl useful from 215 to 255 K)		
5.	Heat or light, including ultraviolet and high-energy radiation		
6.	Electrolytic electron transfer		

or hydroperoxides can also be a component of a redox-induced radical reagent. Benzoyl peroxide reacts with tertiary amines to give benzoyl-tertiary amine and benzoyl radicals (Eq. 2.27) at lower temperatures than the corresponding thermal scission reaction.

Caution: peroxides such as benzoyl peroxide are potentially explosives, especially if slightly impure. Also, metals with two stable oxidation states tend to decompose peroxides.

$$PhC(O)O\text{–}OC(O)Ph + R_3N \longrightarrow R_3NOC(O)Ph + PhC(O)O\bullet \quad (2.27)$$

For a purely inorganic radical reaction, the addition of iron(II) to hydrogen peroxide gives iron(III) plus the hydroxide ion and the hydroxy radical (Eq. 2.28).

$$H_2O_2 + Fe^{2+} \longrightarrow OH^- + \bullet OH + Fe^{3+} \quad (2.28)$$

Two other redox reactions that produce radicals are shown in Table 2.4 (#3). Whereas silver alkyls easily produce radicals (Eq. 2.29), their useful temperature range is well below room temperature. Note that this is a spontaneous redox reaction containing only one reactant. Because the alkyls decompose well below 0 °C, they must be prepared at even lower temperatures to avoid spontaneous decomposition before isolation and storage.

$$Ag(C_2H_5) \longrightarrow Ag + C_2H_5\bullet \quad (2.29)$$

Carraher and Pittman (104) note that a wide variety of vinyl cyclopentadienyl derivatives have been polymerized by radical initiation. They yield Type III metal polymers. The azo catalyst AIBN serves as a good polymerization catalyst for tricarbonyl(vinylcyclopentadienyl)manganese(I) and for ferrocenyl ethyl acrylate and the corresponding methyl acrylate. The reactions are first order in the monomer and half order in the initiator, consistent with the thermal splitting of AIBN into two radicals plus nitrogen, with the radicals subsequently propagating the polymerization. A schematic chain reaction sequence is shown in Figure 2.22. On the other hand, vinylferrocene does not polymerize well with AIBN because of electron transfer termination as shown in Figure 2.23. Also, tricarbonylmethyl(vinylcyclopentadienyl)tungsten(II) does not polymerize well using AIBN.

2.2.1.1 Photopolymerization

Light can also be used to produce radicals for chain polymerizations. For example, ultraviolet light severs bonds of nonmetallic species; for example, polysilanes react with ultraviolet light to produce silyl radicals (Eq. 2.30) that can add to carbon-carbon double bonds and thus become photoinitiators. Although the

Chain reaction for metal-containing monomer*

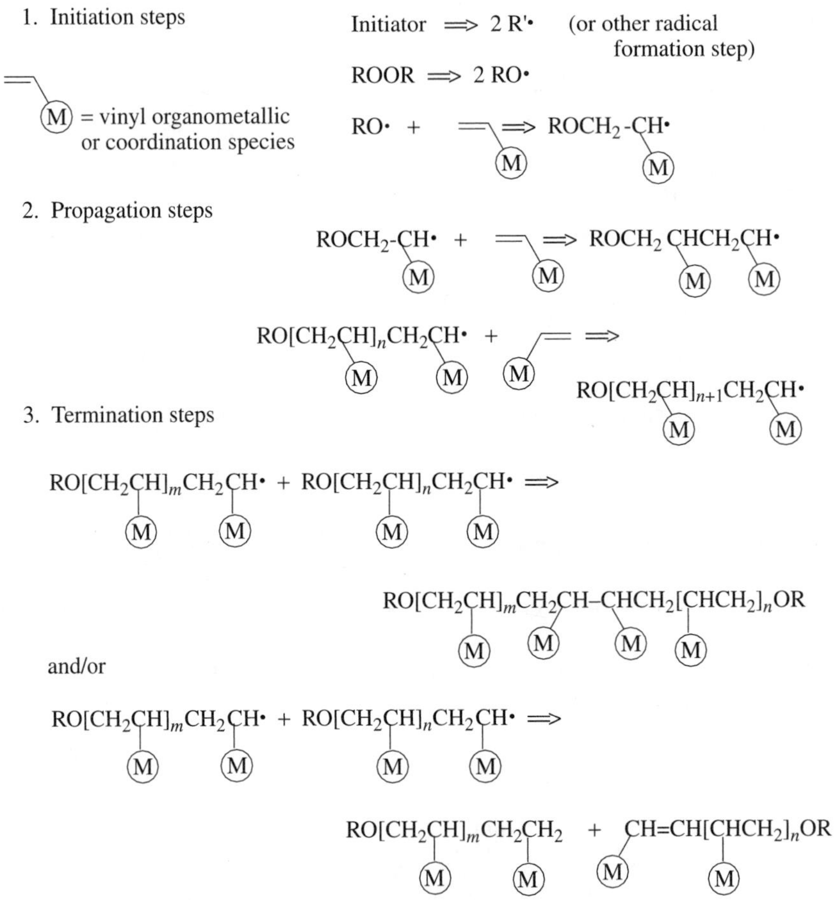

1. Initiation steps

Initiator ⟹ 2 R'• (or other radical formation step)

ROOR ⟹ 2 RO•

(M) = vinyl organometallic or coordination species

RO• + $CH_2{=}CH(M)$ ⟹ $ROCH_2\text{-}CH(M)•$

2. Propagation steps

$ROCH_2\text{-}CH(M)•$ + $CH_2{=}CH(M)$ ⟹ $ROCH_2CH(M)CH_2CH(M)•$

$RO[CH_2CH(M)]_nCH_2CH(M)•$ + $CH_2{=}CH(M)$ ⟹ $RO[CH_2CH(M)]_{n+1}CH_2CH(M)•$

3. Termination steps

$RO[CH_2CH(M)]_mCH_2CH(M)•$ + $RO[CH_2CH(M)]_nCH_2CH(M)•$ ⟹

$RO[CH_2CH(M)]_mCH_2CH(M)\text{–}CH(M)CH_2[CH(M)CH_2]_nOR$

and/or

$RO[CH_2CH(M)]_mCH_2CH(M)•$ + $RO[CH_2CH(M)]_nCH_2CH(M)•$ ⟹

$RO[CH_2CH(M)]_mCH_2CH_2(M)$ + $(M)CH{=}CH[CH(M)CH_2]_nOR$

*Radical polymerization shown, but ionic similar.
Chain transfer and other complications not shown.

Figure 2.22 Schematic chain reaction for a vinyl organometallic or coordination compound.

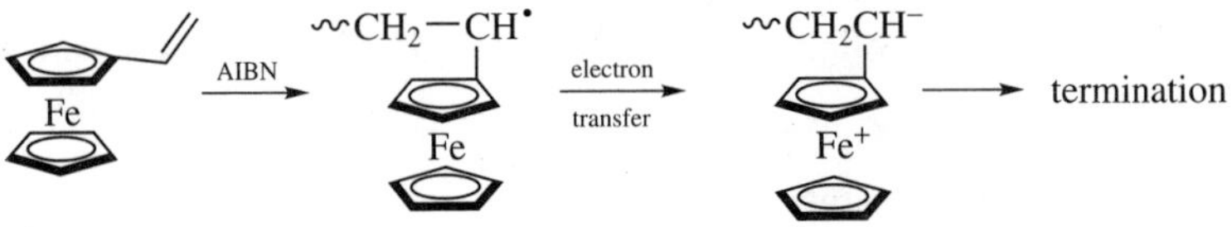

Figure 2.23 The electron-transfer reaction that inhibits ferrocene polymerization with AIBN and other radical initiators.

polysilanes are not very efficient as photoinitiators, they are less susceptible to oxygen inhibition than most other photoinitiators.

$$M\,[\mathrm{Si(CH_3)(C_6H_5)}]_n + \mathrm{h}\nu(254\ \mathrm{nm}) \longrightarrow [\mathrm{Si(CH_3)(C_6H_5)}]_m\bullet + [\mathrm{Si(CH_3)(C_6H_5)}]_{n-m}\bullet \quad (2.30)$$

Transition metal coordination compounds of metals with two or more oxidation states have proven particularly versatile in photopolymerizations of unsaturated organic compounds (38) and should be useful for polymerizing vinyl derivatives of both nonmetal and metal inorganic species. To illustrate, a photoactivated platinum-catalyzed hydrosilylation polymerization of vinyldimethylsilane (Fig. 2.24) has been shown to remain catalytically active for an extended time period. The vinyl and hydride end-groups continue to link for several months, with the weight-average molecular mass ($\overline{M}_W$ increasing from 5500 to 12,300 during that time (39). A ring-opening carbosilane photopolymerization will be discussed below (Section 2.3).

Mesogenic metal Schiff-base derivatives with monoacrylate end groups (Fig. 2.24) can be polymerized photochemically in a melt blend with 3% of a fluorinated diaryltitanocene photoinitiator* that generates radicals for the polymerization (40). The insoluble nature of the products means that the polymerization results were based on optical properties and the evolution of heat

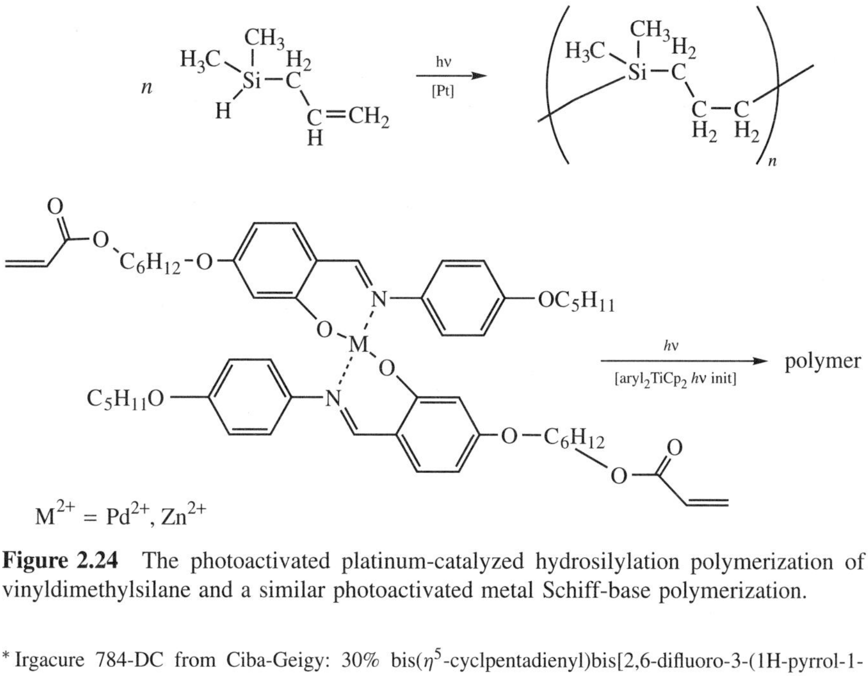

$M^{2+} = Pd^{2+}, Zn^{2+}$

Figure 2.24 The photoactivated platinum-catalyzed hydrosilylation polymerization of vinyldimethylsilane and a similar photoactivated metal Schiff-base polymerization.

* Irgacure 784-DC from Ciba-Geigy: 30% bis(η^5-cyclpentadienyl)bis[2,6-difluoro-3-(1H-pyrrol-1-yl)phenyl]titanium with 70% Dicalite).

relative to polymerizations of analogous organic species. In addition to successful results with palladium and zinc, the copper(II) derivative appeared unreactive and the vanadyl(IV) (VO^{2+}) derivative appeared unstable in solution. (Solutions are used to prepare the blends.) Photochemical polymerization in toluene solution was unsuccessful for all of the metal-containing Schiff-bases.

Classically insoluble polymeric metal-containing polymers were also obtained through photoactivation. The blue of blueprint paper is the polymeric iron(II)/iron(III) cyano polymer called Prussian blue (cf. Fig. 1.16).

2.2.1.2 Electrochemical Polymerization

Finally, electrochemically induced electron transfer can be used in place of a redox reagent or light to meet the same end; that is, to form free radicals capable of inducing polymerization. The electropolymerization of copper(II) and nickel(II) complexes of the tetradentate Schiff-base prepared from 3,4-diaminothiophene and two molecules of salicylaldehyde is a recent example (Fig. 2.25). The reaction is thought to involve oxidative polymerization of radicals formed electrochemically followed by electrochemical reduction to stabilize the product (41, 42). These are similar to earlier reported electro-polymerizations of metal salicylaldimine chelates (43, 44). Polymerization of nickel, copper, and zinc chelates have all been reported, and oxidation leads to an unstable state followed by polymerization with rapid electron transfer. Both sets of experiments have also shown that the polymerization occurs through the para position of the phenolic ring of the salicylaldimine ligand. These products were all intractable.

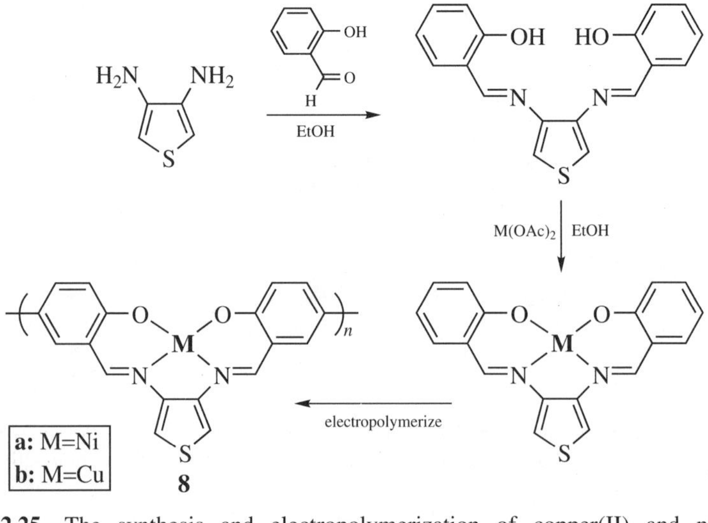

Figure 2.25 The synthesis and electropolymerization of copper(II) and nickel(II) 3,4-bis(salicylaldimino)thiophene Schiff-base chelates (41).

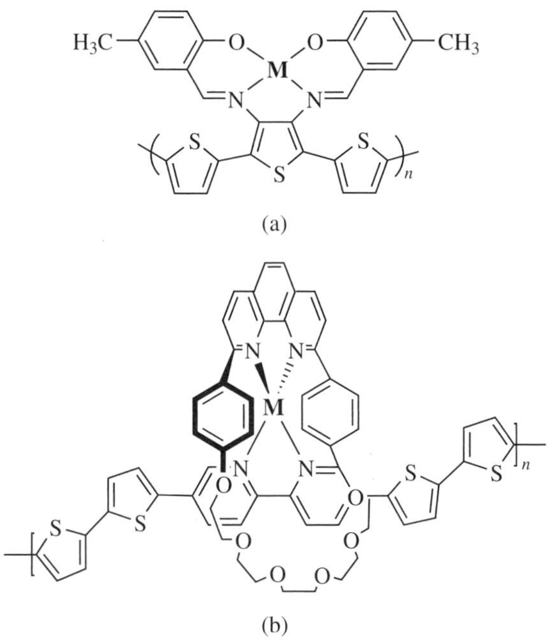

(a)

(b)

Figure 2.26 Examples of metal-containing polymers synthesized electrochemically (42, 45).

Type III metal-containing polymers have also been prepared electrochemically. The 2- position in the thiophene ring provides the polymerization site (41, 42, 45). Polymerization products obtained in this manner are shown in Figure 2.26. Many more examples could be provided; for example, a large number of metal polypyridyl complexes were obtained electrochemically during the 1980s (46), as were several metallated polyporphyrins, polycyclams, polyferrocene, etc (47).

2.2.2 Cationic Polymerizations

Chain polymerizations of unsaturated compounds can often be catalyzed by cationic initiators, either strong protonic acids such as sulfuric acid, perchloric acid, or hydrochloric acid or Lewis acids such as BF_3, $BF_3{\bullet}O(C_2H_5)_2$, BCl_3, $TiCl_4$, Al_2Cl_6, or $SnCl_4$. Water or methanol is a cocatalyst (equimolar or less) with these Lewis acid catalysts, at least for vinyl monomers. Cationic catalysis is favored for monomers with electron-rich or electron-donating groups. The actual catalyst is a proton either from a strong acid (such as H_2SO_4, $HClO_4$, or HCl) or from a Lewis acid plus a cocatalyst of methanol or water at a concentration that is lower than the Lewis acid catalyst. That is, if a catalyst like boron trifluoride etherate is used as the catalyst, the cocatalyst (e.g., methanol) reacts with the catalyst to produce the proton ion pair (Eq. 2.31) needed for the cationic catalysis. The protonated ether is a very strong proton donor. The analogous reaction

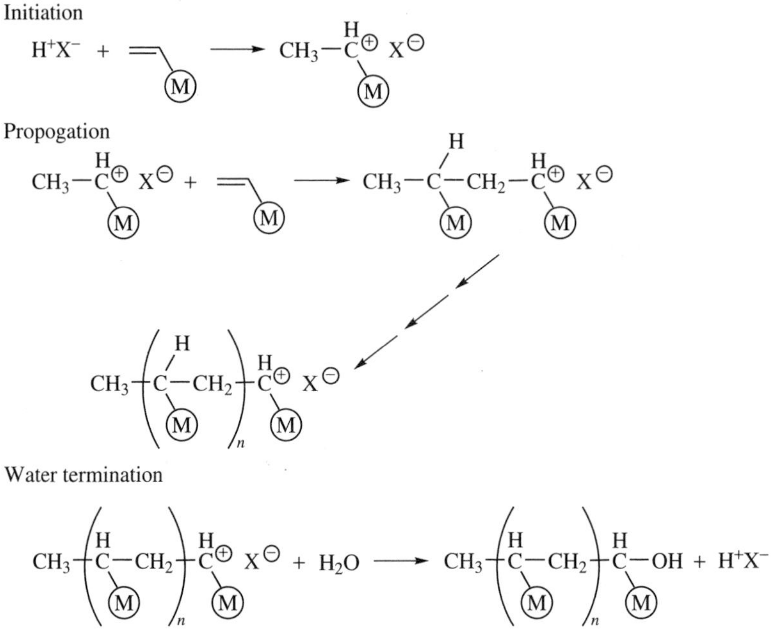

Initiation

$$H^+X^- + CH_2{=}CH(M) \longrightarrow CH_3{-}\overset{H}{C}^{\oplus}(M)\; X^{\ominus}$$

Propogation

$$CH_3{-}\overset{H}{C}^{\oplus}(M)\; X^{\ominus} + CH_2{=}CH(M) \longrightarrow CH_3{-}\overset{H}{C}(M){-}CH_2{-}\overset{H}{C}^{\oplus}(M)\; X^{\ominus}$$

$$\longrightarrow CH_3{-}\left(\overset{H}{C}(M){-}CH_2\right)_n{-}\overset{H}{C}^{\oplus}(M)\; X^{\ominus}$$

Water termination

$$CH_3{-}\left(\overset{H}{C}(M){-}CH_2\right)_n{-}\overset{H}{C}^{\oplus}(M)\; X^{\ominus} + H_2O \longrightarrow CH_3{-}\left(\overset{H}{C}(M){-}CH_2\right)_n{-}\overset{H}{C}(M){-}OH + H^+X^-$$

Figure 2.27 Cationic initiated polymerization—Lewis acid to proton donor transformation discussed in text. Chain transfer not shown.

for a protonic acid in a nonaqueous medium like ether would be very similar, as shown in Eq. 2.32.

$$(C_2H_5)_2O : BF_3 + CH_3OH \longrightarrow (C_2H_5)_2O : H^+(CH_3OBF_3)^- \quad (2.31)$$

$$HCl + (C_2H_5)_2O \longrightarrow (C_2H_5)_2O : H^+Cl^- \quad (2.32)$$

The cationic initiation is thought to involve the addition of the ion pair across the double bond. A carbonium ion is produced, with the proton going to the more electron-rich carbon. The reaction sequence that follows is shown schematically in Figure 2.27, where $H^+ X^-$ is an ion pair, such as those formed in Eqs. 2.31 and 2.32. Addition of monomers to the carbocation is rapid, even at low temperatures.

Theoretically, one chain should be produced per ion pair; however, chain transfer of a proton to another monomer can occur, especially as the temperature is increased. The transferred proton then catalyzes a second chain. The result is a lower molecular mass per chain. Termination can also occur because of traces of water (cf. Fig. 2.27) or, eventually, by total consumption of the monomer.

2.2.2.1 Vinyl Organometallic Polymers

Vinyl ferrocene and other species with electron-rich rings undergo cationic addition polymerizations quite effectively. With boron trifluoride etherate as a catalyst, divinyl ferrocene polymerizes to a product with an average

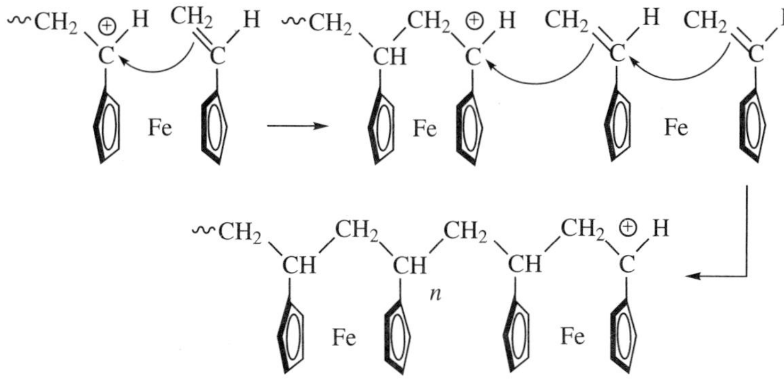

Figure 2.28 The cationic initiated polymerization of divinylferrocene.

molecular mass of 35,000 (see Fig. 2.28). On the other hand, replacement of the $-CH{=}CH_2$ vinyl group of vinylferrocene with a $-C(CF_3){=}CH_2$ group provides a very sluggish polymerization center. Also, cationic initiation of tricarbonylmethyl(vinylcyclopentadienyl)molybdenum(II) is not very effective.

2.2.2.2 Polyphosphazene Synthesis at or near Ambient Temperatures

2.2.2.2.1 Poly(dichlorophosphazene) (48, 49)

The phosphoranimine $Cl_3P{=}NSiMe_3$ reacts with a trace of PCl_5 at room temperature in methylene chloride to produce a polymer with a $\overline{M}_N = 40,000$ and a polymer dispersion index of 1.18 in 4 h. On the basis of monomer studies with PCl_5, the mechanism is thought to involve a cation-induced mechanism, as shown in Eqs. 2.33 and 2.34, where Eq. 2.33 is the initiation step and Eq. 2.34 is the first propagation step.

$$Cl_3P{=}NSiMe_3 + PCl_5(\text{in}CH_2Cl_2) \longrightarrow \{Cl_3P = NPCl_3{}^+\}PCl_6{}^- + Me_3SiCl \qquad (2.33)$$

$$\{Cl_3P{=}NPCl_3{}^+\}PCl_6{}^- + nCl_3P{=}NSiMe_3 \longrightarrow \{Cl_3P{=}N - [PCl_2 = N]_nPCl_3{}^+\}PCl_6{}^- \qquad (2.34)$$

$SbCl_5$ also catalyzes this polymerization, although most other Lewis acids give sluggish reactions at room temperature and broad polymer dispersion indices in boiling CH_2Cl_2.

2.2.2.2.2 Poly(organophosphazenes)

Organic substituted monomers such as $Ph_2ClP{=}NSiMe_3$ and $(CH_3)(C_2H_5)ClP{=}NSiMe_3$ can also be polymerized in a similar fashion near room temperature with a PCl_5 catalyst to form the $[N{=}PPh_2]_n$ and $[N{=}P(CH_3)(C_2H_5)]_n$ polymers, respectively (50). Other species such as $[N{=}PFPh]_n$ have also been prepared by this approach (51). Again, a cation-induced reaction appears to be involved.

$$\mathrm{Ph_2ClP{=}NSiMe_3 + PCl_5(in\ CH_2Cl_2) \longrightarrow \{Ph_2ClP{=}NPCl_3{}^{+}\}PCl_6{}^{-} + Me_3SiCl} \quad (2.35)$$

$$\mathrm{\{Ph_2ClP{=}NPCl_3^{+}\}PCl_6{}^{-}} + n\mathrm{Ph_2ClP{=}NSiMe_3 \longrightarrow \{Ph_2ClP{=}N\text{–}[PCl_2{=}N]_{\mathit{n}}PCl_3^{+}\}PCl_6{}^{-}} \quad (2.36)$$

The large molecular masses obtained by many of these catalyzed reactions are indicative of chain polymerization reactions. These "living" cationic polymerizations can be quenched to control their size. An example of quenching that leads to telechelic di- and triblock polymers is given at the end of this chapter (Section 2.6.2).

2.2.3 Anionic Polymerizations

A variety of anionic initiators have been used in organic reactions and in vinyl inorganic reactions. They include:

Alkali metals in tetrahydrofuran (THF) or liquid ammonia
Alkyl or aryl lithium reagents
Grignard reagents
Aluminum alkyls
Organic radical anions

Monomers with electron-withdrawing groups work best with anionic initiators because the electron-withdrawing groups leave a partial positive charge on at least one carbon atom, and that stabilizes the anion-substrate initiator. For example, see Figure 2.29 for an example of activation by a lithium alkyl.

2.2.3.1 Organometallics

Monomers that are electron rich, such as vinylferrocene, do not polymerize under anionic conditions. However, ferrocenylmethyl acrylate and the methacrylate analogue do polymerize under anionic conditions (Fig. 2.30). The electron-withdrawing oxygen atoms on the acrylate or methacrylate are more influential on the double bond than is the electron-rich ferrocenyl group. Other examples can

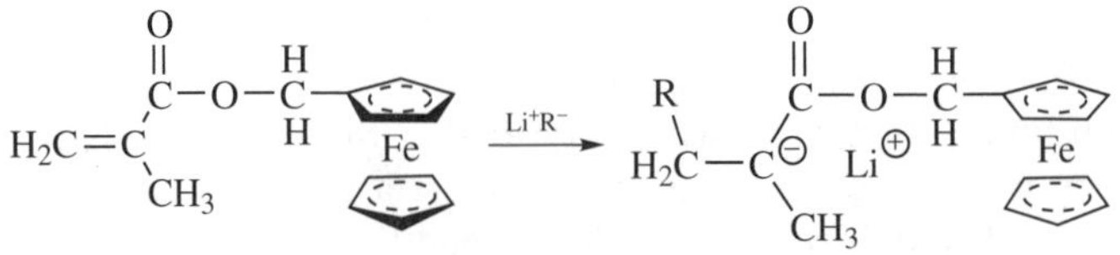

$$\mathrm{H_2C{=}C(CH_3)\text{–}C({=}O)\text{–}O\text{–}CH_2\text{–}Fc \xrightarrow{Li^+R^-} R\text{–}H_2C\text{–}C^{\ominus}(CH_3)\text{–}C({=}O)\text{–}O\text{–}CH_2\text{–}Fc\ \ Li^{\oplus}}$$

Figure 2.29 Lithium alkyl activation of a schematic polymerization reaction.

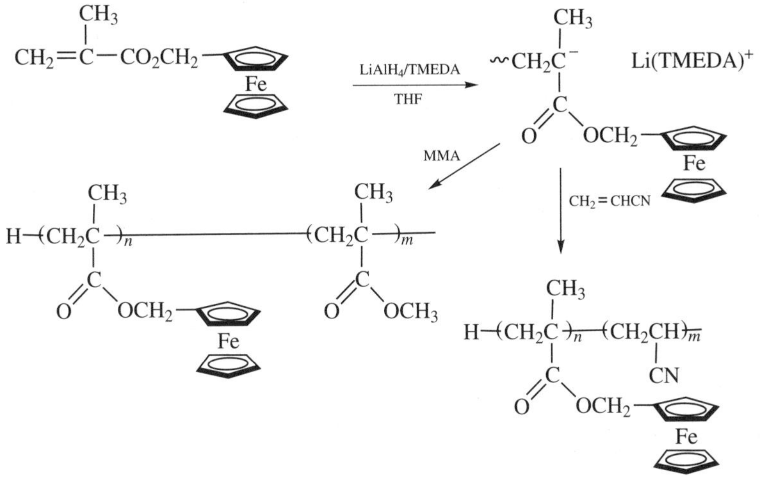

Figure 2.30 Anionic initiation of ferrocenylmethyl methacrylate with lithium aluminum hydride and use to prepare block copolymers with methyl methacrylate (MMA) and acrylonitrile.

be found in detailed reviews. Anionic catalysis of ring-opening polymerizations of silaferrocenes and related organometallics are noted in Section 2.3.

2.2.3.2 Polyphosphazenes

Alkyl and aryl polyphosphazenes cannot be synthesized through the normal ring-opening and substitution approach noted for polyphosphazenes in Section 2.3. However, *N*-silylphosphoranimines, $Me_3SiN{=}PR_2X$, thermally condense to provide poly(dialkylphosphazenes) and poly(alkyl/arylphosphazenes) with the appropriate silylphosphoranimine (52, 53). The reaction is shown in Eq. 2.37, where R is an alkyl group, R′ is an alkyl or aryl group, and X is OCH_2CF_3 or OC_6H_5:

$$n\mathrm{Me_3SiN{=}PRR'X} \longrightarrow \mathrm{Me_3Si[N{=}PR_2]_nX} + (n-1)\mathrm{Me_3SiX} \tag{2.37}$$

Unfortunately, the reactions require high temperatures and long reaction times. Nucleophilic catalysts (e.g., F^-) lowers the temperature somewhat (usually to 150 °C for one day vs. 180–200 °C for several days without a catalyst). The molecular weight distributions indicate that this is a chain reaction rather than a step-growth reaction, even though a trimethylsilyl derivative condenses during the reaction. The lower temperature and time requirements in the presence of nucleophiles, particularly anionic nucleophiles, suggest an anion-induced reaction.

2.3 RING-OPENING POLYMERIZATIONS

Historically, the first inorganic ring-opening polymerization (ROP) belongs to studies with elemental sulfur. The formation of "inorganic rubber" by ring opening of the six-membered dichlorophosphazene trimer has also been well known for at least a century. Whereas both sulfur and the phosphazene were studied during the nineteenth century, the first metal-containing monomers shown to polymerize through ring opening appear to be beryllium polymers first reported at the beginning of the 1960s. However, tractable organometallic polymers through ring-opening polymerizations were unknown before 1992. Rather than give a historical treatment, however, we will follow the precedent used in the rest of this chapter and give the metallic species first.

2.3.1 Metal-Coordination ROP

Kluiber and Lewis (54) showed that metal-coordination polymers, beryllium polymers in particular, could be synthesized by ROP if the precursor complex forms a ring (Fig. 2.7). The interesting point is that both the monomeric and the dimeric beryllium complexes undergo ROP to the same linear product. The resulting amorphous polymers have enough solubility in benzene to obtain viscosity measurements. The polymer with an octyl $(CH_2)_8$ bridge showed properties consistent with a polymer having a molecular mass of about 10,000 or more. The ring-opening method provides beryllium polymers superior to those obtained by the condensation methodology attempted by others (37). The toxicity of beryllium compounds has limited further research in this area. Thus ROP of metal-coordination polymers is a virtually untapped area of synthetic research.

2.3.2 Organometallic ROP

Brandt and Rauchfuss (55) showed that a disulfide-bridged ferrocene polymer could be formed from a trisulfide-bridged monomeric ferrocene plus a tertiary phosphine. The condensation byproduct is the tertiary phosphine sulfide (see Fig. 2.31). Note that this ROP involves the abstraction of a sulfide from the ring and could be considered a condensation (or atom abstraction) ROP.

Organomonosilyl bridged to the two rings of ferrocene {[1]ferrocenophanes} has sufficient ring strain (60–80 kJ/mol) that thermal ROP at 150–180 °C yields polymers of high molecular mass ($\overline{M}_N > 100,000$) (56) (see Fig. 2.32). For R = CH_3, the product is soluble in aromatic or chlorinated hydrocarbon solvents and has a glass-transition temperature of about 33 °C. For R = C_2H_5 to C_6H_{13} and for R = $C_{18}H_{37}$, the polymers are gummy polymeric products that are soluble

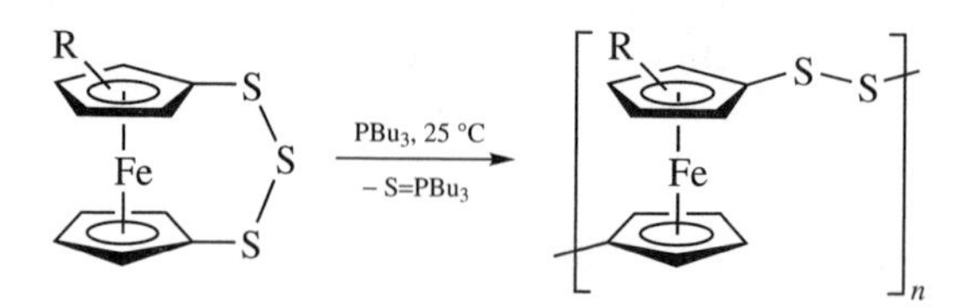

Figure 2.31 The ring-opening polymerization of trisulfide bridged ferrocene catalyzed by tributyl phosphine.

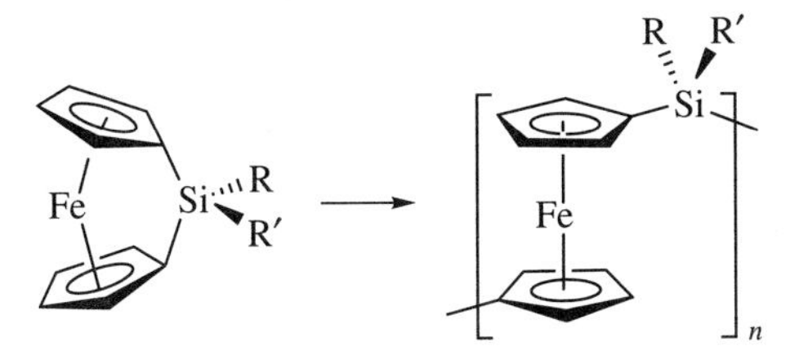

Figure 2.32 The thermal ring-opening polymerization of a silyl-bridged ferrocene.

in hexane. The hexyl derivative has a glass-transition temperature of −26 °C. The polymer produced when R = phenyl exhibits poor solubility in normal organic solvents.

Block copolymers can be prepared thermally from the dimethylsilyl-bridged ferrocene plus $[Si(CH_3)(C_6H_5)]_4$ (see Fig. 2.33). The polysilane segments can be cleaved photochemically with ultraviolet light. These and many other derivatives have been discussed in detail by Ian Manners (57–59), in whose laboratories a vast majority of the successful ring-opening ferrocenyl work has occurred. Thermal ROP methodology works on a wide variety of metallocenes with a variety of bridging groups as noted below.

The analogous GeR_2-bridged ferrocenes, for methyl, ethyl, *n*-butyl, and phenyl R groups, can be synthesized thermally at about 90 °C to provide molecular masses of the order of 100,000. The analogous di-*tert*-butyltin-bridged ferrocenes ring-open thermally at 150 °C to produce a polymer with $\overline{M}_N \approx 80{,}000$, and analogous PR-bridged ferrocenes also undergo ROP. Ethylene

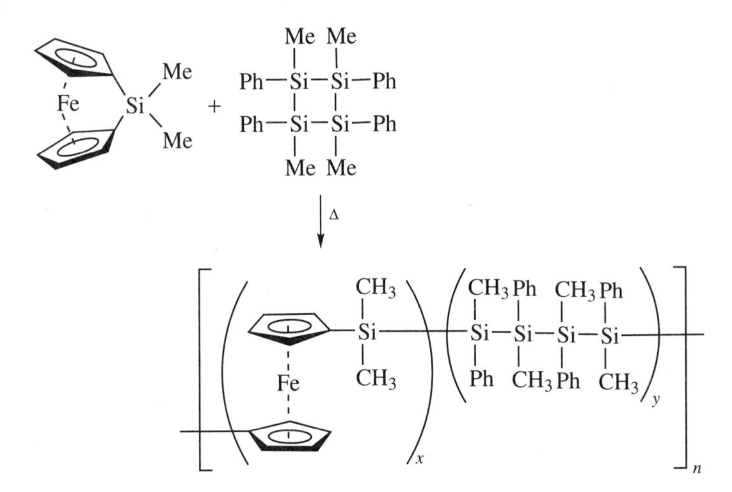

Figure 2.33 A dual thermal ring-opening polymerization reaction between a silyl-bridged ferrocene and a cyclic silane.

–CR_2–CR_2–bridged ferrocenes can be polymerized around 300 °C. When R = H, the polymer is insoluble, but when R = methyl a soluble polymer results. The analogous ruthenocene has more ring strain and undergoes thermal ROP at 220 °C. A sizable number of other substituted ferrocenes and ruthenocenes have also been studied (57, 58).

2.3.2.1 Catalyzed ROP Reactions of Bridged Ferrocenes

ROP of strained-ring organometallics is not limited to thermal reactivity. Anionic initiators, such as *n*-BuLi, allow the synthesis of "living" poly(ferrocenylsilane) derivatives in THF at room temperature with controlled molecular weights and narrow polydispersities. However, pure reactants and solvents are necessary. Block copolymers of 1,1′-dimethylsilane-bridged ferrocene and $[(CH_3)_2SiO]_3$, both of which have ring strain, are possible using this anionic ROP living polymerization method (see Fig. 2.34). Block copolymers of poly(ferrocene) derivatives and organic polymers provide redox-active domains in solid-state spheres, cylinders, films, and micelles (60).

Transition metal-catalyzed ROP of organosilyl-bridged [1]ferrocenophanes provides another method of synthesizing polymers and block copolymers without the necessity of an ultrapure monomer. The mechanism or mechanism of the catalytic reactions using rhodium(I), palladium(II), and platinum(II) as well as palladium(0) and platinum(0) has been undergoing investigation for some time, and the current indication is that the catalysts are colloidal noble metals (61).

SnR_2 bridged ferrocenophanes do not undergo platinum-catalyzed ROP but can be catalyzed by bases such as amines (62). In fact, impure products of R_2Sn-bridged [1]ferrocenophane from preparations that involve amines can undergo slow ROP even at room temperature (62). Thus it is possible that the so-called thermal ROP reactions noted above are catalyzed by traces of amines left in other ferrocenophanes as well (62).

Hydrophilic water-soluble neutral and cationic poly(ferrocenylsilanes) of high molecular mass have been synthesized at room temperature by

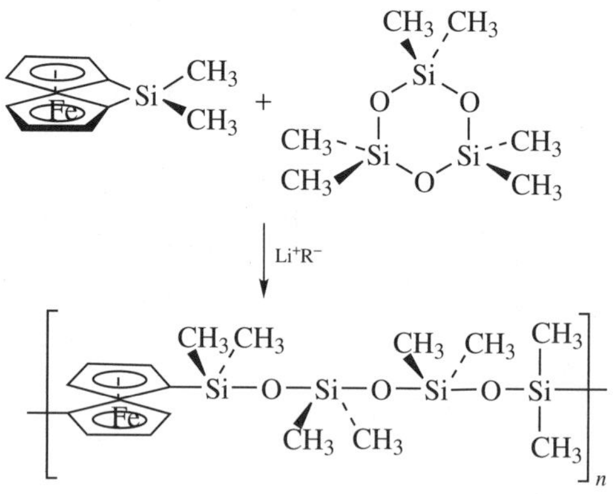

Figure 2.34 Anionic ring-opening polymerization of a silyl-bridged ferrocene.

transition metal-catalyzed ring-opening catalysis of chloromethylsila-bridged [1]ferrocenophane followed by chloro replacement with polyethylene glycol methyl ester or oxoethyl-*N,N*-dimethylamine groups. Quaternization of the latter with methyl iodide provides soluble polyelectrolytes. The polyethylene glycol derivative with $\overline{M}_n \approx 190{,}000$ is also quite soluble (63).

2.3.3 Main Group ROP (32, 57, 64)

2.3.3.1 Polyphosphazenes

"Inorganic rubber," or poly(dichlorophosphazene), has been known for over a century. The synthesis of $(NPCl_2)_n$ typically involves the thermal ring opening of $(NPCl_2)_3$ at greater than 200 °C in high-boiling organic solvents, such as one of the di- or trichlorobenzenes, or neat at about 250 °C (Fig. 2.35). Careful reaction control is required to obtain a soluble linear polymer. Lewis acid catalysts, for example, anhydrous aluminum chloride, catalyze this ROP. Less time or a lower temperature is needed. $Cl_3P{=}NP(O)Cl_2$ thermolysis also produces $(NPCl_2)_n$ ($n \geq 1000$).

A sizable number of polyphosphazene derivatives have been synthesized from poly(dichlorophosphazene) plus a suitable nucleophilic reagent, for example, the sodium salt of an alcohol or phenol, as shown in Eq. 2.38. Similarly, primary or secondary amines can provide other derivatives as indicated in Eqs. 2.39 and 2.40. See Table 1.3 for a comparison of the properties of these various

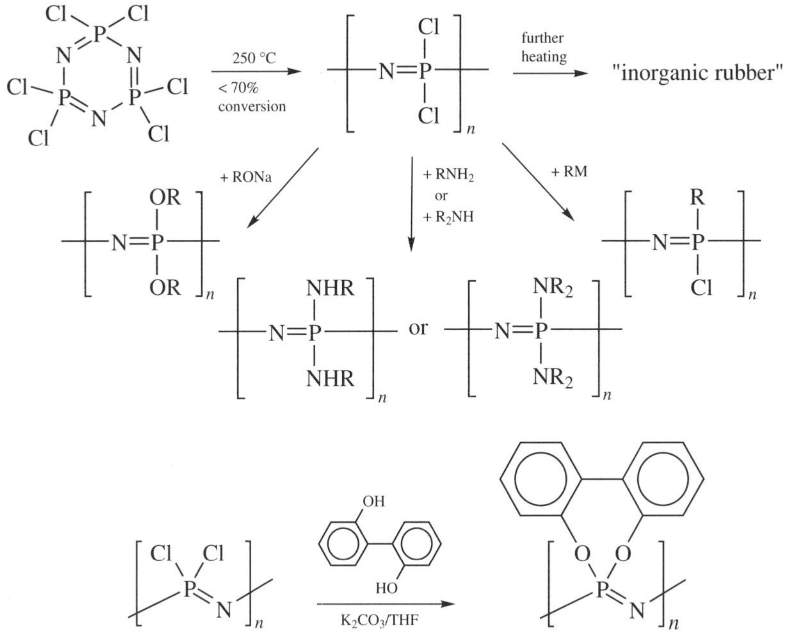

Figure 2.35 A trimeric dichlorophosphazene ring-opening polymerization.

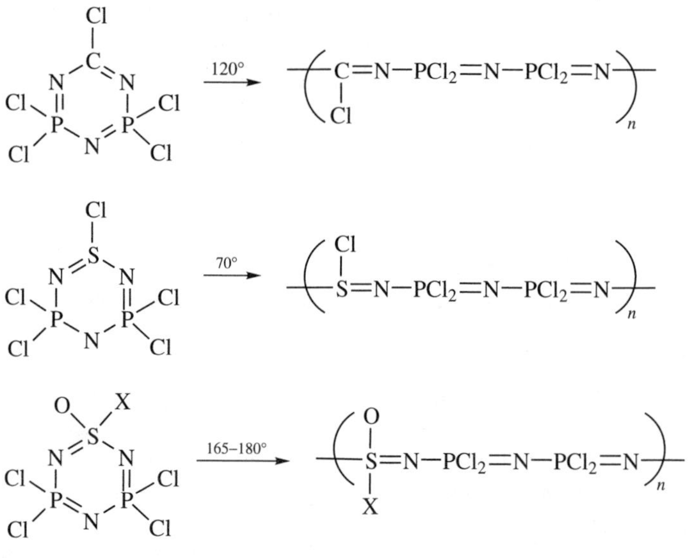

$X = Cl, F$

Figure 2.36 Ring-opening polymerizations of several substituted trimeric phosphazenes, specifically, rings in which CCl, $S^{IV}Cl$, $S^{VI}OCl$, and $S^{VI}OF$ replace one of the $P^{V}Cl_2$ groups (where the Roman numeral represents the formal oxidation state of the atom in the ring).

polyphosphazene derivatives. Note that the materials vary from elastomers to thermal plastics simply by different chloro replacement reactions.

$$(N{=}PCl_2)_n + 2n\text{NaOR} \longrightarrow [N{=}P(OR)_2]_n + 2n\text{NaCl} \quad (2.38)$$

$$(N{=}PCl_2)_n + 2n\text{RNH}_2 \longrightarrow [N{=}P(NHR)_2]_n + 2n\text{HCl} \quad (2.39)$$

$$(N{=}PCl_2)_n + 2n\text{R}_2\text{NH} \longrightarrow [N{=}P(NR_2)_2]_n + 2n\text{HCl} \quad (2.40)$$

A number of related polymers can also be synthesized by ROPs. (see Fig. 2.36). The temperatures for such reactions can be reduced with appropriate catalysts. For example, cyclic thionylphosphazenes undergo room temperature ROP in the presence of gallium(III) chloride (65).

2.3.3.2 Polysilanes

The anionic ROP of tetrameric $(R_2Si)_4$ or $(RR'Si)_4$ (where R or R′ = H, alkyl, aryl, etc.) provides long chain polymers, at least one of which has a practical use as a precursor of β-silicon carbide. The tetrameric ring compound can be synthesized by reductive coupling of the dimer, $R_2Si{-}SiR_2$. This and the analogous reductive coupling polymerization reaction are noted Section 2.4. The anionic induced ROP reaction can be simply written as

$$n/4(\mathrm{RR'Si})_4 \longrightarrow (\mathrm{RR'Si})_n \tag{2.41}$$

where R and R′ are typically methyl and phenyl groups. The octaphenyl is too sterically hindered to react, and the octamethyl produces an intractable polymer.

2.3.3.3 Polysiloxanes

ROP is now a common method for the commercial synthesis of polysiloxanes. Typically, the first step is a controlled hydrolysis of a dichlorodialkylsilane, a dichloroalkylarylsilane, or a dichloroalkylhydrosilane to form cyclosiloxane trimers and/or tetramers. The cyclosiloxane ring formation is favored by acidic catalysts. The ROP of this cyclosiloxane can then be conducted thermally or with an anionic or a cationic catalyst. Once again, anionic catalysis of siloxane ROPs provides controlled polymerization with low polydispersity (Fig. 2.37).

An example of the synthesis of a nonsymmetrical polysiloxane by ROP is shown in Figure 2.38. When ether solutions of tetramethyldisiloxane-1,3-diol and dichlorodimethylsilane are mixed in the presence of triethylamine, the cyclic 1-hydrido-1,3,3,5,5-pentamethylcyclotrisiloxane is formed and separated by fractional distillation. The cyclic siloxane in tetrahydrofuran ring opens quite rapidly at −78 °C with the anionic catalyst dilithium diphenylsilanediolate. The resulting poly(1-hydrido-1,3,3,5,5-pentamethylsiloxane) has a $\overline{M}_N$ of over 50,000 with a polydispersity of 1.59 based on gel permeation chromatography and polystyrene standards.* From NMR results, the polymer appears to be highly regular, suggesting that the anionic attack is exclusively on the hydridosilicon atoms (66).

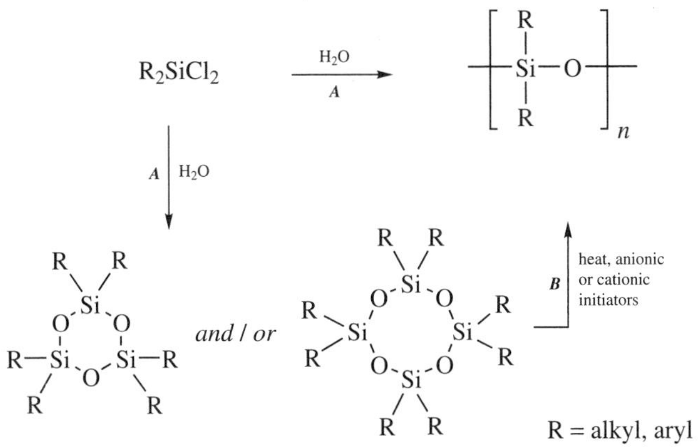

Figure 2.37 Schematic ring and chain siloxane formation through silane hydrolysis (A) followed by siloxane ring-opening polymerization (B).

* See Section 3.2.1 for information on this technique.

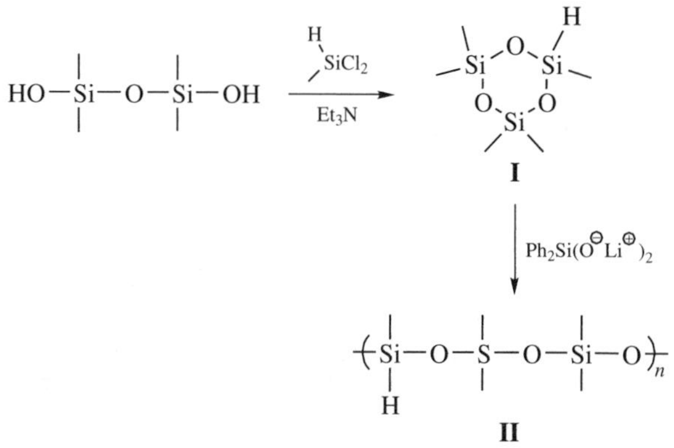

HO—Si—O—Si—OH →(HSiCl₂ / Et₃N) I →($Ph_2Si(O^{\ominus}Li^{\oplus})_2$) II

Figure 2.38 Preparation of an unsymmetrical but highly regular polysiloxane using anionic ring-opening catalysis (66).

2.3.3.4 Polycarbosilanes

Whereas polycarbosilanes were classically prepared by either reductive coupling (cf. Section 2.4) or by the thermal rearrangement of poly(dimethylsilane) at about 400 °C (cf. Eq. 2.42),

$$[\text{-Si(CH}_3)_2\text{-}]_n + \text{heat}(\sim 400\,^\circ\text{C}) \longrightarrow [\text{-SiH(CH}_3)\text{-CH}_2\text{-}]_n \qquad (2.42)$$

Interrante and coworkers (67–72) have perfected the ROP of cyclic SiCSiC ring compounds (73) and substitution reactions as shown in Figure 2.39. The dichlorosilane groups are reduced with lithium aluminum hydride to an alternating silylene-methylene chain. Poly(silylene-methylene) $[SiH_2CH_2]_n$ has a melting temperature of about 25 °C and a glass transition temperature of about −135 °C, is air stable, and dissolves readily in common organic solvents. With a chloromethylsilylene group a variety of other species can be formed (Fig. 2.39). The $[Si(CH_3)(OR)CH_2]_n$ derivatives vary in both glass transition temperatures and hydrolytic sensitivity depending on the R group (that includes ethyl, 1,1,1-trifluoroethyl, acetyl, and phenyl). Polycarbodifluorosilane $[SiF_2CH_2]_n$ and $[Si(CH_3)(CN)CH_2]_n$ have also been synthesized by appropriate substitution reactions (74, 75). A number of other laboratories have also reported ring-opening carbosilane polymerization reactions as detailed in a recent review (64).

Even with the H_2PtCl_6 catalyst, temperatures of 80–100 °C or extended reflux is required for solvents that boil below that temperature range. Thus the photocatalyzed ROP of a disilacyclobutate with platinum(II) acetylacetonate at room temperature (76) (Fig. 2.40) is encouraging. Compelling evidence for both homogeneous and heterogeneous paths has been presented, although the mechanistic details are somewhat uncertain (76).

2.3.3.5 Polythiazyl

Polythiazyl [or poly(sulfur nitride)] is another interesting inorganic polymer that is synthesized via a ROP reaction (77). The polymer is synthesized from the

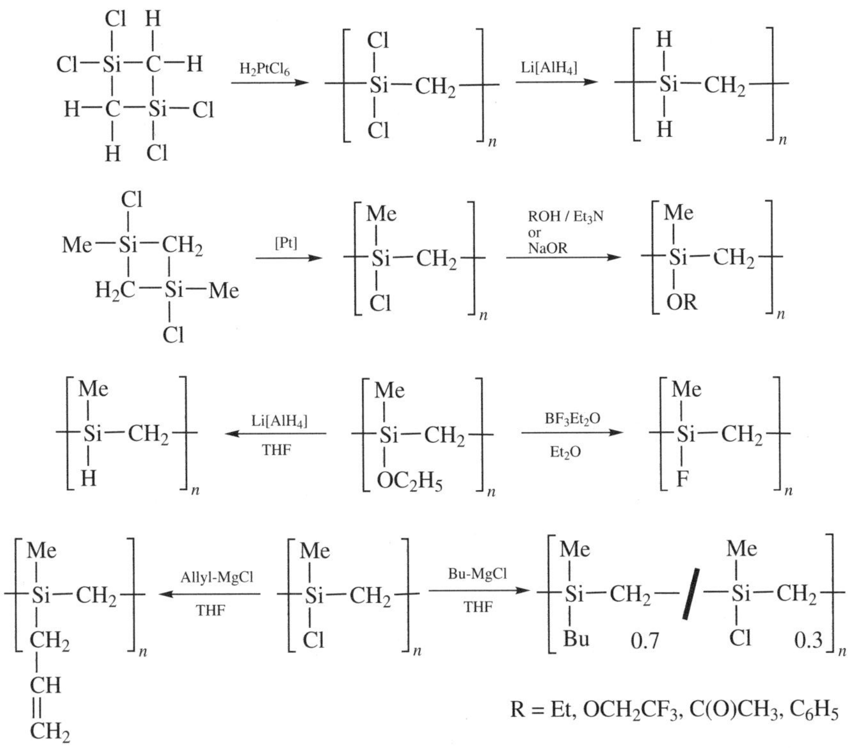

$R = Et, OCH_2CF_3, C(O)CH_3, C_6H_5$

Figure 2.39 Carbosilane ring-opening syntheses.

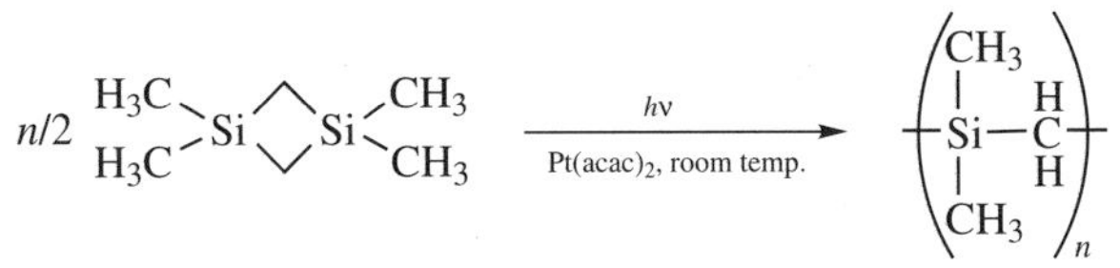

Figure 2.40 The photocatalyzed ring-opening polymerization of a carbosilane.

$(SN)_2$ alternating four-member ring dimer. The dimer is itself synthesized from the cyclic alternating sulfur nitride tetramer $(SN)_4$, which can be synthesized in a number of ways: $S_2Cl_2 + NH_3(l)$ or $SF_4 + NH_3(l)$ or $S_8 + NH_3(l)$. The tetramer is an unstable orange-yellow crystalline solid with a melting point of 178 °C. The conversion of the tetramer to the dimer that is used in the ROP obviously also involves a ring-opening reaction of tetramer vapor ($t_{subl} = 85\,°C$ at 0.01 torr) catalyzed with hot silver wool at 200–300 °C. The dimer is also very unstable (potentially explosive). It is condensed on a liquid nitrogen-cooled cold finger and purified to a colorless crystalline material by sublimation *in vacuo* at 25 °C to another trap at 0 °C. The solid-state polymerization occurs at 25 °C (3 days) plus 75 °C for 2 h. The final product is a gold-metallic diamagnetic polymer that conducts electrons; in fact, it is a superconductor at very low temperatures (≤ 0.33 K). (During the course of the polymerization a blue-black paramagnetic material is formed.) Unfortunately, the polymeric product is insoluble in all solvents with which it does not react, and it also slowly reacts with both air and water.

$$n/4(\text{SNSNSNSN})_{\text{ring}} \xrightarrow{200^{\circ}/0.01\text{torr}} n/2(\text{SNSN})_{\text{ring}} \xrightarrow{25^{\circ}} (\text{–S=N–})_n \qquad (2.43)$$

2.4 REDUCTIVE COUPLING AND OTHER REDOX POLYMERIZATION REACTIONS

2.4.1 Reductive Coupling

The Wurtz reductive coupling reaction between a dichlorosilane and sodium metal for the production of **polysilanes** or polysilylenes should be more forgiving of inexact stoichiometry than a step-growth condensation because all of the coupled molecules are identical. In fact, polysilane molecules with molecular masses of greater than 10^6 have been prepared in this manner. The reaction produces NaCl as well as the polysilane (as shown simplistically in Eq. 2.44); thus this reaction is a condensation polymerization, but not of the step-growth type. For the reaction to proceed at a reasonable rate, elevated temperatures and rapid mixing are required. These conditions help disperse the sodium. The temperature can be reduced through the use of

(1) Ultrasonic radiation to disperse the sodium and provide constant fresh surfaces,
(2) sodium-potassium alloy that is a liquid at room temperature, or
(3) a crown ether in an organic solvent to provide solutions of Na(crown ether)$^+$Na$^-$ (see Fig. 2.41 for typical reactions).

$$n\text{R}_2\text{SiCl}_2 + 2n\text{Na} \longrightarrow (\text{SiR}_2)_n + 2n\text{NaCl} \qquad (2.44)$$

However, the polymerization is not a simple reaction. Even though up to 20% of the monomer is converted to a polymer with an average molecular mass of 10^6, a multitude of products are formed including silane rings and a bimodal polymer distribution that includes a cluster of short polymer chains as well as the very long polymer chains noted above (cf, Exercise 2.7). The actual yield of polymeric silane is quite low by this method; thus other synthetic procedures are also under development (78).

Polygermanes by the Wurtz synthetic route have provided $(R_2Ge)_n$ polymers with $\overline{M}_N$ of greater than 500,000 (79, 80). Alternate non-Wurtz methods are also being developed (81), but none has provided such high-molecular-mass materials (cf. Exercise 2.8). Sigma electron delocalization is more pronounced in polygermanes than in polysilanes.

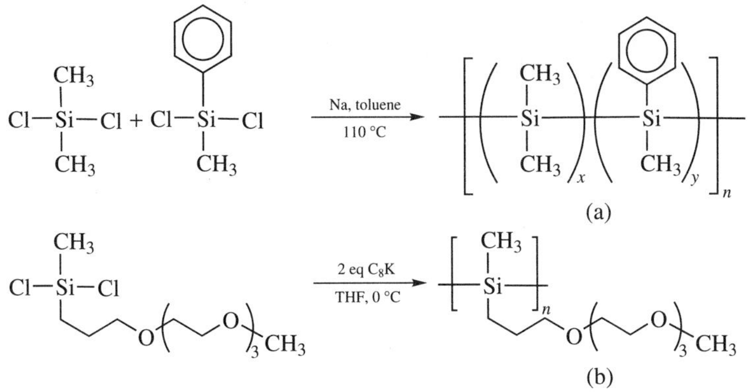

Figure 2.41 Wurtz and modified Wurtz coupling reactions in the synthesis of soluble polysilanes: (a) is soluble in organic solvents and (b) is soluble in water even though it is nonionic.

Polystannanes have also been synthesized using the Wurtz synthetic route with dibutyldichlorotin(IV) with a 15-crown-5-sodium catalyzed reaction (82). An improved Wurtz synthesis by Devylder and coworkers (83) has provided polystannanes with $\overline{M}_N \approx 10^6$ as estimated by gel permeation chromatography.* Although comparing a polystannane with polystyrene does not lead to good precision in molecular mass measurements, this recent polystannane is undoubtedly a high polymer.

Polycarbosilanes have also been prepared by reductive coupling. Two examples are shown in Figure 2.42. One, in which magnesium metal is used as the reductant, leads to a carbosilane with alternating aromatic carbon rings and silanes with an organic R group and an ethoxo (ethoxy) group with $\overline{M}_N \approx 6000$ to 18,000 (84). The ethoxo group can be metathetically replaced with fluoro, chloro, hydrido, or another substituent as shown in the figure. The other example builds a carbosilane with four backbone silane groups per methylene backbone unit (85).

Among the methods attempted for the synthesis of **polyferrocene**, the reductive coupling of 1,1′-diiodoferrocene with a stoichiometric quantity of magnesium leads to a short polymer ($\overline{M}_N \leq 5000$) (see Fig. 2.43). Even so, this reductive coupling is as good as any other synthetic method for synthesizing the unsubstituted polyferrocene.

Reductive coupling may also be involved in the synthesis of the soluble **ruthenium**(II) polymers (Fig. 2.5) from ruthenium(III) monomers discussed previously. In the second paper of their synthesis (19) Kelch and Rehahn note the addition of a reducing agent such as ethylmorpholine to ensure complete reduction to ruthenium(II) during the final stages of the polymerization. No reducing agent was noted in the original communication (20), in which average degrees of polymerization were of the order of 10 or 15; therefore, the reduction must be fairly spontaneous.

* See Section 3.2.1 for information on this technique.

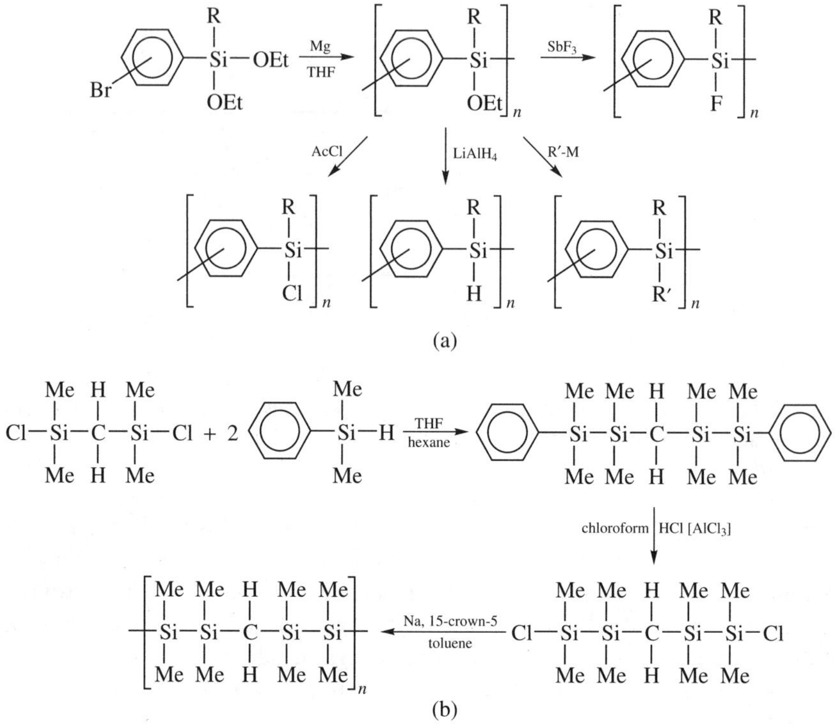

(a)

(b)

Figure 2.42 Reductive coupling reactions in the synthesis of polycarbosilanes—reactions of the ethoxy group of (a) provides at least four other polycarbosilanes. The monomer synthesis for polymer (b) is also shown.

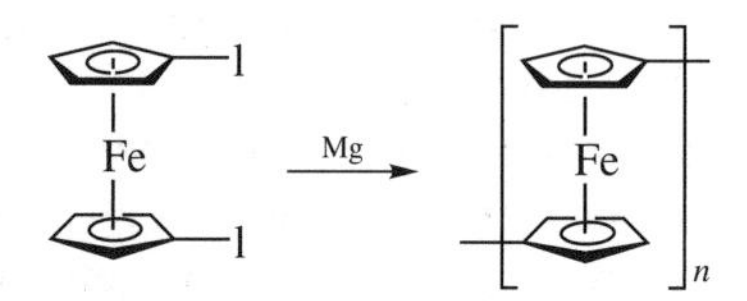

Figure 2.43 Ferrocene polymer by reductive coupling.

2.4.2 Oxidative Addition Polymerizations

Oxidative addition polymerizations have also been used to synthesize metal-containing polymers (86). The **tungsten**(IV) coordination polymer noted earlier in this chapter (Fig. 2.15), where a seven-coordinate tungsten(II) monomer is oxidized by a quinone (quinoxalinedione) that becomes a diolato ligand (quinoxalinediolato) (23), is just one example. The general oxidative reaction scheme for quinones and metal ions is shown in Figure 2.44. Several soluble polymers of **germanium** ($\overline{M}_N \sim 10^5$ based on gel permeation chromatography* relative to

* See Section 3.2.1 for information on this technique.

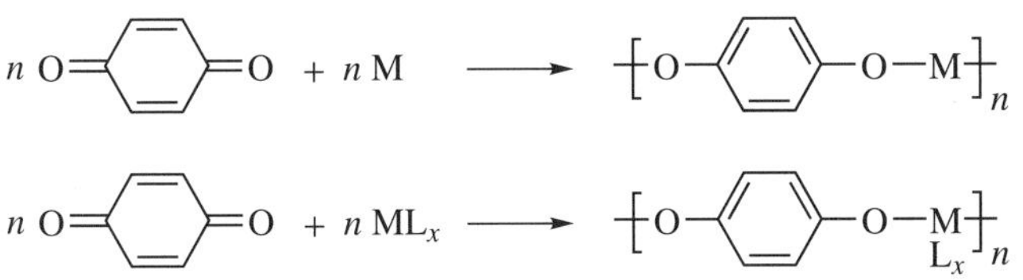

$$n\,O{=}C_6H_4{=}O + n\,M \longrightarrow \left[O{-}C_6H_4{-}O{-}M \right]_n$$

$$n\,O{=}C_6H_4{=}O + n\,ML_x \longrightarrow \left[O{-}C_6H_4{-}O{-}M(L_x) \right]_n$$

Figure 2.44 Oxidative addition reaction scheme for quinones and metal atoms, ions, or coordination or organometallic species.

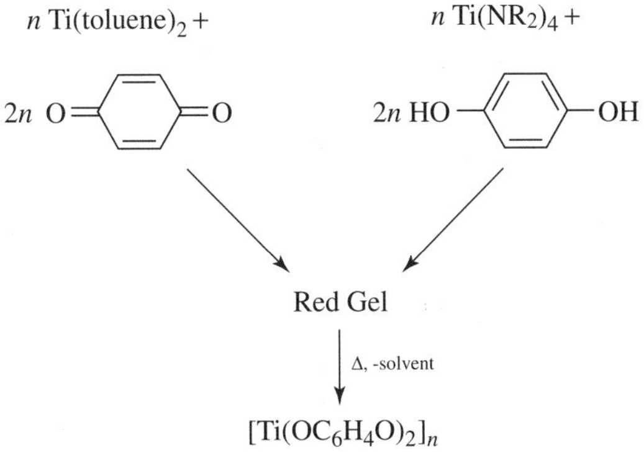

n Ti(toluene)$_2$ + 2n O=C$_6$H$_4$=O → Red Gel ← n Ti(NR$_2$)$_4$ + 2n HO–C$_6$H$_4$–OH

Red Gel —(Δ, -solvent)→ $[Ti(OC_6H_4O)_2]_n$

Figure 2.45 Oxidative addition and condensation polymerization pathways for the synthesis of $[Ti(OC_6H_4O)_2]_n$ (86).

polystyrene in chloroform) have been obtained by reactions of $(R_2N)_2Ge^{II}$, where $R = Si(CH_3)_3$, with 1,4-benzoquinone and several substituted benzoquinones (87). Difficultly soluble polymers of **tin** have been obtained through the reaction of R_2Sn^{II}, where $R = (CH_3)_2NCH_2CH_2–$ or $CH_3COCH{=}C(CH_3)–$, with benzoquinone, although they have not been well characterized. Similar reactions of benzoquinone with stannous β-diketonates provide insoluble polymers that are unstable in moist air.

Titanium(IV) polymers from Ti(toluene)$_2$ and quinones involve a four-electron change or 2 quinones/titanium. This reaction is compared with the tetraamidotitanium(IV) reaction with two moles of 1,4-hydroquinone in Figure 2.45. Identical insoluble 3-D network $[Ti(OC_6H_4O)_2]_n$ polymers result in both cases. With 2,5-dihydroxy-1,4-benzoquinone and Ti(toluene)$_2$, a polymer results that is stable to water (86). The resulting 1,2,4,5-tetraolate bridging ligand can chelate as it bridges and improve the hydrolytic stability.

2.5 CONDENSATION (DESOLVATION) OLIGOMERIZATIONS/ POLYMERIZATIONS

In addition to the step-growth condensation polymerizations noted in Section 2.1, a number of other inorganic condensation reactions are known. Most of them do not lead to soluble linear inorganic polymeric species, but a brief accounting of them seems appropriate.

2.5.1 Cationic Aggregations

The simplest examples of cationic aggregations are the formation of colloidal hydrous hydroxides and oxides through condensation. These reactions start as spontaneous hydrolysis reactions in acidic solution with small, highly charged metal ions, such as iron(III):

$$[M(H_2O)_n]^{m+} + H_2O \longrightarrow [M(H_2O)_{n-1}(OH)]^{m-1} + H_3O^+ \tag{2.45}$$

or by the addition of base in larger or lower-charged ions:

$$[M(H_2O)_n]^{m+} + OH^- \longrightarrow [M(H_2O)_{n-1}(OH)]^{m-1} + H_2O \tag{2.46}$$

Olation (the condensation of hydroxo ligands) follows:

$$2[M(H_2O)_{n-1}(OH)]^{m-1} \longrightarrow [(H_2O)_{n-2}M(\mu - OH)_2M(H_2O)_{n-2}]^{2(m-1)} \tag{2.47}$$

In addition to the (μ-OH)$_2$ double bridge, where μ is the symbol used to designate bridging ligands, (μ-OH)$_3$ and (μ-OH)$_2$ bridges can be formed, where μ_3 represents a bridging group coordinated to three metal ions, as in the $(\mu\text{-OH})_4M_4$ cubane structures.

The olation can continue to a polymeric hydrous hydroxide, $M(OH)_m{\bullet}XH_2O$ or through oxolation to a hydrous oxide, $MO_{m/2}{\bullet}YH_2O$ or $M_2O_m{\bullet}ZH_2O$, for even and odd m values, respectively. Sometimes the olations and oxolations stop at small species as in $[M_4(\mu\text{-OH})_8(H_2O)_{16}]^{8+}$, where $M = Zr^{IV}$ or Hf^{IV}. $[M^{II}(\mu_4\text{–}O)(OH)_6]$ and $[M_3^{III}O(OH)_6(H_2O)_3]^+$ species are known for several divalent and trivalent metal ions, but most of them are metastable unless stabilized by other species or by the electronic structure of the metal ion.

2.5.2 Anionic Aggregations

Both transition metal and nonmetals form anionic species in high oxidation states in basic solutions. On the addition of acid these anions, such as chromate, $CrO_4{}^{2-}$, the analogous molybdate and tungstate ions, vanadate, $VO_4{}^{3-}$, silicate, $SiO_4{}^{4-}$, phosphate, $PO_4{}^{3-}$, etc., aggregate. Using chromium as an example, the condensation products are $Cr_2O_7{}^{2-}$, $Cr_3O_{10}{}^{2-}$, etc., leading eventually to CrO_3. Vanadate goes through small aggregates to a specific $V_{10}O_{28}{}^{6-}$ ion (and protonated forms such as $H_2V_{10}O_{28}{}^{4-}$) on the way to V_2O_5 that then dissolves on even further acidification to $VO_2{}^+$. Polymolybdates, polytungstates, polysilicates, and polyphosphates are all well known. Details can be found in any standard inorganic chemistry textbook.

2.5.3 Desolvation at Elevated Temperature

Polymeric oxides are often formed from oxoacids or hydrous hydroxides by heating. For example, As_2O_5 can be prepared by the dehydration of crystalline H_3AsO_4 at about 200 °C. This oxide consists of equal numbers of corner-shared AsO_4 tetrahedra and AsO_6 octahedra to provide cross-linked strands with tubular cavities. The cubic form of MnO is formed by heating hydrated $Mn(OH)_2$. A number of other examples could be given.

However, desolvation is not limited to water removal. Silicon nitride can be prepared by the thermal reaction shown in Eq. 2.48:

$$3x[Si(NH_2)_4] \longrightarrow [Si_3N_4]_x + 8xNH_3 \tag{2.48}$$

The essence of much of the preceramic chemistry of the past few years involves synthesizing new inorganic species (polymers, clusters, etc.) that have the right composition so that the removal of small molecules from the species provides a ceramic not easily obtained by the normal ceramic "brute force" methods.

Sol-gel techniques can provide thin layers of polymeric species using volatilization of small molecules as part of the process. The antireflective coating of alumina or titania can be applied through sol-gel techniques and the water can be removed by heating or baking the item to which the gel has been applied. The preparation of cadmium sulfide layers for photocells is more complex. The layers are prepared by spraying an ammoniacal aqueous solution of cadmium chloride and thiourea onto a substrate surface. The thiourea hydrolyzes to produce the sulfide needed to produce the cadmium sulfide film. The water is removed by heating. Further baking at up to 500 °C ensures the decomposition of the ammonium chloride to gaseous HCl and NH_3. Thus the desolvation includes an acid and a base as well as water and carbon dioxide (cf. Eqs. 2.49 and 2.50).

$$CdCl_2(aq) + (NH_2)_2CS(aq) + 2H_2O \longrightarrow CdS(s) + CO_2(g) + 2NH_4Cl(aq) \tag{2.49}$$

$$NH_4Cl(aq) + \text{heat} \longrightarrow NH_3(g) + HCl(g) + (H_2O)(g) \tag{2.50}$$

2.5.4 Solvolysis-Desolvation Reactions

The best-known examples of a combination of solvolysis followed by desolvation to produce a polymeric material come from phosphorus chemistry. Phosphorus(V) chloride can be partially hydrolyzed by water to $HOP(O)Cl_2$ followed by dehydrohalogenation to $[PO_2Cl]_n$. Alternatively, phosphorus(V) chloride can undergo partial ammonolysis with ammonium chloride followed by dehydrohalogenation to a phosphazene. These reactions are shown in Eqs. 2.51 and 2.52.

$$nPCl_5 + 2nH_2O \longrightarrow \begin{matrix} n\ HOP(O)Cl_2 \\ +3n\ HCl \end{matrix} \longrightarrow \begin{matrix} [PO_2Cl]_n \\ +n\ HCl \end{matrix} \tag{2.51}$$

$$nPCl_5 + nNH_4Cl \longrightarrow (\text{partialammonolysis}) \longrightarrow [PNCl_2]_n + 4nHCl \tag{2.52}$$

2.6 MISCELLANEOUS SYNTHESIS COMMENTS

2.6.1 Solubility

Alkyl groups tend to promote solubility in organic solvents, whereas phenyl groups tend to decrease solubility. An interesting example of the effects of alkyl side groups in promoting solubility and the inverse effect of phenyl groups in chain polymers is well documented in the palladium-catalyzed condensation reactions between bis(halophenyl)ferrocene derivatives and aryldiboronic acids as shown in Figure 2.46. For the case in which the boronic acid is attached directly to the benzene ring with the alkyl side groups ($y = 0$), the average degree of polymerization that is attained for these amorphous polymers is about 55. The glass transition temperature is lowered from about 80 °C for R = n-hexyl to about 20 °C for R = n-dodecyl. The average degree of polymerization that is attained for similar polymers is only about 40 when $y = 1$ and 10 when $y = 2$. Aromatic stacking appears to predominate as the phenylene chain increases in length and limits the number of units that combine before precipitation, after which chain length increases are minimal.

Conversely, solubility in polar solvents is enhanced with polar groups. For example, poly(silaferrocenes) are made soluble, even in water, through the use of polyether side groups, as noted in Section 2.3.2. Metal-coordination polymers seem to be most soluble in polar organic solvents such as DMSO, NMP, dimethyl acetamide, and dimethyl formamide.

On the other hand, if polymers of labile metal ions are desired, for example, tetrahedral bis(diimines) of silver(I) or copper(I), even protecting the ions with bulky substituents on the diimines is not sufficient because of the lability.

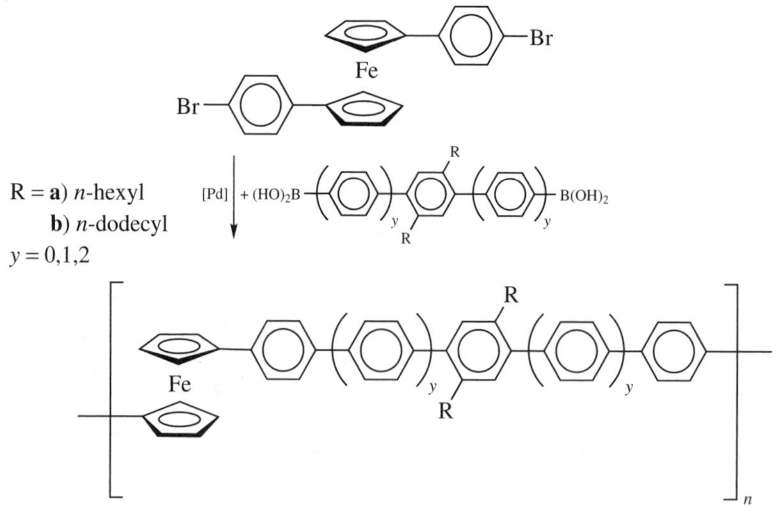

Figure 2.46 Palladium-catalyzed condensation reactions between a bis(halophenyl)ferrocene derivative and aryldiboronic acids (64).

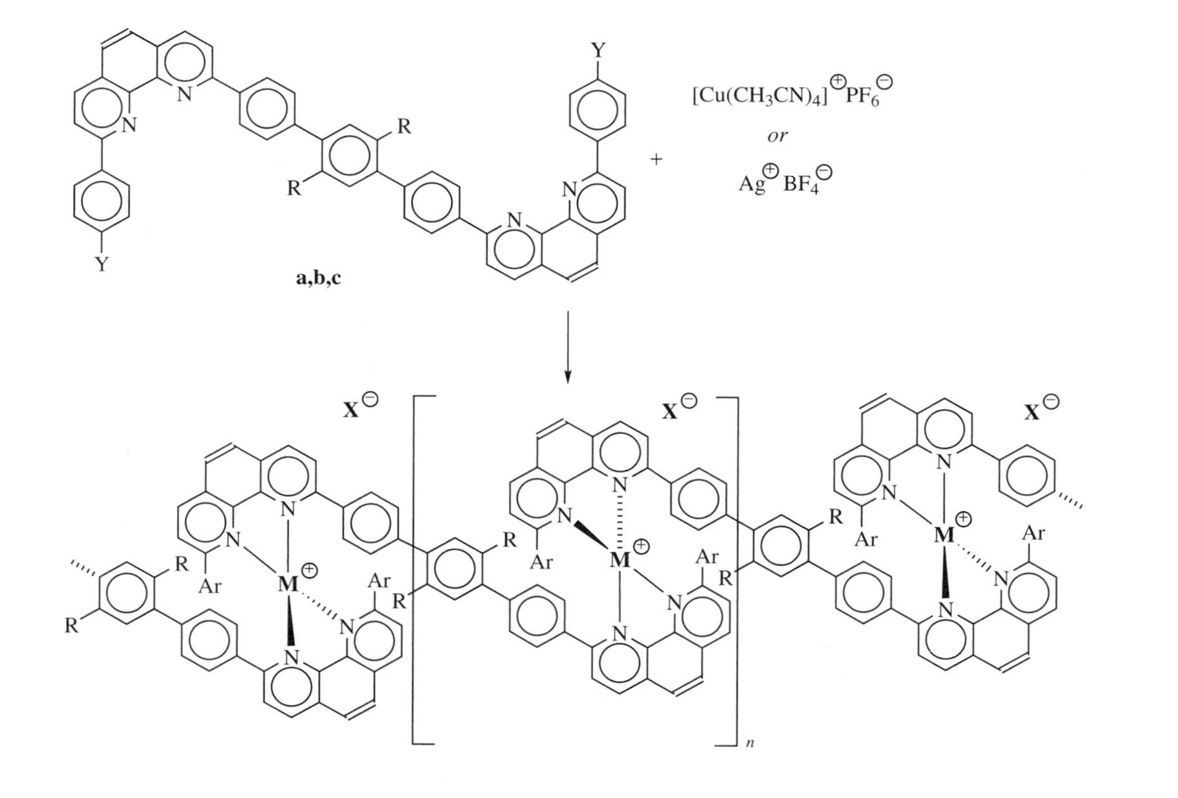

Figure 2.47 Tetrahedrally coordinated bis(diimine) coordination polymers (64). $R = C_6H_{13}$, Y = (a)H, (b) Cl, (c) OCH_3, $Ar = C_6H_4$-Y.

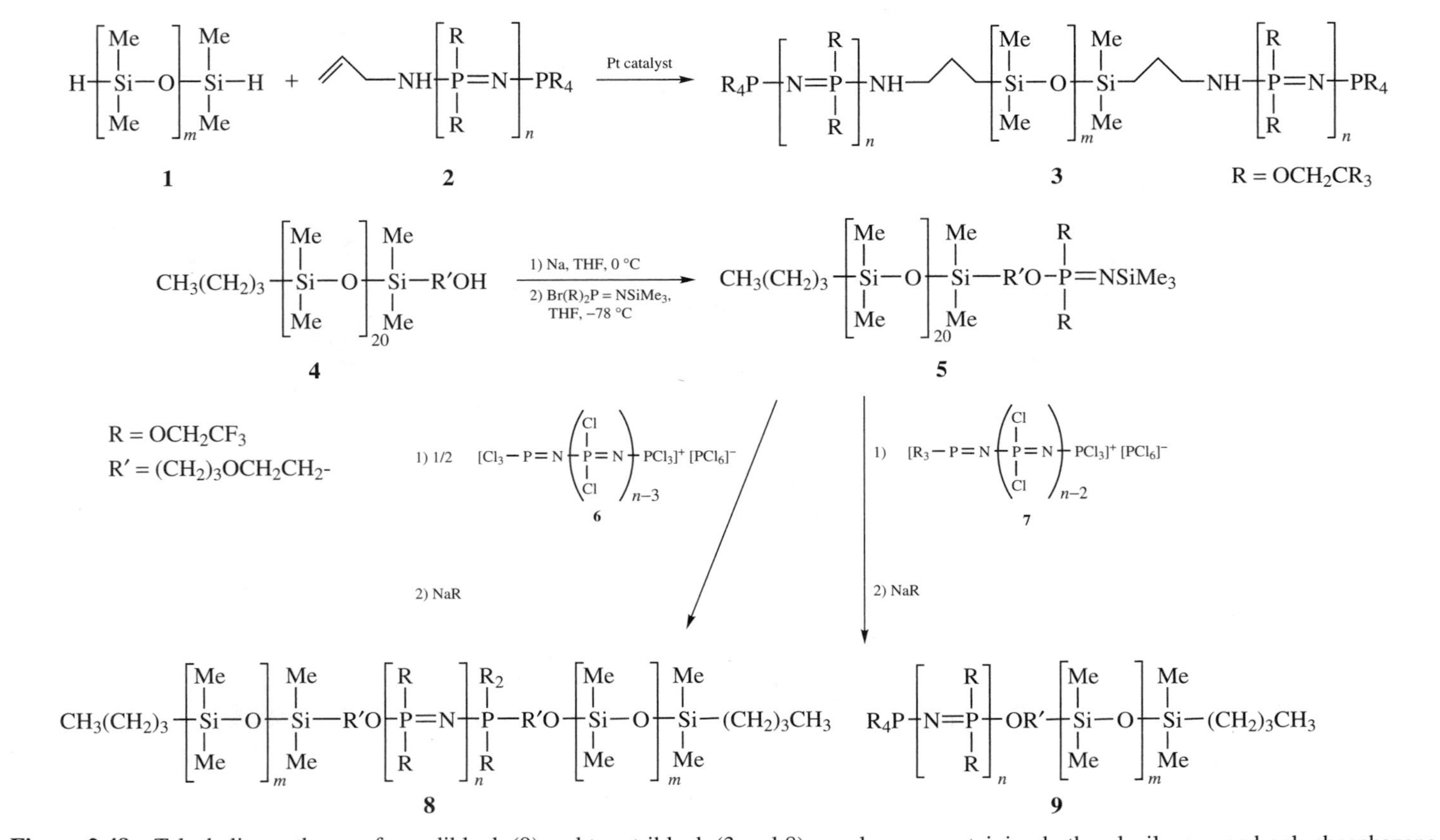

Figure 2.48 Telechelic syntheses of one diblock (9) and two triblock (3 and 8) copolymers containing both polysiloxane and polyphosphazene blocks (reprinted with permission from Ref. 90; © 1999 American Chemical Society).

However, a combination of bulky side groups plus dissolution in pure halogenated hydrocarbons or acetone provides soluble coordination polymers that are stable in the solid state for months and in solutions of the aforementioned solvents (see Fig. 2.47) (64, 88). Although such polymers must be kept away from polar solvents and water, they could provide an entry to solid state devices that can be protected from the "elements."

2.6.2 Telechelic Polymers

Block copolymers are often synthesized with **telechelic** polymers. Telechelic polymers are polymers, generally of moderate molar masses, that have a functional group at both ends of each polymer chain. If both functional groups are identical, the **homotelechelic** polymer can be combined with two equivalents of a monofunctional polymer chain to form an A-B-A-type triblock copolymer. **Heterotelechelic** polymers have different functional groups on the two ends of each chain (89).

Homotelechelic dihydride-terminated siloxanes have been coupled with heterotelechelic polyphosphazenes containing a vinyl terminal group using a divinylsiloxane/platinum catalyst. A phosphazene/siloxane/phosphazene triblock polymer of controlled size results (Fig. 2.48). Other phosphazene/siloxane di- and tri-block copolymers have also been synthesized using this telechelic approach (see Fig. 2.48) (90).

Telechelic synthetic methods have also been used to prepare other phosphazene copolymers (91, 92) and siloxane-containing copolymers, including siloxane/organic copolymers, (93–95), as well as a telechelic "polycarbosiloxane" (96), actually an organic copolymer containing disiloxane units (cf. Exercise 2.9). Telechelic syntheses of metal-containing polymers have also been claimed (97, 98).

2.6.3 Catalyzed Dehydrogenation Reactions

Although a thermal dehydrogenation reaction of borazine was noted above, this synthetic chapter would be incomplete without discussion of the catalytic dehydrogenation reactions that have been used for synthesizing polysilanes, polygermanes, and polystannanes (99–103). Although these reactions have not provided high-molecular-mass species, the syntheses seem more controlled than the Wurtz reactions. Trends in delocalization have been obtained, with the polystannanes being the most delocalized. A polystannane polymer with a $\overline{M}_N$ of 22,000 has extended sigma delocalization into the near ultraviolet ($\lambda_{max} \approx 385$ nm), and when doped with arsenic pentafluoride, its electrical conductivity was up to 0.3 S cm^{-1} (103).

REFERENCES

1. Rosen, S. L. in *Kirk-Othmer Encycl. Chem. Tech.*; 4th ed., Kroschwitz, J. I. and Howe-Grant, M., Ed., Wiley-Interscience: New York, 1996; Vol. 19, pp 881–904.

2. Anonymous in *Encycl. Polymer Sci. Engr.*; Mark, H. F., Bikales, N. M., Overberger, G. O., Menges, G. and Kroschwitz, J. I., Ed., Wiley-Interscience: New York, 1989; Vol. 15, pp 625–631.
3. Cotton, F. A., Wilkinson, G. *Advanced Inorganic Chemistry*; Wiley-Interscience: New York, 1988.
4. Joyner, R. D., Kenney, M. E. *Inorg. Chem.* 1962, **1**, 717.
5. Archer, R. D. *Coord. Chem. Rev.* 1993, **128**, 49.
6. Metz, J., Pawlowski, G., Hanack, M. *Z. Naturforsch.* 1983, **38B**, 378.
7. Archer, R. D., Hardiman, C. J., Kim, K. S., Grandbois, E. R., Goldstein, M. in *Metal-containing polymeric systems*; Sheats, J. E., Carraher, C. E., Jr. and Pittman, C. U., Jr., Ed., Plenum Press: New York, 1985, pp 355–366.
8. Carraher, C. E., Jr. *Macromolecules* 1971, **4**, 360.
9. Neuse, E. W. in *Organometallic Polymers*; Carraher, C. E., Jr., Sheats, J. E. and Pittman, C. U., Jr., Ed., Academic Press: New York, 1978, pp 95–100.
10. Pittman, C. U., Jr. *Makromol. Chem.* 1974, **175**, 1427.
11. Neuse, E. W., Bednarik, L. *Macromolecules* 1979, **12**, 187.
12. Archer, R. D., Chen, H., Cronin, J. A., Palmer, S. M. in *Metal-Containing Polymeric Materials*; Pittman, C. U., Jr., Carraher, C. E., Jr., Zeldin, M., Sheats, J. E. and Culbertson, B. M., Ed., Plenum Press: New York, 1996, pp 81–91.
13. Pittman, C. U., Jr., Carraher, C. E., Jr., Sheats, J. E., Timken, M. D., Zeldin, M. in *Inorganic and Metal-Containing Polymeric Materials*; Sheats, J. E., Carraher, C. E., Jr., Pittman, C. U., Jr., Zeldin, M. and Currell, B., Ed., Plenum Press: NY, 1990, pp 1–27.
14. Chen, H., Archer, R. D. *Macromolecules* 1995, **28**, 1609.
15. Chen, H., Archer, R. D. *Macromolecules* 1996, **29**, 1957.
16. Carraher, C. E., Jr., Sheats, J. E. *Makromol. Chem.* 1973, **166**, 23.
17. Basolo, F., Pearson, R. G. *Mechanisms of Inorganic Reactions*; 2nd ed., Wiley: New York, 1967.
18. Taube, H. *Chem. Rev.* 1952, **50**, 69.
19. Kelch, S., Rehahn, M. *Macromolecules* 1997, **30**, 6185.
20. Knapp, R., Schott, A., Rehahn, M. *Macromolecules* 1996, **29**, 478.
21. Chang, C.-H., Archer, R. D. *Korean J. Chem.* 1990, **34**, 85.
22. Donahue, C. J., Archer, R. D. *J. Am. Chem. Soc.* 1977, **99**, 6613.
23. Archer, R. D., Batchelet, W. H., Illingsworth, M. L. *J. Macromol. Sci.-Chem.* 1981, **A16**, 261.
24. Archer, R. D., Illingsworth, M. L., Rau, D. N., Hardiman, C. J. *Macromolecules* 1985, **18**, 1371.
25. Whitmarsh, C. K., Interrante, L. V. *Organometallics* 1991, **10**, 1337.
26. Rushkin, I. L., Shen, Q., Lehman, S. E., Interrante, L. V. *Macromolecules* 1997, **30**, 3141.
27. Kellogg, G. E., Gaudiello, J. G. in *Inorganic Materials*; Bruce, D. W. and O'Hare, D., Ed., John Wiley & Sons: Chichester, 1992, pp 354–404.
28. Friedel, C., Crafts, J. M. *Ann. Chim. Phys.* 1866, **[4] 9**, 5.
29. Ladenburg, A. *Ann.* 1872, **164**, 311, reported in Merck Index, 9th ed.

30. Hani, R., Lenz, R. W. in *Silicon-based polymer science: a comprehensive resource*; Zeigler, J. M. and Fearon, F. W. G., Ed., American Chemical Society: Washington, DC, 1989; No. 224 Advances in Chemistry Series, pp 741–52.
31. Senshu, K., Furuzono, T., Koshizaki, N., al., e. *Macromolecules* 1997, **30**, 4421.
32. Allcock, H. R., Lampe, F. W. *Contemporary Polymer Chemistry*; 2nd ed., Prentice-Hall: Engelwood Cliffs, NJ, 1990.
33. McCarthy, D. W., Mark, J. E., Schaefer, D. W. *J. Polym. Sci. Part B Polym. Phys.* 1998, **36**, 1167.
34. Szafran, Z., Pike, R. M., Singh, M. M. *Microscale Inorganic Chemistry: A Comprehensive Laboratory Experience*; Wiley: New York, 1991.
35. Carraher, C. E., Jr., Scherubel, G. A. *Makromol. Chem.* 1972, **160**, 259.
36. Roy, A. K., Burns, G. T., Grigoras, S., Lie, G. C. in *Inorganic and Organometallic Polymers II*; Wisian-Neilson, P., Allcock, H. R. and Wynne, K. J., Ed., American Chemical Society: Washington, DC, 1994, pp Chapter 26.
37. Klein, R. M., Bailar, J. C., Jr. *Inorg. Chem.* 1963, **2**, 1190.
38. Hennig, H., Rehorek, D., Archer, R. D. *Coord. Chem. Rev.* 1985, **61**, 1–53 (especially 26–30).
39. Fry, B. E., Neckers, D. C. *Macromolecules* 1996, **29**, 5306.
40. Cano, M., Oriol, L., Piñol, M., Serrano, J. L. *Chem. Mater.* 1999, **11**, 94.
41. Reddinger, J. L., Reynolds, J. R. *Macromolecules* 1997, **30**, 673.
42. Reddinger, J. L., Reynolds, J. R. *Synthetic Metals* 1997, **84**, 225.
43. Goldsby, K. A., Blaho, J. K., Hoferkamp, L. A. *Polyhedron* 1989, **8**, 113.
44. Hoferkamp, L. A., Goldsby, K. A. *Chem. Mater.* 1989, **1**, 348.
45. Zhu, S. S., Carroll, P. J., Swager, T. M. *J. Am. Chem. Soc.* 1996, **118**, 8713.
46. Deronzier, A., Moutet, J.-C. *Acc. Chem. Res.* 1989, **22**, 249, refs. 8–12.
47. Deronzier, A., Moutet, J.-C. *Acc. Chem. Res.* 1989, **22**, 249, refs. 13–18.
48. Allcock, H. R., Crane, C. A., Morrissey, C. T., Nelson, J. M., Reeves, S. D., Honeyman, C. H., Manners, I. *Macromolecules* 1996, **29**, 7740.
49. Honeyman, C. H., Manners, I., Morrissey, C. T., Allcock, H. R. *J. Am. Chem. Soc.* 1995, **117**, 7035.
50. Allcock, H. R., Nelson, J. M., Reeves, S. D., Honeyman, C. H., Manners, I. *Macromolecules* 1997, **30**, 50.
51. Nelson, J. M., Allcock, H. R., Manners, I. *Macromolecules* 1997, **30**, 3191.
52. Neilson, R. H., Jinkerson, D. L., Kucera, W. R., Longlet, J. J., Samuel, R. C., Wood, C. E. in *Inorganic and Organometallic Polymers* II; Wisian-Neilson, P., Allcock, H. R. and Wynne, K. J., Ed., American Chemical Society: Washington, DC, 1994, Chapter 18.
53. Wisian-Neilson, P. in *Encyclopedia of Inorganic Chemistry*; King, R. B., Ed., John Wiley & Sons: Chichester, 1994; Vol. 6, pp 3371–89.
54. Kluiber, R. W., Lewis, J. W. *J. Am. Chem. Soc.* 1960, **82**, 5777.
55. Brandt, P. F., Rauchfuss, T. B. *J. Am. Chem. Soc.* 1992, **114**, 1926.
56. Foucher, D. A., Tang, B. Z., Manners, I. *J. Am. Chem. Soc.* 1992, **114**, 6246.
57. Manners, I. *Ann. Rep. Prog. Chem., Sect. A, Inorg. Chem.* 1991–1997, **88–94**, 77,93,103,131,127,129,603, respectively.

58. Nguyen, P., Gómez-Elipe, P., Manners, I. *Chem. Rev.* 1999, **99**, 1515.
59. Manners, I. *Ang. Chem. Intl. Ed. Engl.* 1996, **35**, 1602.
60. Massey, J. A., Power, K. N., Manners, I., Winnik, M. A. *Adv. Mater.* 1998, **10**, 1559.
61. Temple, K., Jakle, F., Lough, A. J., Sheridan, J. B., Manners, I. *Polymer Preprints* 2000, **41**, 429.
62. Manners, I., $FcSnR_2$ ROP discussion.
63. Power-Billard, K. N., Manners, I. *Macromolecules* 2000, **33**, 26.
64. Rehahn, M. *Acta Polymer.* 1998, **49**, 201.
65. Peckham, T. J., Massey, J., Gates, D. P., Manners, I. *Phosphorus Sulfur Silicon & Related Elem.* 1999, **146**, 217.
66. Paulasaari, J. K., Weber, W. P. *Macromolecules* 1999, **32**, 6574.
67. Wu, H.-J., Interrante, L. V. *Chem. Mater.* 1989, **1**, 564.
68. Wu, H.-J., Interrante, L. V. *Macromolecules* 1992, **25**, 1849.
69. Interrante, L. V., Wu, J. J., Apple, T., Shen, Q., Ziemann, B., Narsavage, D. M. *J. Am. Chem. Soc.* 1994, **116**, 2085.
70. Rushkin, I. L., Interrante, L. V. *Macromolecules* 1995, **28**, 5160.
71. Rushkin, I. L., Interrante, L. V. *Macromolecules* 1996, **29**, 3123, 5784.
72. Shen, Q. H., Interrante, L. V. *Macromolecules* 1996, **29**, 5788.
73. Knoth, W. H., Jr., *Carbosilane ring-opening polymerization*, U. S. Patent no. 2,850,514, 1958.
74. Lienhard, M., Rushkin, I., Verdecia, G., Wiegand, C., Apple, T., Interrante, L. V. *J. Am. Chem. Soc.* 1997, **119**, 12020.
75. Lienhard, M., Wiegand, C., Apple, T., Interrante, L. V. *Polymer Preprints* 2000, **41**, 570.
76. Wu, X., Neckers, D. C. *Macromolecules* 1999, **32**, 6003.
77. Labes, M. M., Love, P., Nichols, L. F. *Chem. Rev.* 1979, **79**, 1.
78. Wisian-Neilson, P., Allcock, H. R., Wynne, K. J. *Inorganic and Organometallic Polymers II*; American Chemical Society: Washington, DC, 1994.
79. Trefonas, P., West, R. *J. Polym. Sci., Polym. Chem. Ed.* 1985, **23**, 2099.
80. Miller, R. D., Sooriyakumaran, R. *J. Polym. Sci., Polym. Chem. Ed.* 1987, **25**, 111.
81. Mochida, K., Chiba, H. *J. Organomet. Chem.* 1994, **473**, 45.
82. Zou, W. K., Yang, N.-L. *Polym. Preprints* 1992, **33(2)**, 188.
83. Devylder, N., Hill, M., Molloy, K. C., Price, G. J. *J. Chem. Soc., Chem. Commun.* 1996, 711.
84. Ohshita, J., Yamashita, A., Hiraoka, T., Shinpo, A., Kunai, A., Ishikawa, M. *Macromolecules* 1997, **30**, 1540.
85. Isaka, H. *Macromolecules* 1997, **30**, 344.
86. Burch, R. R. *Chem. Mater.* 1990, **2** , 633.
87. Kobayashi, S., Iwata, S., Abe, M., Shoda, S. *J. Am. Chem. Soc.* 1990, **112**, 1625.
88. Velten, U., Rehahn, M. *J. Chem. Soc., Chem. Commun.* 1996, 2639.
89. Colombani, D., Chaumont, P. *Acta Polym.* 1998, **49**, 225.
90. Prange, R., Allcock, H. R. *Macromolecules* 1999, **32**, 6390, and references therein.

91. Inoue, K., Negayama, S., Itaya, T., Sugiyama, M. *Macromol. Rapid Commun.* 1997, **18**, 225.
92. Miyata, K., Watanabe, Y., Itaya, T., Tanigaki, T., Inoue, K. *Macromolecules* 1996, **29**, 3694.
93. Brzezinska, K. R., Wagener, K. B., Burns, G. T. *J Polym Sci A-Polym Chem* 1999, **37**, 849.
94. Yoon, S. C., Ratner, B. D., Ivan, B., Kennedy, J. P. *Macromolecules* 1994, **27**, 1548.
95. Nishida, H., Yamane, H., Kimura, Y., Kitao, T. *Kobunshi Ronbunshu* 1995, **52**, 25, Web of Science citation.
96. Smith, D. W., Wagener, K. B. *Macromolecules* 1993, **26**, 1633.
97. Matsuda, H. Polym Advan. Technol. 1997, **8**, 616, Web of Science citation.
98. Liu, G., White, B., VancsoSzmercsanyi, I., Vancso, G. J. *J. Polym. Sci. Part B-Polym. Phys.* 1996, **34**, 277, Web of Science citation.
99. Aitken, C. T., Harrod, J. F., Samuel, E. *J. Organomet. Chem.* 1985, **279**, C11.
100. Aitken, C. T., Harrod, J. F., Samuel, E. *J. Am. Chem. Soc.* 1986, **108**, 4059.
101. Harrod, J. F. in *Inorganic and Organometallic Polymers*; Zeldin, M., Wynne, K. J. and Allcock, H. R., Ed., American Chemical Society: Washington, 1988; Vol. 360 ACS Symposium Series, pp 89–100.
102. Imori, T., Tilley, T. D. *J. Chem. Soc., Chem. Commun.* 1993, 1607.
103. Imori, T., Lu, V., Cai, H., Tilley, T. D. *J. Am. Chem. Soc.* 1995, **117**, 9931.
104. Carraher, C. E., Jr., Pittman, C. U., Jr. in *Metal-containing Polymeric Systems*; Sheats, J. E., Carraher, C. E., Jr. and Pittman, C. U., Jr., Ed., Plenum Press: New York, 1985, pp 1–42.

EXERCISES

2.1. To synthesize metal-containing polymers that are less apt to come apart in solution, compile lists of metal ions (including an example of a suitable species) that

a. have 6-coordinate octahedral complexes that are inert to ligand exchange;

b. have 4-coordinate planar complexes that are inert to ligand exchange; and

c. have cyclopentadienyl organometallic derivatives that are inert to ligand exchange.

2.2. Postulate reasons for the fact that the sodium polyelectrolytes shown in Figure 2.13 are more soluble than analogous polyelectrolytes with **either** smaller cations (Li^+ and H^+) **or** larger cations (K^+ and Cs^+).

2.3. The reaction between tetrapyrido[3,2-a:2′,3′-c:3″,2″-h:2‴,3‴-j]phenazine and $Ru(bpy)Cl_3$ produces a soluble linear polymer of ruthenium(II), even in the absence of added reducing agents. Because ruthenium is reduced during the polymerization reaction, something must be oxidized. Suggest an oxidation product for this reaction and justify your answer.

2.4. **a.** Show how the coordination of an aldehyde to a metal ion provides the charge density required on the carbonyl carbon for nucleophilic attack by an amine.

b. Explain how the lability of these metal ions is important for the reaction to proceed to completion.

2.5. **a.** Sketch the oligomerization reaction between phosgene ($COCl_2$) and $Li_2C_2B_{10}H_{10}$.

b. Suggest possible reasons as to why this reaction stopped as an oligomer.

c. From your knowledge of organic chemistry or a good organic chemistry textbook, suggest alternate organic reagents that might be used for a polymerization with $Li_2C_2B_{10}H_{10}$.

2.6. Write the chemical reactions involved in the condensation of Cl_3SiCH_2Cl with Mg in diethyl ether and the subsequent reduction with lithium aluminum hydride to produce the polymeric carbosilane silaethylene: $[SiH_2CH_2]_n$.

2.7. Write balanced reactions related to the production of polysilane including end groups and byproducts for

a. rings (assume 8-membered rings) and

b. linear polymers formed by the reactions represented in Eq. 2.44.

c. Write a balanced equation for the subsequent destruction of excess sodium metal from the ends of the linear polymers.

2.8. **a.** One alternate route to polygermanes uses GeI_2 and alkyl Grignard reagents, although the molecular masses obtained are of the order of 10^3-10^4 (81). Write out a possible polymerization reaction including bonding **and** classify the polymerization reaction from among the types presented in this chapter.

b. Another route to polygermanes uses GeI_2 and organolithium reagents (81). Again, the molecular masses obtained are modest. Write out a possible polymerization reaction and classify the polymerization reaction as in part a.

2.9. Sketch the structures of the reactants and the telechelic ABA copolymer that results from the condensation reaction between $Cl(CH_3)_2Si(CH_2)_4[CH{=}CH(CH_2)_6]_yCH{=}CH(CH_2)_4Si(CH_3)_2Cl$ and 2HO $[Si(CH_3)_2O]_xSi(CH_3)_3$ (93). {**Caution**: the structure of the copolymer shown in the table of contents of the journal is incorrect.}

CHAPTER 3

INORGANIC POLYMER CHARACTERIZATION

It is one thing to synthesize a material that you think is a polymer and quite another to prove that a polymeric product has actually been synthesized. Normally soluble synthetic polymers have a distribution of molecular masses; thus the "nonsporting method" of molecular structure by X-ray crystallography is not appropriate for polymer characterization.* This chapter explores a variety of methods of characterizing the molecular masses of inorganic polymers after exploring the typical molecular mass distributions that are obtained during polymer synthesis. The determination of molecular mass by the methods used for small molecules is typically inappropriate for large polymers, although such methods can be used in conjunction with physical measurements that allow estimates or actual molecular mass measurements. Conversely, some of the physical methods used for polymers become unreliable at low molecular masses.

Other methods of characterization that are important for inorganic polymers include a variety of thermal, spectroscopic, and viscoelastic methodologies. To keep this chapter to a reasonable length, an introduction to each of these techniques is provided along with references where more details can be found.

* Crystalline polymeric structures, which are held together by ionic interactions and do not have a distribution of molecular masses, are not in the same class as the synthetic polymers considered in this book. Typically, the crystalline materials are either insoluble or degrade into nonmacromolecular species on dissolution.

Inorganic and Organometallic Polymers, by Ronald D. Archer
ISBN 0-471-24187-3

3.1 AVERAGE MOLECULAR MASSES AND DEGREES OF POLYMERIZATION

Whereas chemical molecules typically have a definite **molecular mass** (or classically, molecular weight), soluble macromolecules have a range of molecular masses within every synthesis. Thus an **average molecular mass**, $\overline{M}$, is more appropriate than a simple molecular mass.

These average molecular masses (or weights) have distributions that provide different results depending on the nature of the measurement. Two common and important averages are the **number-average** ($\overline{M}_N$) and the **weight-average** ($\overline{M}_W$) molecular masses, which are defined as follows:

$$\overline{M}_N = (N_1M_1 + N_2M_2 + \cdots)/(N_1 + N_2 + \cdots) = \sum_i N_iM_i / \sum_i N_i \tag{3.1}$$

$$\overline{M}_W = (N_1M_1^2 + N_2M_2^2 + \cdots)/(N_1M_1 + N_2M_2 + \cdots) = \sum_i N_iM_i^2 / \sum_i N_iM_i \tag{3.2}$$

where N_i is the number of molecules that is an i-mer (i.e., an oligomer or polymer with i repeating units) and M_i is the mass of the i-mer. Although these two ($\overline{M}_N$ and $\overline{M}_W$) are the most commonly quoted averages, other averages of molecular mass also exist and will be noted as needed.

As noted in the Chapter 2, the **average degree of polymerization** ($\overline{DP}$) for a two-component step polymerization is almost twice the average number of repeating units ($\overline{n}$), specifically, $\overline{DP} = 2\overline{n} - 1$. Also, for perfect stoichiometry

$$\overline{DP} = 1/(1 - \rho) \tag{3.3}$$

where ρ is the **extent of reaction** [i.e., the fraction of the molecules whose functional groups have reacted to become part of a polymer (or oligomer) chain, cf. Table 2.2]. The degree of polymerization is also dependent on the **stoichiometric ratio of reactants** (r) as noted in Chapter 2 (Table 2.2). For the rest of this discussion we will assume that the stoichiometric ratio is exactly 1:1.

For a siloxane condensation reaction in a nonaqueous solvent, the reaction would be:

$$n\mathrm{SiR_2Cl_2} + n\mathrm{H_2O} \rightarrow \mathrm{Cl[R_2SiO]}_n\mathrm{H} + 2n - 1\mathrm{HCl} \tag{3.4}$$

Assuming that the reaction is first order in the dialkyldichlorosilane and first order in water, the rate equation is

$$\frac{-d[\mathrm{SiR_2Cl_2}]}{dt} = k_2[\mathrm{SiR_2Cl_2}][\mathrm{H_2O}] \tag{3.5}$$

If $[\mathrm{SiR_2Cl_2}] = [\mathrm{H_2O}]$, as is necessary for a high-molecular-mass polymerization, then

$$\frac{-d[\mathrm{SiR_2Cl_2}]}{dt} = k_2[\mathrm{SiR_2Cl_2}]^2 \tag{3.6}$$

Separation of the variables and integration from $t = 0$ to $t = t$ gives the usual second-order rate expression

$$\frac{1}{[SiR_2Cl_2]} = \frac{1}{[SiR_2Cl_2]_0} + k_2t \tag{3.7}$$

However, by definition ρ is the fraction of reactive groups that have reacted, that is,

$$\rho = 1 - [SiR_2Cl_2]/[SiR_2Cl_2]_0 \tag{3.8}$$

or by rearrangement:

$$[SiR_2Cl_2] = [SiR_2Cl_2]_0(1 - \rho) \tag{3.9}$$

Substituting Eq. 3.9 in Eq. 3.7:

$$\frac{1}{[SiR_2Cl_2]_0(1 - \rho)} = \frac{1}{[SiR_2Cl_2]_0} + k_2t \tag{3.10}$$

and multiplying through by $[SiR_2Cl_2]_0$ gives

$$1/(1 - \rho) = 1 + [SiR_2Cl_2]_0k_2t \tag{3.11}$$

(cf. Exercise 3.1).

A graph of this relationship is shown schematically in Figure 3.1, where $1/(1 - \rho)$ is plotted as a function of time. Note that this plot is equivalent to plotting $\overline{DP}$ as a function of time (cf. Eq. 3.3).

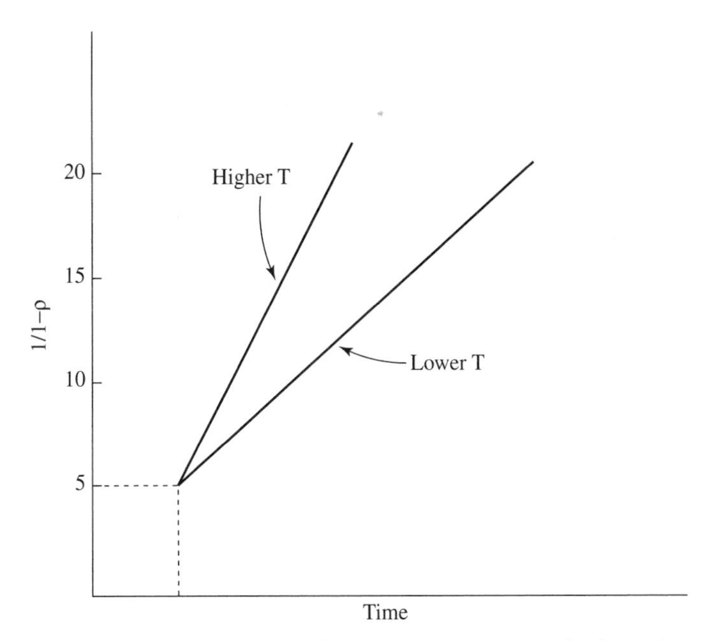

Figure 3.1 Plot of $1/(1 - \rho)$ vs. time for a simple step polymerization, where $\rho =$ the extent of reaction.

Reactions that require an acid or base catalyst may have a different time dependency. Specifically, an esterification reaction (acid catalyzed) that has no acid added (other than the acid being esterified) is third order in the carboxylic acid. Linear plots analogous to Figure 3.1 for this third-order reaction are time vs. $1/(1-\rho^2$ plots. On the other hand, added acid causes a second-order dependence on the carboxylic acid, and the time dependency reverts to that shown in Figure 3.1 (1). These graphs are based on the assumption that the reaction rates of oligomers are fairly invariant with size (after the first few oligomerization steps). This assumption has been shown to be true for a number of model condensation systems (2). This generality is logical because the rates of virtually all step polymerization reactions are much slower than diffusion rates (1).

Another aspect of interest is the effect that the extent of reaction ρ has on polymerization mass distributions. The distributions are shown in Figures 3.2 to 3.4. Figure 3.2 shows the effect that ρ has on the number-average molecular mass using Eq. 3.12.

$$\overline{M}_N = [\mathrm{M}_0/(1-\rho)] + \Sigma(\text{endgroups}) \tag{3.12}$$

This equation is based on

1. The average degree of polymerization $\overline{DP}$ being equal to $1/(1-\rho)$ (Eq. 3.3),
2. the number-average molecular mass $\overline{M}_{\mathrm{N}}$ being the mass of the monomer (M_0) times $\overline{DP}$, with M_0 being the average of the two components for a two-component condensation or the mass of a one-component condensation, and

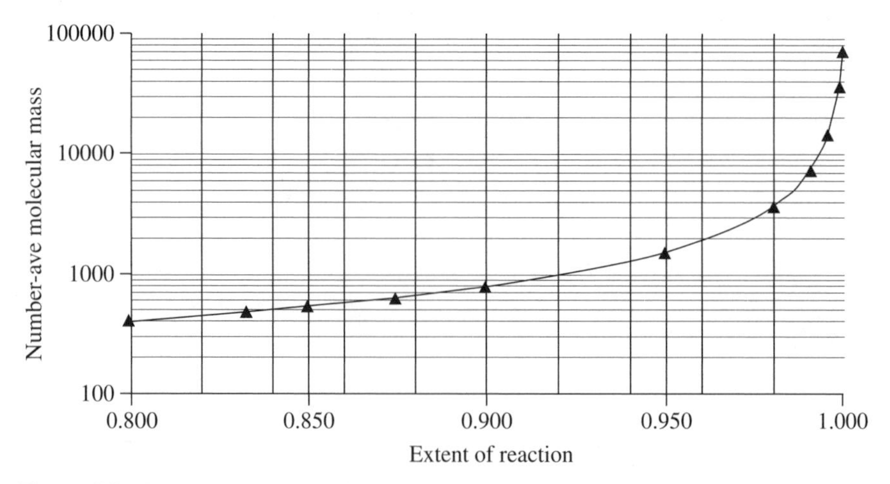

Figure 3.2 Number-average molecular weight vs. extent of reaction for a simple step polymerization reaction.

3. Σ (end groups) will be 18 if water is the condensation product, 36.5 if HCl is the condensation product, etc. That is, the average polymer molecule has one of each type of end group, which is true for good stoichiometry.

Note that a very large extent of reaction ($\rho > 0.99$) is necessary to obtain an average molecular mass above 10,000.

Figure 3.3 shows the logarithm of the most probable number of molecules vs. mass for step condensation reactions with ρ values of 0.90 and 0.99 based on a standard statistical evaluation assuming that the reactivity of the functional groups are invariant with size, that is, according to Flory (3):

$$N_i/N_o = (1-\rho)^2\rho^{(i-1)} \tag{3.13}$$

where N_i = the number of molecules with i structural units (an i-mer) and N_0 = the total number of molecules. Note that at very large extents of reaction, the actual number of small molecules is still quite large.

Multiplying the probabilities for the selected ρ values for each i value (Fig. 3.3) by i:

$$W_i = i(N_i/N_o) \tag{3.14}$$

leads to the weight-fraction graph (Fig. 3.4). Such a weight-fraction graph of a polymer is related to the $\overline{M}_N$ for the polymer; that is,

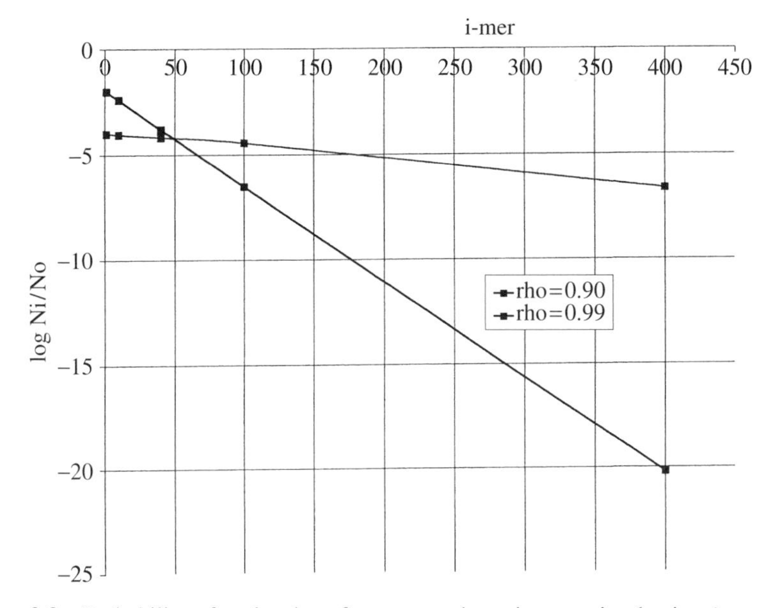

Figure 3.3 Probability of molecules of a step condensation reaction having i structural units (an i-mer) at extent of reaction (ρ) values of 0.90 and 0.99, where N_i = number of i-mers and N_o = total number of molecules.

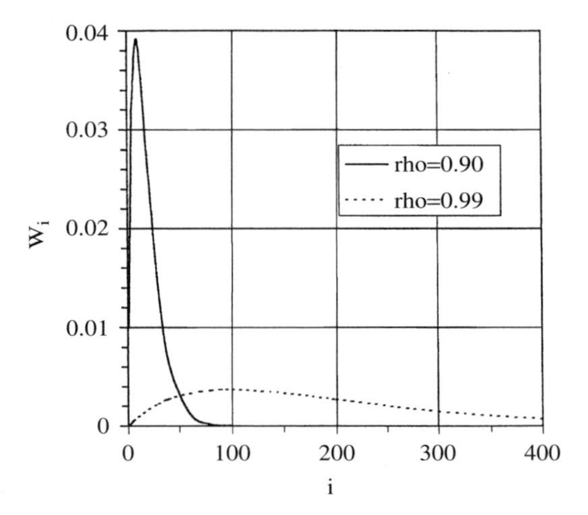

Figure 3.4 Weight-fraction graph of step polymerization using the data shown in Figure 3.3.

1. $\overline{M}_{\mathrm{N}}$ is proportional to the sum of the individual N_i values multiplied by the masses of the i-mers (see Eq. 3.1), and
2. the i-mer mass is i times the monomer mass, M_0, or the average monomer mass for a two-component system.

Thus Figure 3.4 graphs the individual i-mer components, except that they have not been multiplied by the monomer mass, M_0.

Also, using similar probability arguments for the weight-average molecular mass of a step polymerization, the result is

$$\overline{M}_{\mathrm{W}} = (1 + \rho)/(1 - \rho) \tag{3.15}$$

if the minor contributions from the end groups are neglected. And because the number-average relationship (Eq. 3.12 without the minor end-group contribution) is

$$\overline{M}_{\mathrm{N}} = 1/(1 - \rho) \tag{3.16}$$

the weight-average molecular mass is

$$\overline{M}_{\mathrm{W}} = \overline{M}_{\mathrm{N}}(1 + \rho) \tag{3.17}$$

Note that for a reaction that approaches 100%, the weight-average molecular mass for a step polymerization becomes *double* the number-average molecular mass (cf. Exercise 3.2). When comparing molecular mass results from different research papers. it is important to note the average that has been determined in each case. Also, as will become evident from the discussion of the methods of determining these averages, the method or methods used are also critical in evaluation of molecular mass results.

3.2 METHODS OF CHARACTERIZING AVERAGE MOLECULAR MASSES

3.2.1 Gel Permeation Chromatography

Gel permeation chromatography (GPC) or **size exclusion chromatography** (SEC) is a common technique that can be used to evaluate the average molecular mass (both $\overline{M}_N$ and $\overline{M}_W$) as well as the mass distribution of a polymer. A schematic representation of a GPC instrument is shown in Figure 3.5. The separation is based on the fact that larger molecules are unable to diffuse into the small pores of the cross-linked gel column and thus they pass through the column more rapidly than the smaller molecules. The smaller molecules spend time in the small pores on their way through the column. In the range of molecular masses for which the column is useful, the time on the column is approximately proportional to the inverse of the logarithm of the molecular masses of the molecules. See Figure 3.6 for an example of a GPC column calibration curve. Normally, a series of "monodisperse" polystyrene samples (dissolved in the solvent that is to be used with the unknown polymer samples) is used to calibrate the GPC column. The GPC columns may be listed according to the range of molecular masses for which they are effective or the pore size of the column. Columns of different porosity can be used in tandem to improve the resolution and the molecular mass range that can be analyzed. An extended-range mixed-bed column is a newer alternative for broad molecular mass determinations. Table 3.1 provides a sampling of

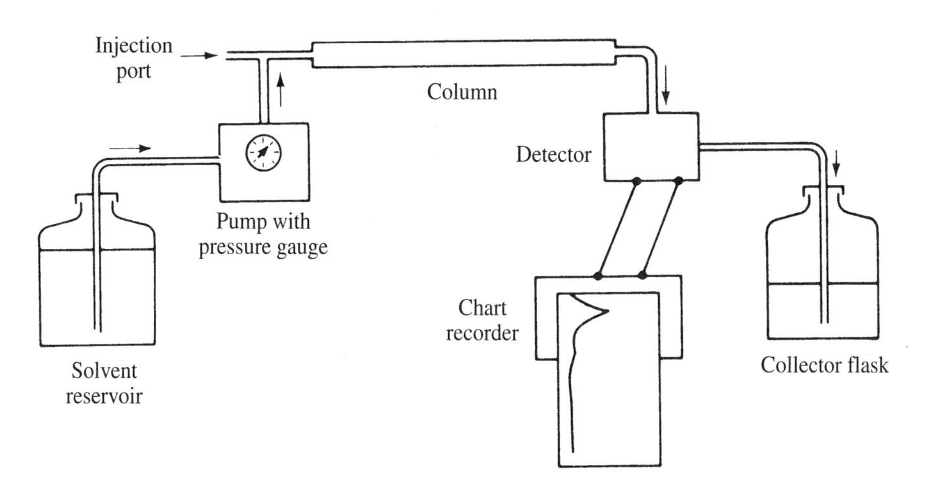

Figure 3.5 Schematic diagram of a gel permeation chromatograph for GPC measurements. From p. 396, *Contemporary Polymer Chemistry*, 2/E by Allcock/Lampe, © 1990. Adapted by permission of Prentice-Hall, Inc., Upper Saddle River, NJ.

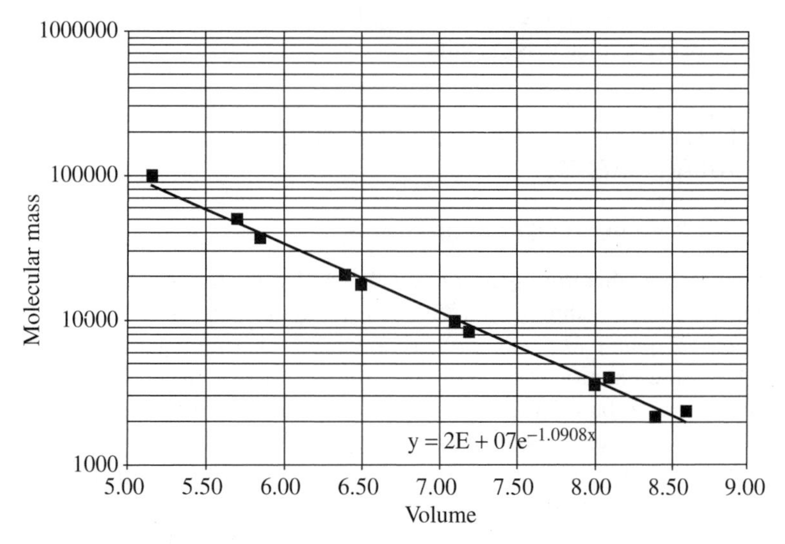

Figure 3.6 Example of a gel permeation chromatography column calibration curve — a 10^3 Å Ultrastyragel® column using monodisperse polystyrene standards in NMP. (Calibration by C. J. Hardiman).

TABLE 3.1 Molecular Mass Ranges Accessible Through a Series of High-Resolution Styragel® GPC Columns Provided by Waters Corporation[a].

Column	Effective Molecular Mass
HR 1	100–5,000
HR 2	500–20,000
HR 3	500–30,000
HR 4	5,000–600,000
HR 5	50,000–4,000,000
HR 6	200,000–10,000,000
HR 4E	50–100,000
HR 5E	2,000–4,000,000

[a] Syragel is Waters' trademark for their styrene divinylbenzene particle column packings.

high-resolution columns available for "low molecular weight analysis" by one manufacturer. Note that by choosing an appropriate column almost any molecular mass range can be determined (at least on a relative basis). Because the columns discussed so far are gels, they must be kept in a solvent matrix *at all times*. Although columns are typically available from suppliers in several solvents, the change to another solvent requires extensive flushing of the column with the new solvent. It is also important to note that the gels are only swelled sufficiently for use with solvents in which the noncross-linked gel substrate would be soluble. For the normal cross-linked polystyrene columns, this allows solvents as diverse as tetrahydrofuran (THF), benzene, toluene and many other substituted phenyl derivatives, dimethyl formamide (DMF) and dimethyl acetamide (DMA), *N*-methylpyrrolidone (NMP), chloroform, fluorinated alcohols, etc.

For solvents that cannot maintain the gel nature of cross-linked polystyrene (water, alcohols, etc.) porous glass has been used as a size exclusion chromatographic column. Unfortunately, adsorption often occurs on such polar stationary substrates. If adsorption is severe, elution is sometimes impossible. Even with less severe adsorption, differential absorption may distort the GPC results.

The detector is typically a refractive index sensor, although a detector can be an ultraviolet monochromatic or photodiode array sensor, a micro viscometer or micro light-scattering unit. The pump must provide constant reproducible pressures, and a thermostated column provides improved reproducibility.

Although both $\overline{M}_{\mathrm{N}}$ and $\overline{M}_{\mathrm{W}}$ can be evaluated by analyzing the GPC curves, the method is empirical and the standard may or may not be suitable for replicating molecular masses for specific polymers of interest. Because the solvent volume for elution (or the time required to traverse a specific GPC column) is actually related to size, a random-coil polymer appears to be smaller than a rigid-rod polymer of the same molecular mass. Polymer confirmations (e.g., the amount of coiling of the polymer chains) are typically solvent dependent; therefore, it is essential to use the same solvent for measurements as had been used in the standardization of the column.

As noted in Section 3.1, the number-average molecular mass ($\overline{M}_{\mathrm{N}}$) is based on the number of molecules of each mass times the mass summed over all mass values and divided by the total number of molecules. Assuming that the change in refractive index relative to the solvent is proportional to number of units (i) of each individual molecule (i-mer) passing through the detector, the area of a plot of the change in the refractive index vs. retention-volume is integrated. The retention-volume value at 1/2 of the area under the curve is converted to the corresponding molecular mass based on calibration standards. This value is the number-average molecular mass ($\overline{M}_{\mathrm{N}}$). A similar procedure is used to determine the value of the weight-average value ($\overline{M}_{\mathrm{W}}$) except that each point on the plot must be multiplied by the molecular mass of that retention volume. The same concept also holds for ultraviolet or visible light detectors because the absorption produced by each molecule should be proportional to the number of chromophores in each molecule (i.e., proportional to i for each i-mer).

The number-average molecular mass ($\overline{M}_{\mathrm{N}}$) values obtained from the GPC results are quite reasonable and are very good if the column has been calibrated with polymers of known molecular masses and structures similar to the masses to be determined. These $\overline{M}_{\mathrm{N}}$ values can be improved even further through viscosity measurements and universal calibration (as noted below).

The precision of weight-average molecular mass ($\overline{M}_{\mathrm{W}}$) is often poor for polymers with broad distributions. The high-molecular-mass frontal region has an undue influence on the results if the baseline is not perfectly flat. Small

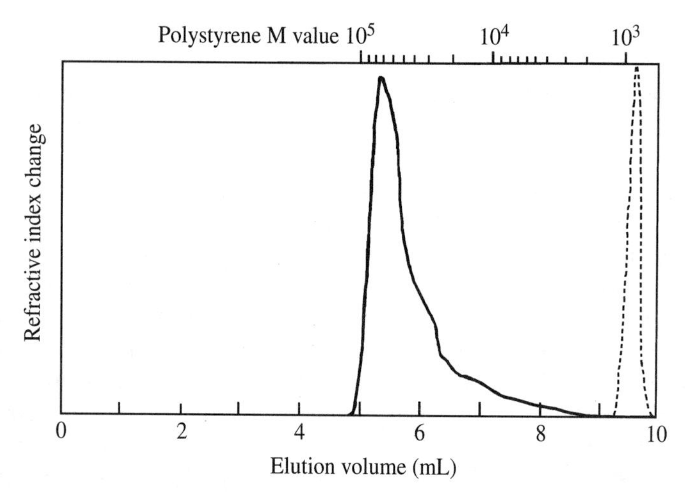

Figure 3.7 Gel permeation chromatogram of a metal coordination polymer prepared by step polymerization—poly(1,2,4,5-tetraaminobenzenezirconium(IV)) in NMP with a 10^3 Å Ultrastyragel® column. (chromatogram by C. J. Hardiman).

molecules in nonfractionated polymer samples can influence both number-average and weight-average molecular mass results if these molecules affect the detector. In fact, one paper (4) notes that GPC $\overline{M}_N$ values are comparable to membrane osmometry $\overline{M}_N$ values only if the portion of the GPC curve representing mass values below 10,000 is eliminated. Of course, this difference is due to the fact that low-molecular-mass species pass through the membrane used in the osmometry. Because low-molecular-mass oligomers are always present in step polymerization syntheses before fractionation, this difference is logical and does not actually invalidate the GPC results.

A typical GPC result for a polymer made by step polymerization is shown in Figure 3.7. This particular chromatogram resulted from the chromatographic analysis of a metal coordination polymer in NMP with a Styragel(r) column (cf. Exercise 3.3).

3.2.1.1 Nonideality

Up to this point, the discussion of gel permeation chromatography has assumed fairly simple polymers that are soluble in solvents suitable for GPC measurements. However, complications can arise in the evaluation of newly synthesized polymeric materials.

The most common problem with metal-containing polymers is finding a solvent that will dissolve the polymer sufficiently for molecular mass measurements in solution without degrading the column. The present author has found that NMP is quite suitable for linear metal-containing polymers. Dimethyl sulfoxide (DMSO) is too harsh on standard GPC columns. DMA and DMF have also been used for coordination polymers by other authors. Naturally, the solvent must also be compatible with the column being used as noted above in the discussion of GPC columns.

Another common problem is a GPC curve that is polymodal (i.e., the curve has more than one peak). If one of the peaks is fairly sharp and occurs where a trimer, tetramer, or hexamer would occur, ring formation is probably occurring. Small rings of siloxanes and phosphazenes are quite common and are actually used for ring-opening polymerizations as noted in Chapter 2. Actually, two or three different rings can be formed, and therefore the GPC curve can be even more complicated.

Also, if the two end groups of a linear polymer interact quite differently with the solvent being used for the GPC evaluation, the solvent volume required for a polymer of the same mass to traverse the column can be end group-dependent. Because approximately 25% of the molecules in a condensation polymer have the more polar end group on both ends, 50% have one of each end group, and 25% have the less polar end group on both ends, bimodal or trimodal curves can occur.

Other problems common to all types of chromatography include column overloading and the adsorption of polymers on the GPC columns. Overloading is not much of a problem given the low solubility of most metal-containing polymers, but it can be a problem with some polysiloxanes. Instructions come with most columns that indicate the loading levels that can be tolerated by the column. The adsorption problem has already been noted in the discussion of column materials above.

Despite the problems noted above, GPC measurements are very useful in determining the extent of polymerization that has occurred in a step polymerization reaction. Together with other molecular mass measurements, the shapes of the GPC curves provide valuable information to the synthetic inorganic polymer chemist.

More sophisticated multidimensional and multidetector chromatographic methods have been developed to obtain better molecular mass information on polymers that may be branched and/or have different functionalities and copolymers that may be random, block, or graft in nature. The reader is referred to a review paper by Trathnigg (5) for more details on both these more sophisticated methods and for a more thorough evaluation of the pitfalls of standard GPC (or SEC in the noted article).

3.2.2 Viscosity

Early in the twentieth century, Einstein noted that the increased viscosity of solutions containing large particles is directly related to the volume fraction of the particles. Because the size of a polymeric molecule is related to the mass of the polymer and viscosity measurements are relatively easy to do and reproduce, viscosity measurements have become a prominent method of evaluating the size and mass of polymeric species. The simple, inexpensive equipment necessary for viscosity measurements also contributes to the popularity of the method.

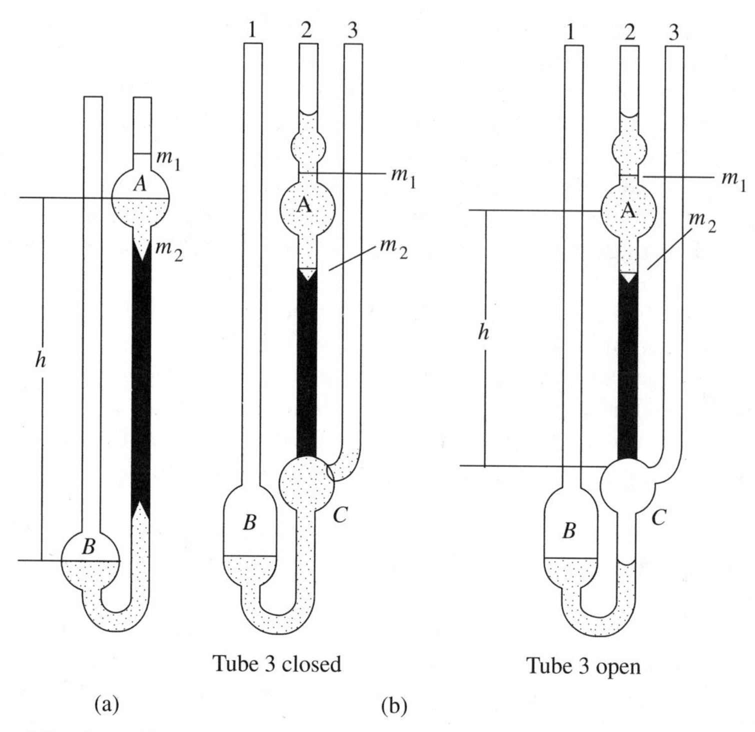

Figure 3.8 Ostwald (a) and Ubbelohde (b) viscometers (reprinted in part with permission from Ubbelohde, L. *Ind. Engr. Chem., Anal. Ed.*, 1937, **9**, 85; © American Chemical Society).

An **Ubbelohde** or **Ostwald viscometer** (Fig. 3.8), a constant-temperature bath, and a stop watch are all that are needed for making viscosity measurements on soluble inorganic polymers. The time required for a solvent or solution in a viscometer to drop from one etched mark to a second mark depends on the average driving force for liquid flow. This force is based on the average difference in the heights of the top and bottom of the flowing liquid column. With an Ubbelohde viscometer, exact volumes of solution or solvent are not required in the viscometer to obtain a constant average height from one measurement to another — the simpler Ostwald viscometer requires exact volumes for reliable measurements. The viscosity of a liquid is quite temperature sensitive, so viscosity measurements must be made in a constant-temperature bath for reliable results. With a good stopwatch, most individuals can obtain reproducible times to within a few one-hundredths of a second. Viscometers with different capillary bores allow solutions in solvents with quite different viscosities to be measured in a reasonable time frame (1–2 min), although that may seem like a long time for measurements that can be reproduced to a small fraction of a second. However, the difference in times for the solvent and dilute solutions is quite small unless the polymers are very large.

Some of the equations used to determine the **intrinsic viscosity** (the viscosity as zero concentration) of inorganic polymer solutions are as follows:

$$\eta_r = \eta/\eta_0 = t/t_0 \tag{3.18}$$

where η_r = **relative viscosity** (a unitless ratio), η is the measured viscosity, η_0 is the viscosity of the solvent, t is the measured time for the solution in the viscometer, and t_0 is the reference time for the solvent under same conditions. This is related to the **specific viscosity**, η_{sp}, which is also unitless:

$$\eta_{sp} = (\eta - \eta_0)/\eta_0 = \eta_r - 1 \tag{3.19}$$

The specific viscosity divided by the concentration, or η_{sp}/c, is called the **reduced viscosity** or reduced specific viscosity. Extrapolation of the reduced viscosity to zero concentration leads to the **intrinsic viscosity**, $[\eta]_i$. That is,

$$\lim_{c \to 0}(\eta_{sp}/c) = [\eta]_i \tag{3.20}$$

Thus a plot of specific viscosity vs. concentration can be used to obtain the extrapolated $[\eta]_i$ value; however, a better method is noted below. The extrapolation eliminates the concentration effects, but because there are first- and second-order concentration effects, the viscosities should be measured at concentrations low enough to minimize the second-order effects but high enough to get reasonable differences between t and t_0.

An alternate definition (6) of intrinsic viscosity is

$$\lim_{c \to 0}(\ln \eta_r/c) = [\eta]_i \tag{3.21}$$

A series expansion of the natural logarithm of the relative viscosity demonstrates the equivalence between Eqs. 3.20 and 3.21 (2). The latter definition (Eq. 3.21) implies that a plot of the natural logarithm of the relative viscosity divided by the concentration versus the concentration and extrapolated to zero concentration should also provide the intrinsic viscosity. In fact, this $\ln \eta_r/c$ vs. c plot is the preferred plot because the change in $\ln \eta_r/c$ is less than the change in η_{sp}/c for the same concentration range (2). Thus the precision of the intrinsic viscosity obtained by extrapolation is improved with a $\ln \eta_r/c$ vs. c plot. A $\ln \eta_r/c$ vs. c plot of a fractionated sample of a metal-containing polymer is shown in Figure 3.9.

Note that both the specific and the intrinsic viscosity values have the units of reciprocal concentration and are currently reported in units of cubic centimeters per gram (cm^3/g). However, the units of deciliter per gram (dL/g) were commonly used until quite recently. One dL/g is equivalent to 100 cm^3/g; thus the newer cm^3/g values appear to be two orders of magnitude larger than the earlier dL/g values found in the earlier polymer science literature.

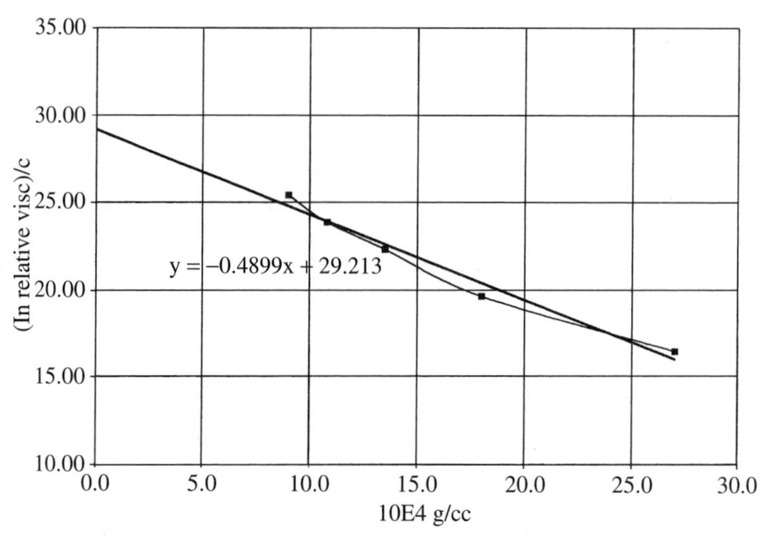

Figure 3.9 An (ln η_r)/c vs. concentration plot for a fractionated sample of a cerium Schiff-base polymer in NMP at 30 °C. (Results obtained by H. Chen.)

Viscosity measurements on fractionated polymers can provide good estimates of molecular masses. However, like GPC, the viscosity method is not an absolute method of molecular mass determination. Large randomly coiled polymer molecules manifest a somewhat different effect on the viscosity than rigid-rod molecules, and polyelectrolytes pose even further problems. However, viscosity provides an inexpensive method for obtaining approximate molecular mass measurements after calibration with standards of similar polymers. Extrapolation of viscosity measurements with oligomers and low-molecular-mass polymers to higher-molecular-mass polymers of the same type is another approach for obtaining quite satisfactory molecular mass results (7).

For a fractionated polymer, a plot of the logarithms of the intrinsic viscosities of each fraction vs. the logarithms of their molecular masses (determined by other means) is linear. Equation 3.22 is the logarithmic version of the Mark–Houwink equation (Eq. 3.23).

$$\log[\eta]_i = \log K + a \log M_i \tag{3.22}$$

$$[\eta]_i = KM_i^a \tag{3.23}$$

K and a are constants specific to an individual polymer in a specific solvent at a specific temperature and are only valid if obtained on well-fractionated samples of the polymer, where M_i is the molecular mass of the fraction. An example of a Mark–Houwink plot for a fractionated metal-containing polymer is shown in Figure 3.10. The constants are typical of well-behaved random-coil polymers.

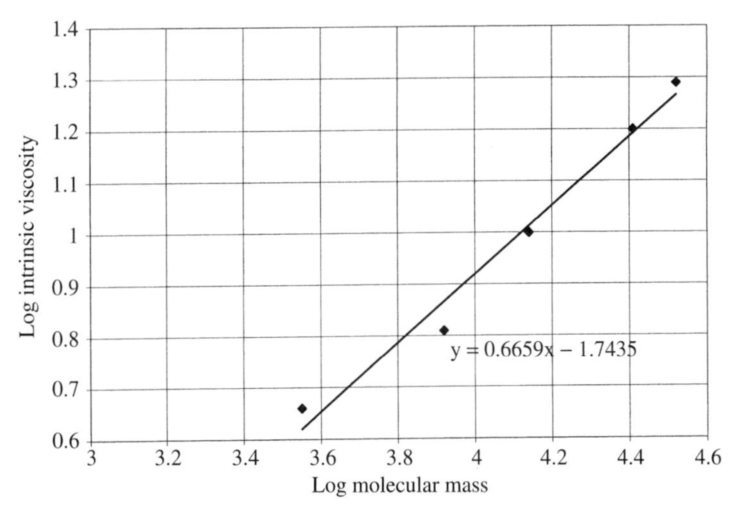

Figure 3.10 A Mark-Houwink plot of a fractionated zirconium Schiff-base polymer in NMP with molecular mass from end-group analysis—with $K = 1.8 \times 10^{-2}$ and $a = 0.67$. (Results obtained by B. Wang.)

The range of a values observed for organic and inorganic polymers is from 0.50 to a little less than 1.00. A value of 0.50 is anticipated for a polymer in a good theta solvent. (A solvent is a theta solvent for a polymer at the temperature at which the polymer-polymer, the polymer-solvent, and the solvent-solvent interactions have comparable free energies).

K values have been tabulated for a large number of organic polymers in a variety of solvents. The values range from about 1.7×10^{-3} cm^3/g for gelatin in water to 3.0×10^{-1} cm^3/g for poly(vinyl alcohol) in water.

Caution: K values are unit dependent; that is, the dL/g values in the older literature must be converted to cm^3/g units.

The molecular mass determined by viscometry is termed $\overline{M}_V$. It can be shown (1,2) that $\overline{M}_V$ is closer to $\overline{M}_W$ than to $\overline{M}_N$. For highly fractionated polymers the values are all close together, but for the product of a nonfractionated step-polymerization reaction, this difference is quite significant. As noted earlier in this chapter, $\overline{M}_W$ is approximately twice the value of $\overline{M}_N$ for a step-polymerization reaction product. For such a situation $\overline{M}_V$ will range from 1.75 $\overline{M}_N$ (or about 0.88 $\overline{M}_W$) for $a = 0.50$ to almost 2.0 $\overline{M}_N$ when a approaches 1.0, at which point $\overline{M}_V = \overline{M}_W$.

As noted in the Chapter 2, stoichiometry variation is used to obtain maximum molecular masses in step-growth syntheses. Viscosity values can be used to ascertain whether the proper stoichiometry is being used. Typical results

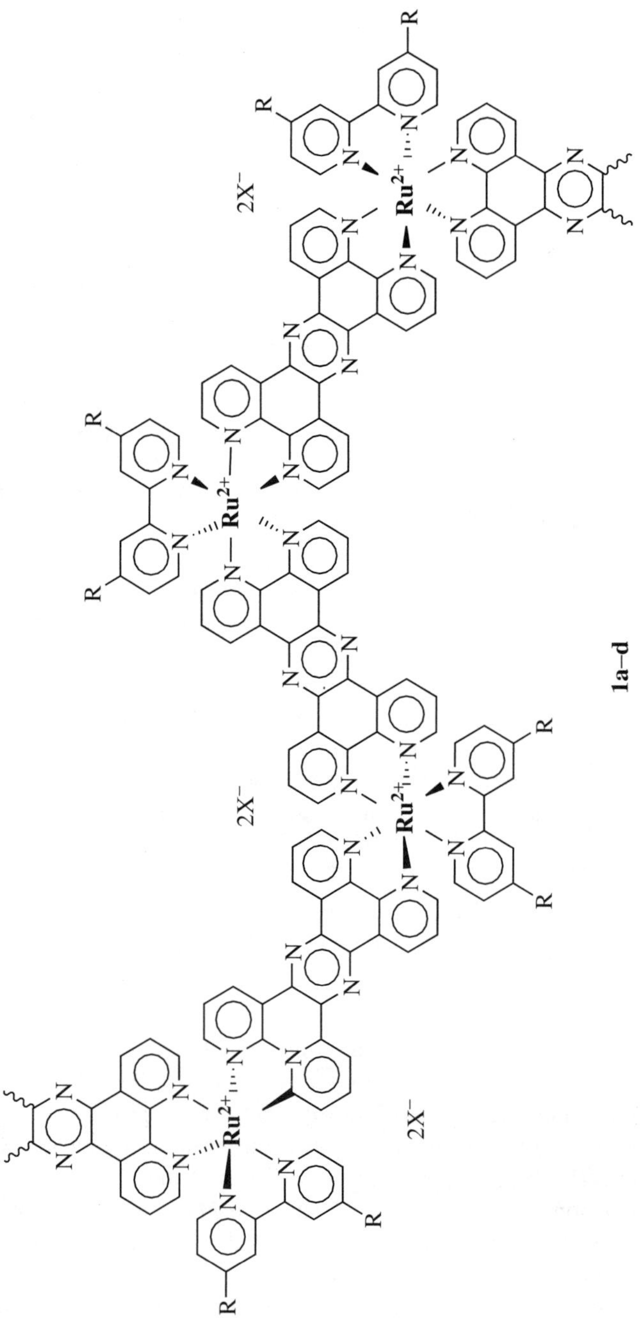

$2X^-$ Ru^{2+}

1a–d

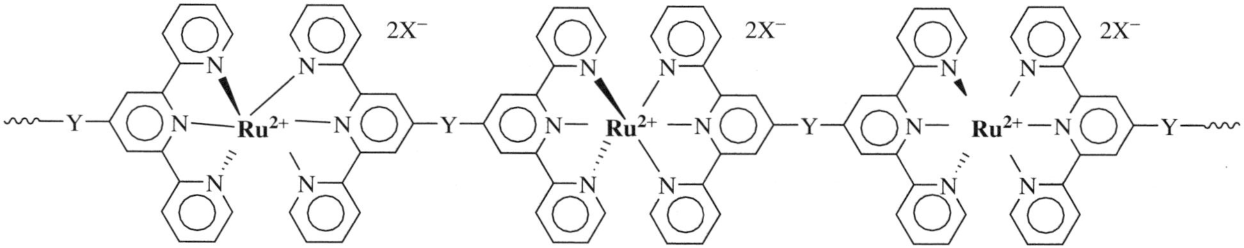

2

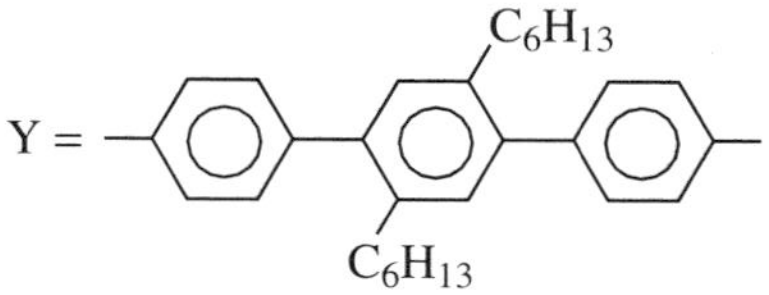

Y = [p-phenylene–(2,5-di-C_6H_{13}-phenylene)–p-phenylene]

R = a : H

b : CH_3

c : C_6H_5

d : $CH_2-C_{10}H_{15}$

X^- : Cl^-, BF_4^-, PF_6^-,...

Figure 3.11 Ruthenium polyelectrolytes, both normal (**1a–d**) and rigid rod (**2**) (reprinted with permission from Kelch and Rehahn, *Macromolecules*, 1999, **32**, 5818; © 1999 American Chemical Society).

TABLE 3.2 Degrees of Polymerization vs. Stoichiometry for a Cerium(IV) Schiff-Base Polymer[a].

Ce^{IV} : $tsdb^{4-}$ Stoichiometry	η_i (cm^3/g)	$\overline{M}_N$	$\overline{DP}$
0.990:1.000	6.7	7900	19
0.995:1.000	12.6	18,100	45
1.000:1.000	15.1	23,100	58
1.000:0.995	10.2	13,800	34
1.000:0.990	6.4	7500	18

[a] $tsdb^{4-}$ = anion of N,N′,N″,N‴tetrasalicylidene-3,3′-diaminobenzidine; the $\overline{M}_N$ values are based on the Mark-Houwink parameters for this $[Ce(tsdb)]_n$ polymer using NMR end-group calibration.

are shown in Table 3.2 for one of the cerium(IV) Schiff-base polymers. A Mark–Houwink plot and the appropriate constants had been developed using NMR end-group molecular mass values; thus, the term $\overline{M}_N$ is used, even though viscosity values are being used. Naturally, the $\overline{M}_N$ values could be multiplied by 1.88 (based on the a value of 0.76) to give the appropriate $\overline{M}_V$ values.

Polyelectrolytes, especially rodlike metal-containing polymers, provide much better viscosity concentration plots for extrapolation to zero concentration when measured under high-ionic-strength conditions. Kelch and Rehahn (8) found dramatic improvement in viscosity results using 0.02 M NH_4PF_6 in DMA for rodlike terpyridine-based ruthenium(II) coordination polyelectrolytes (Fig. 3.11) [compare Fig. 3.12 (added ionic strength) with Fig. 3.13]. Note that a direct comparison of viscosities between rigid rod and coiled polyelectrolytes is not possible even under the high-ionic-strength conditions.

Uncapped polysiloxanes, polysilanes, and zirconium Schiff-base polymers can couple with silica and change the bore of the capillary of a new viscometer. To ensure that no problem like this has occurred, the solvent t_0 should be checked before and after measurements of the polymer solutions. A small amount of undissolved polymer can manifest a similar effect. Prefiltration of the solutions used in the viscosity measurements can minimize this latter problem.

3.2.3 Universal Calibration

Even though both GPC and viscosity measurements only provide relative molecular mass results (as has been emphasized in the past two sections of this chapter), when they are used together improved molecular mass accuracy can be obtained. Plotting the logarithms of the products of the intrinsic viscosities times the molecular masses ($\log\{[\eta]M_i\}$) vs. the elution volumes on a specific GPC column for monodisperse or well-fractionated polymers with known molecular masses provides a single curve, even if a wide variety of polymers are included. Naturally, a common solvent and a constant temperature are required for this plot to work. A classical example that includes poly(phenyl siloxane) along with a

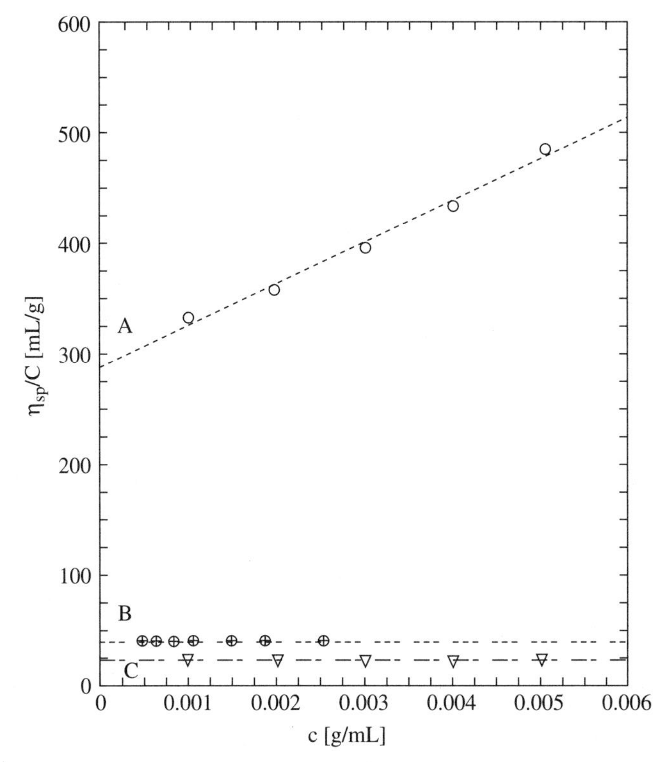

Figure 3.12 Viscosity measurements of ruthenium polyelectrolytes (Fig. 3.11) in 0.02 M ionic strength dimethyl acetamide: A: O = rigid-rod ruthenium polyelectrolyte (**2a**) with $\overline{DP} \geq 30$; B: ⊕ = same rigid-rod ruthenium polyelectrolyte (**2a**) with <10; and C: ∇ = coiled ruthenium polyelectrolyte (**1d**) with $\overline{DP} \geq 30$ (reprinted with permission from Kelch and Rehahn, *Macromolecules*, 1999, **32**, 5818; © 1999 American Chemical Society).

TABLE 3.3 Chromium Polymers: Number-Average Molecular-Weight Estimates.

Polymer	Viscosity Estimates[a]	GPC Estimates[b]	Univ. Calib. Estimates[c]
$[Cr(acacCHO)(acac_2S)]_n$	≥14,000	≥14,800	≥15,800
$[Cr(acacCHO)(acac_2S_2)]_n$	≥15,000	≥17,800	≥21,000
$[Cr(tfac)(acac_2S)]_n$	*d*	≥16,300	*d*
$[Cr(tfac)(acac_2S_2)]_n$	≥17,000	≥17,000	≥15,100

[a]Based on intrinsic viscosities relative to a log-log plot of the intrinsic viscosities vs. the NMR estimates for analogous cobalt polymers.

[b]Based on retention volumes relative to a plot of the GPC retention volumes vs. the logarithms of the NMR estimates for analogous cobalt polymers.

[c]Number-average molecular weights based on the universal calibration method {plot of retention volume vs. $\log(M \times [\eta]_i)$}.

[d]Not comparable — different solvent used for viscosity measurements.

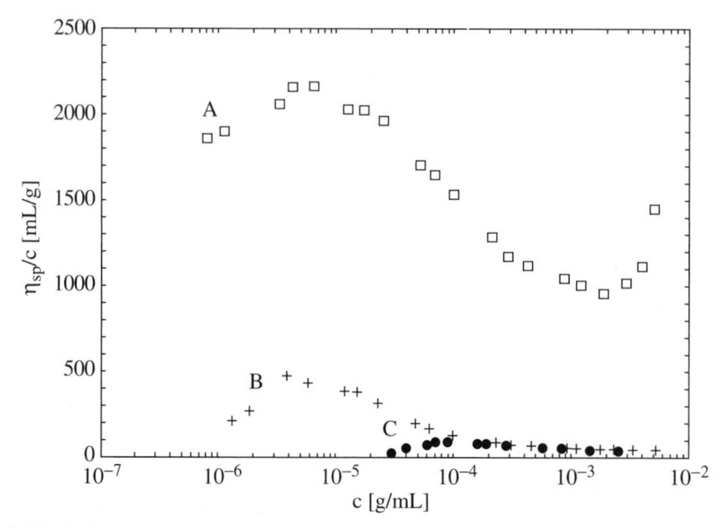

Figure 3.13 Viscosity measurements analogous to Figure 3.12 in neat dimethyl acetamide: *A*: □ = rigid-rod ruthenium polyelectrolyte (**2a**) with $\overline{DP} \geq 30$; B: + = same rigid-rod ruthenium polyelectrolyte (**2a**) with $\overline{DP} < 10$; C: • coiled ruthenium polyelectrolyte (**1a**) with $\overline{DP} \geq 30$ (reprinted with permission from Kelch and Rehahn, *Macromolecules*, 1999, **32**, 5818; © 1999 American Chemical Society).

number of organic polymers including polystyrene, poly(methyl methacrylate), butadiene, and poly(vinyl chloride) is shown in Figure 3.14. Thus, at a given elution volume on the same GPC column, the product of the intrinsic viscosity with the molecular mass is a constant provided all measurements are made in the same solvent at the same temperature. A detailed mathematical analysis of the method has been provided by Allcock (1). This method can be used to determine the molecular mass of other polymers quite accurately provided:

1. No chemical interactions exist between the polymers and the GPC column gel (or porous glass).*
2. Rigid-rod polymers are not compared with normal random-coil polymers.
3. The measurements are conducted at dilute enough concentrations that interactions between individual polymer molecules are negligible.

The present author has used universal calibration to provide estimates of molecular mass for chromium(III) coordination polymers based on NMR end-group molecular mass estimates for cobalt(III) polymers together with the GPC and viscosity measurements on both sets of complexes. The results are shown in Table 3.3.

* Of course, if the polymer interacts with glass, problems with viscosity measurements can also occur as noted in Section 3.2.2.

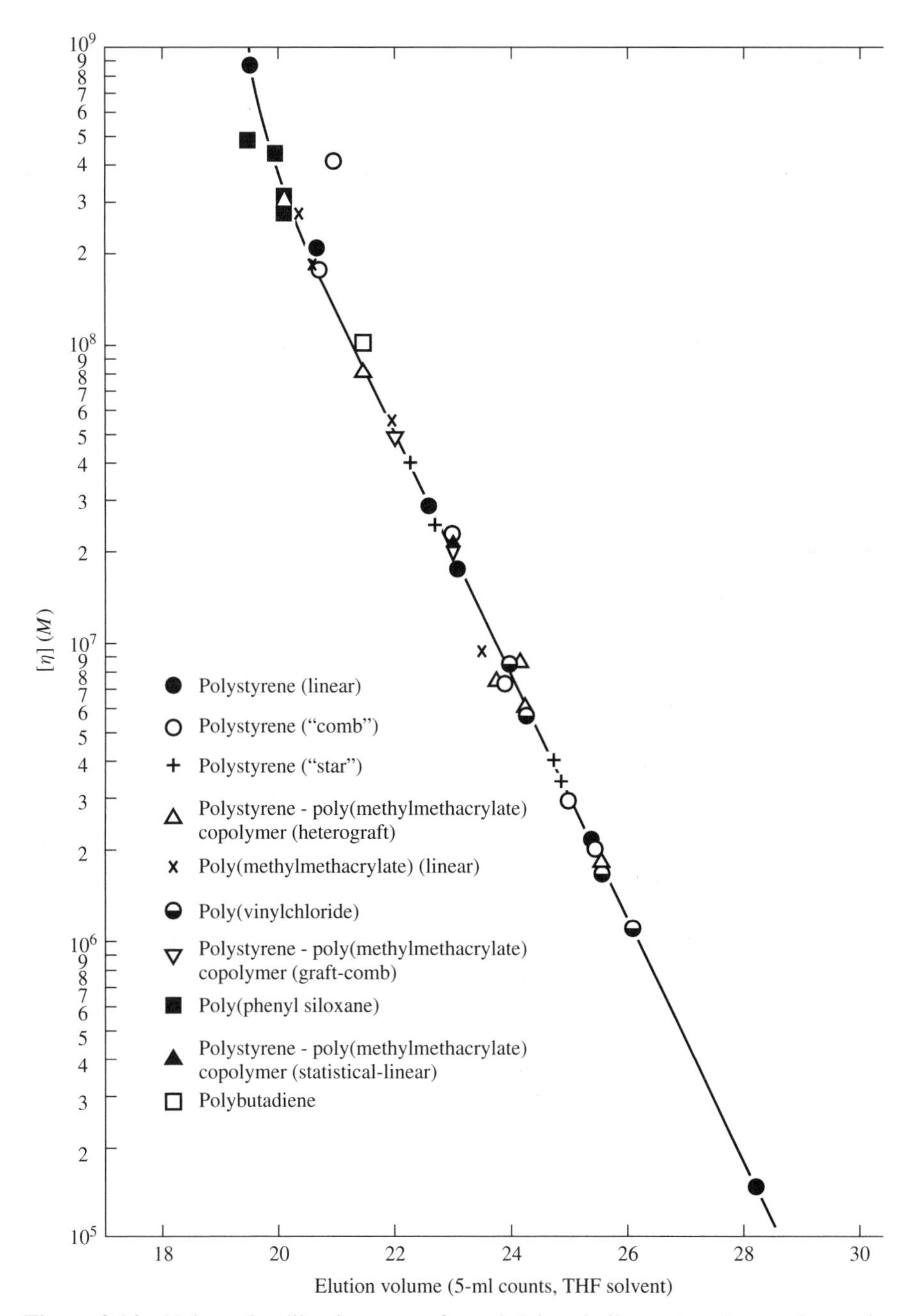

Figure 3.14 Universal calibration curve for poly(phenyl siloxane) and several organic polymers from Grubisic, Z.; Rempp, P.; Benoit, H. *Polymer Lett.*, 1967, **5**, 753.

3.2.4 Light Scattering for Absolute Molecular Mass and Size Measurements

Light scattering provides an "absolute" measure of the average molecular mass, specifically the weight-average molecular mass $\overline{M}_W$, of dilute solutions of polymers of fairly high molecular masses. One important equation that has been developed for this purpose is:

$$\lim_{c \to 0}(Hc/\tau) = 1/\overline{M}_W \tag{3.24}$$

where

$H = (32\pi^3/3\lambda^4 N_0)\tilde{n}_0^2(\tilde{n} - \tilde{n}_0/c)^2$
c = concentration in g/cm^3
τ = turbidity = I'_s/I_0
λ = wavelength
N_0 = Avogadro's number
$\tilde{n}_0$ = refractive index of the solvent
$\tilde{n}$ = refractive index of the solution
I'_s = total scattered intensity per unit path length
I_0 = incident light intensity

The light scattered by each polymer molecule contributes to the turbidity. However, the turbidity is normally too small to be measured at the low concentrations necessary to avoid extensive interparticle interactions that would further complicate the interpretation of the results. Instead, the turbidity can be determined from the Rayleigh ratio, R_θ. The Rayleigh ratio is the fraction of the incident light that is scattered at an angle θ according to the equation:

$$R_\theta = i_\theta r^2/I_0 \tag{3.25}$$

where

i_θ = excess intensity of scattered light per unit volume of solution relative to the solvent at angle θ
r = distance of measurement relative to center of cell
I_0 = incident light intensity as above.

Turbidity is then

$$\tau = (16\pi/3)R_\theta/(1 + \cos^2\theta) \tag{3.26}$$

and Eq. 3.24 becomes

$$\lim_{\substack{c \to 0 \\ \theta \to 0}} (\mathrm{Kc}/\mathrm{R}_\theta)(1 + \cos^2\theta) = 1/\overline{M}_\mathrm{W} \quad (3.27)$$

where $\mathrm{K} = \mathrm{H}(3/16\pi)$.

For modest-molecular-mass polymers, only a simple concentration extrapolation (to zero) should be necessary to eliminate **interparticle** interference. As the molecular mass increases, **intraparticle** interference becomes important and requires both a concentration and an angle extrapolation (to zero). The scattering functions are related to $\sin^2(\theta/2)$ so the extrapolation plot (Zimm plot) (9) is a plot of $\mathrm{Kc}/\mathrm{R}_\theta$ vs. $\boldsymbol{k}\mathbf{c} + \sin^2(\theta/2)$ as shown in Figure 3.15, where $\boldsymbol{k}$ is an arbitrary constant to provide an extrapolative grid (cf. Exercises 3.4 and 3.5).

A schematic of a laser light-scattering photometer for making such measurements is shown in Figure 3.16.

Because polymer molecules vary in structure in solution, corrections must be made that take account of the dissymmetry expected for the polymer in a given solvent. The dissymmetry can be estimated from the intrinsic viscosity in the same solvent and approximate molecular mass of the polymer. Details can be found elsewhere (1, 2).

Other corrections must also be made for polymers that absorb part of the light being passed through the solution (10). Thus, light scattering has not been used extensively with metal coordination and organometallic polymers because of extensive absorption in the visible region where light scattering measurements are normally made.

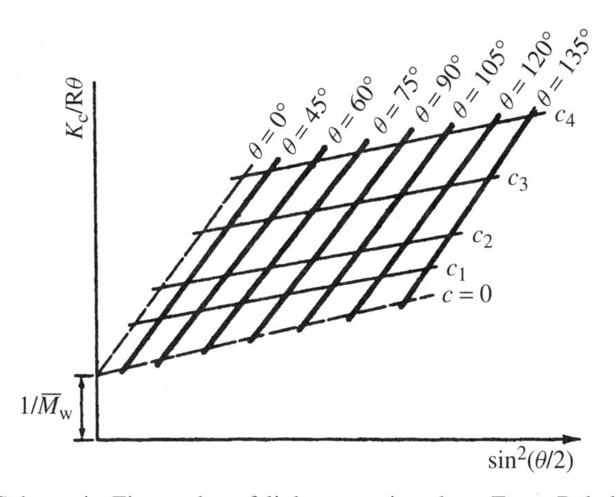

Figure 3.15 Schematic Zimm plot of light-scattering data. From Rabek, *Experimental Methods in Polymer Chemistry*, Wiley: Chichester, UK, 1980, p 196.

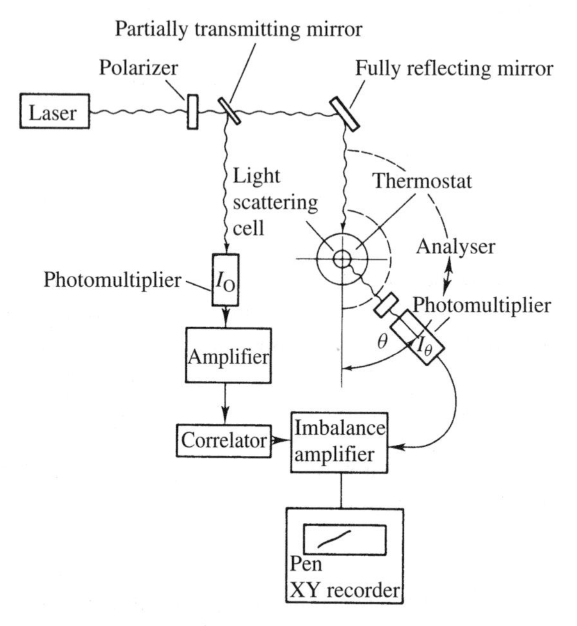

Figure 3.16 Schematic of a light-scattering photometer with a laser source. From Rabek, *Experimental Methods in Polymer Chemistry*, Wiley: Chichester, UK, 1980, p 199.

3.2.5 Colligative Properties (Vapor Pressure Lowering, Boiling Point Elevation, Melting Point Lowering, and Osmotic Pressure)

Of the various colligative properties of solutions, only osmotic pressure enhancement (membrane osmometry) is sensitive enough to allow the evaluation of the molecular masses of large inorganic polymers that have limited solubility. Vapor pressure lowering osmometry, boiling point elevation (ebulliometry), and freezing point depression (cryoscopy) have been claimed to be useful for organic polymers of up to or greater than 100,000. However, the solubilities of inorganic polymers limit the use of vapor pressure osmometry, boiling point elevation, and melting point lowering to molecular masses of about 10,000 or less.

Membrane osmometry involves the nonequilibrium that exists when a semipermeable membrane separates a pure solvent from a solution in that solvent (Fig. 3.17). In the classic static equilibrium method, a net flow of solvent passes through the membrane into the solution until the hydrostatic pressure produces equilibrium. The osmotic pressure, $\pi = \rho g \Delta h$ (where $\rho =$ solvent density in g/cm^3, $g =$ gravitational force, and with Δh as shown in Fig. 3.17) should provide an absolute measure of $\overline{M}_N$. The reduced osmotic pressure, π/c, where c is concentration in g/L, is related to the number-average molecular mass using the approximate van't Hoff expression at the zero concentration limit:

$$\lim_{c \to 0}(\pi/c) = RT/\overline{M}_N \tag{3.28}$$

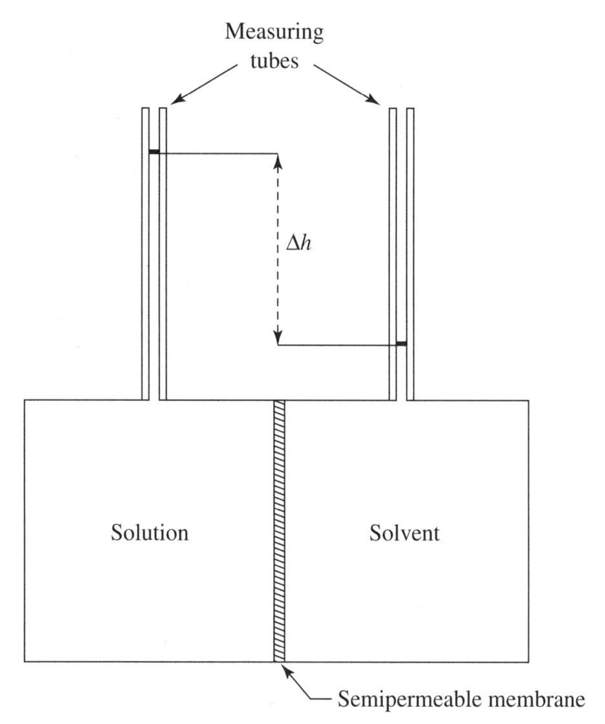

Figure 3.17 Schematic of a membrane osmometer. From p. 340, *Contemporary Polymer Chemistry*, 2/E by Allcock/Lampe, © 1990. Adapted by permission of Prentice-Hall, Inc., Upper Saddle River, NJ.

where R is 0.0821 L atm $mol^{-1}K^{-1}$. The virial equation is dominated by a c^2 term so that extrapolation is best done by plotting $(\pi/c)^{1/2}$ versus c. The sensitivity of the method is such that a 0.001 molal solution supports a column of liquid of the order of 20 mmHg or 270 mm of water. In general, polymers of up to 1,000,000 number-average molecular mass can be measured by this technique, but as noted above, the solubility of the polymer in the solvent being used may limit the range.

Number-average molecular masses determined by membrane osmometry are often larger than the values obtained with gel permeation chromatography, especially with step polymerizations that always have a sizable amount of low-molecular-mass material. The lower-molecular-mass species pass through the semipermeable membrane and leave only the higher-mass species to be measured. Fractionation of the polymer by precipitation of the higher-molecular-mass material before the mass measurements are made avoids this problem. [Alternatively, it has been suggested that comparable results occur if the portion of the polymer below a mass of 10,000 is ignored in calculating the molecular mass from the gel permeation chromatographic curve (4).] Obviously, the demarcation would be dependent on the membrane being used.

Vapor-phase osmometry has been used for molecular mass measurements of polymers using the commercial instrumentation that is used to evaluate the molecular masses of monomeric molecules (11). Drops of solution and solvent are measured with thermistors in a temperature-controlled cavity that is saturated with vapors of the solvent. A solution has a lower vapor pressure, and the heat of condensation causes the temperature of the drop to rise until the solution vapor pressure is increased and equal to that of the solvent. In actuality, this does not happen as a state of equilibrium is reached when the heat losses from the drops become equal to the heat of condensation of the solvent. The increase in temperature is directly proportional to the molality of the solute. A schematic of a vapor-phase osmometer is shown in Figure 3.18. The temperature is controlled to about 1/1000 °C. The difference in temperature is measured with a circuit that measures resistance changes (ΔR) quite precisely. Starting with a species of known molecular mass, a calibration curve of the molality (m) vs. ΔR is constructed and then the unknown molecular mass is determined from the ΔR

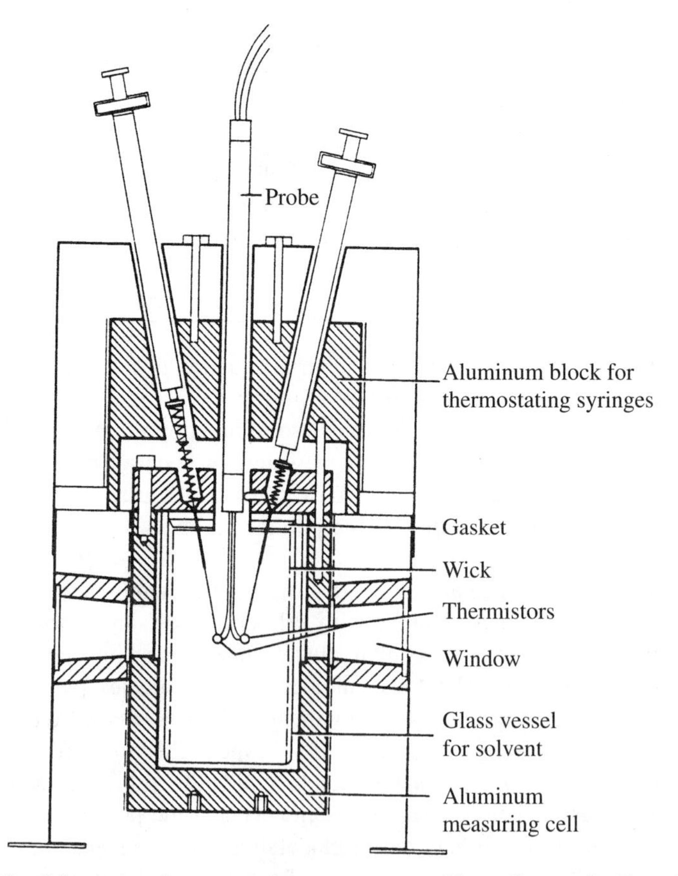

Figure 3.18 Schematic of a vapor-phase osmometer. From Cooper in *Encycl. Polym. Sci. Engr.*, 1987, **10**, 1.

that it produces. If the plot for the known species is linear:

$$\Delta R = \mathrm{a} + \mathrm{b}m \tag{3.29}$$

the number-average molecular mass of the unknown polymer can be calculated from Eq. 3.30:

$$\overline{M}_{\mathrm{N}} = \mathrm{bc}/(\Delta R - \mathrm{a}) \tag{3.30}$$

where c = g solute/1000 g solvent.

This method is quite good for organic polymers that have high solubility in volatile solvents, but it has been less useful for most inorganic polymers. For higher-molecular-mass solutes, the concentration dependence becomes important and extrapolation to zero concentration is also necessary in this technique. In these cases the best plots are usually $(\Delta R/\mathrm{c})^{1/2}$ vs. c extrapolated to zero.

Details and references on the use of ebulliometry and cryoscopy for molecular mass measurements of polymers can be found in the article by Cooper (11).

3.2.6 End-Group Analyses

A number of methods have been developed for the end-group analysis of polymers (11). Specific examples of end-group analyses of inorganic polymers include radioactivity measurements of hot-atom fragments appended to end groups, spectroscopic evaluation of at least one end group or modified end group, and titration of end groups. The limitations depend on the sensitivity of the method relative to the mass of the polymer being measured. The use of radioactive tracer measurements was quite popular several decades ago, is a very sensitive method, and can be used for insoluble polymers (11). Titration methods are possible for end groups that are acidic or basic; however, the sensitivity of the method is often quite low except for polymers with high solubility and/or moderate molecular mass. Likewise, spectroscopic methods are often not sensitive enough for high-molecular-mass species. However, sometimes a group can be substituted on the end group of a polymer chain that increases the sensitivity of the method, for example, the use of *t*-butyl groups to enhance the sensitivity of proton magnetic resonance spectroscopy as noted below.

The polymeric molecules that result from an A–A plus B–B step polymerization would be expected to have an equal number of A ends and B ends. This is the case for exact stoichiometry where 50% of the polymeric product molecules have one A end and one B end, 25% have two A ends, and 25% have two B ends. However, for slightly nonstoichiometric ratios the end-group ratios are *not* 1:1. The mathematical expression for the general case is shown in Eq. 3.31.

$$(1 - \rho)\mathrm{N}_{\mathrm{A}}^{0} + (1 - \mathrm{r}\rho)\mathrm{N}_{\mathrm{B}}^{0} = \mathrm{N}_{\mathrm{ends}} = 2 \times (\text{number of molecules}) \tag{3.31}$$

for reactant ratio $\mathrm{r} = \mathrm{N}_{\mathrm{A}}^{0}/\mathrm{N}_{\mathrm{B}}^{0} \leq 1.00$ where $\mathrm{N}_{\mathrm{A}}^{0}$ = the number of A reactive ends initially, $\mathrm{N}_{\mathrm{B}}^{0}$ = the number of B reactive ends initially, and the extent of reaction

$\rho \leq 1.00$. Initially, each A–A molecule has 2 N_A ends and each B–B molecule has 2 N_B ends.

The approximate values of A ends to B ends for extents of reaction from 0.98 to 1.00 and reactant ratios of 0.98 to 1.00 are shown in Table 3.4. The values calculated using Eq. 3.31 have been rounded; for example, for $\rho = 0.98$ and r = 0.98 the calculated results are 0.0200 N_A^0 and 0.0396 N_B^0, which have been rounded to 0.02 N_A^0 and 0.04 N_B^0. The results for other values of ρ and r can easily be determined using Eq. 3.31. From the results shown in Table 3.4 it should be evident that determining the number-average molecular mass ($\overline{M}_N$) of a polymer by using just one of the end groups can cause a significant error in the results. Because A is the reactant that is in short supply (except when r = 1.00), fewer A end groups exist on the polymeric molecules and molecular mass determinations based on A end groups would be higher than the true value. Conversely, molecular mass determinations based on B end groups would be too low. Table 3.5 provides a comparison of the degrees of polymerization obtained for the same extents of reaction and reaction ratios that were provided in Table 3.4. In Table 3.5, the calculated $\overline{M}_N$ values based on the total number of end groups are compared with the values obtained based on the number of A and B end groups from Eq. 3.31.

The use of end-group analysis is a valuable adjunct to other methods of molecular mass determination, especially when new classes of polymers are being investigated. For example, with zirconium chelate polymers (Fig. 2.16), even

TABLE 3.4 End Groups for Different Reactant Ratios and Extents of Reaction[a].

Extent of Reaction, ρ ↓	Reactant Ratio, r 1.00	0.99	0.98
1.00	0.00 N_A^0; 0.00 N_B^0	0.00 N_A^0; 0.01 N_B^0	0.00 N_A^0; 0.02 N_B^0
0.99	0.01 N_A^0; 0.01 N_B^0	0.01 N_A^0; $\underline{0.02}$ N_B^0	0.01 N_A^0; $\underline{0.03}$ N_B^0
0.98	0.02 N_A^0; 0.02 N_B^0	0.02 N_A^0; $\underline{0.03}$ N_B^0	0.02 N_A^0; $\underline{0.04}$ N_B^0

[a] $(1-\rho)N_A^0 + (1-r\rho)N_B^0 = N_{ends} = 2 \times$ (number of molecules) where N_A^0 and N_B^0 are the number of reactive ends initially (when $\rho = 0$); $r = N_A^0/N_B^0$; the underlined 0.02, 0.03, and 0.04 values are actually 0.0199, 0.0298, and 0.0396, respectively.

TABLE 3.5 Degrees of Polymerization—Actual and Based on End-Group Analyses.

Extent of Reaction, ρ ↓	Reactant Ratio, r 1.00	0.99	0.98
1.00	$\infty(\infty; \infty)$[a]	200 (∞; 100)	100 (∞; 50)
0.99	100 (100; 100)	67 (100; 50)	50 (100; 33)
0.98	50 (50; 50)	40 (50; 32)	33 (50; 25)

[a] The first number in each cell is the calculated degree of polymerization for the extent of reaction and reactant ratio; the first number in the parentheses is the result provided by an A end group analysis; and the second number in the parentheses is the result provided by a B end group analysis.

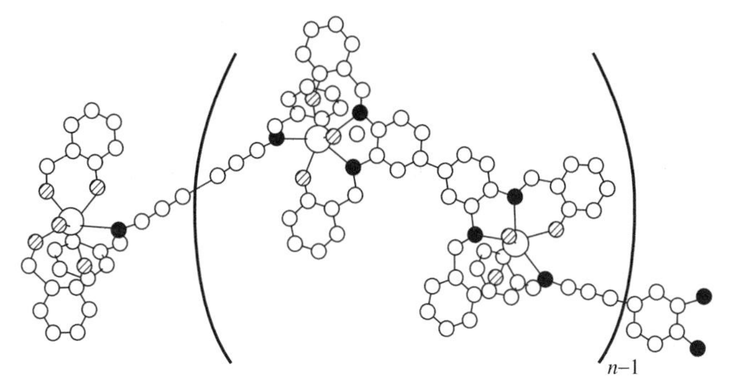

Figure 3.19 A polymer chain of *catena*-[poly(N,N′,N″,N‴-tetrasalicylidene-3,3′,4,4′-tetraaminobiphenyl)zirconium(IV)] showing the end groups — one end group is a diamine and the other end group is a zirconium(IV) ion with two unreacted salicylaldehydato ligands *per average chain*. The black balls represent nitrogen atoms, the striped balls represent oxygen atoms, the small white balls represent carbon atoms, and the large white balls represent zirconium atoms. The portion of the ligands below the planes and the hydrogen atoms are not shown to simplify the diagram. Each zirconium ion is eight-coordinate.

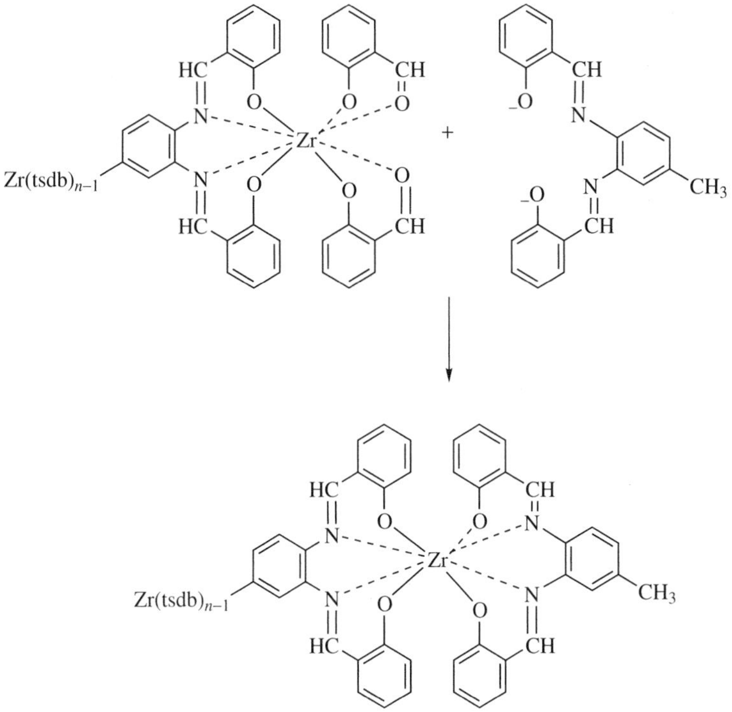

Figure 3.20 Replacement of the salicylaldehydato ligands with the dianion of N,N′-bis(salicylidene)-1,2-diamino-4-toluene provides polymer chains with one toluene methyl group per chain *on average*.

though the gel permeation and viscosity results appeared to provide evidence for long polymer chains, the use of NMR spectroscopy provided confirmation of the results (12). A short polymer chain of poly(*N*,*N*′,*N*",*N*‴-tetrasalicylidene-3,3′-diaminobenzidine)zirconium(IV) is displayed in Figure 3.19 showing the two types of end groups. One end group is a diamine, and the other end group is a zirconium(IV) ion with two unreacted salicylaldehydato ligands. Addition of *N*,*N*′-bis(salicylidene)-1,2-diamino-4-toluene (Fig. 3.20) provides a polymer chain with one toluene methyl group per chain on average. The proton-NMR spectrum in d_6-DMSO (Fig. 3.21) indicates a degree of polymerization of about 90; that is, $1.368/0.0297 = 46$ repeating units/average chain, which is equivalent to an average degree of polymerization of 91 for the step polymerization ($\overline{DP} = 2n - 1$). This high degree of polymerization requires reaction under the best conditions (several hours at 70 °C) and partial fractionation that dissolves the low-molecular-mass materials in a solvent such as tetrahydrofuran.

In later studies, *N*,*N*′-bis(4-*tert*-butylsalicylidene)-1,2-diaminobenzene (Fig. 3.22) was used to end cap metal Schiff-base polymer and polyelectrolyte chains (13). This end-capping reagent provides 18 protons and provides improved precision in the results.

Although the present author is unaware of any use of radioactive labeling for inorganic polymers, radiolabeled sulfur-35 from a bisulfite initiator has been used to determine the number-average molecular masses of Teflon samples (14). The use of this technique with insoluble polymers suggests that it could also be used by inorganic polymer chemists.

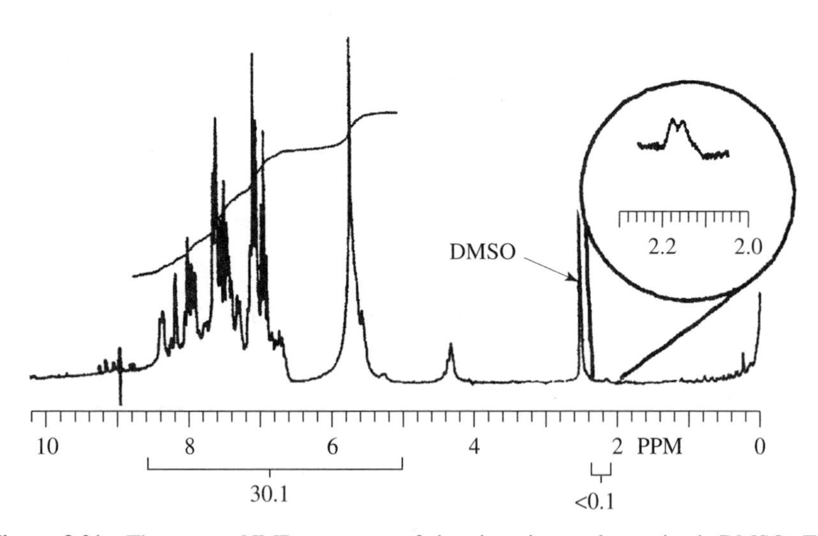

Figure 3.21 The proton-NMR spectrum of the zirconium polymer in d_6-DMSO. The end-group methyl at ~2.2 ppm integrates as 0.0890 (0.0297/proton/toluene end group) relative to the 22 aromatic protons per repeating unit that integrate as 30.1 (or 1.368/proton/repeating unit).

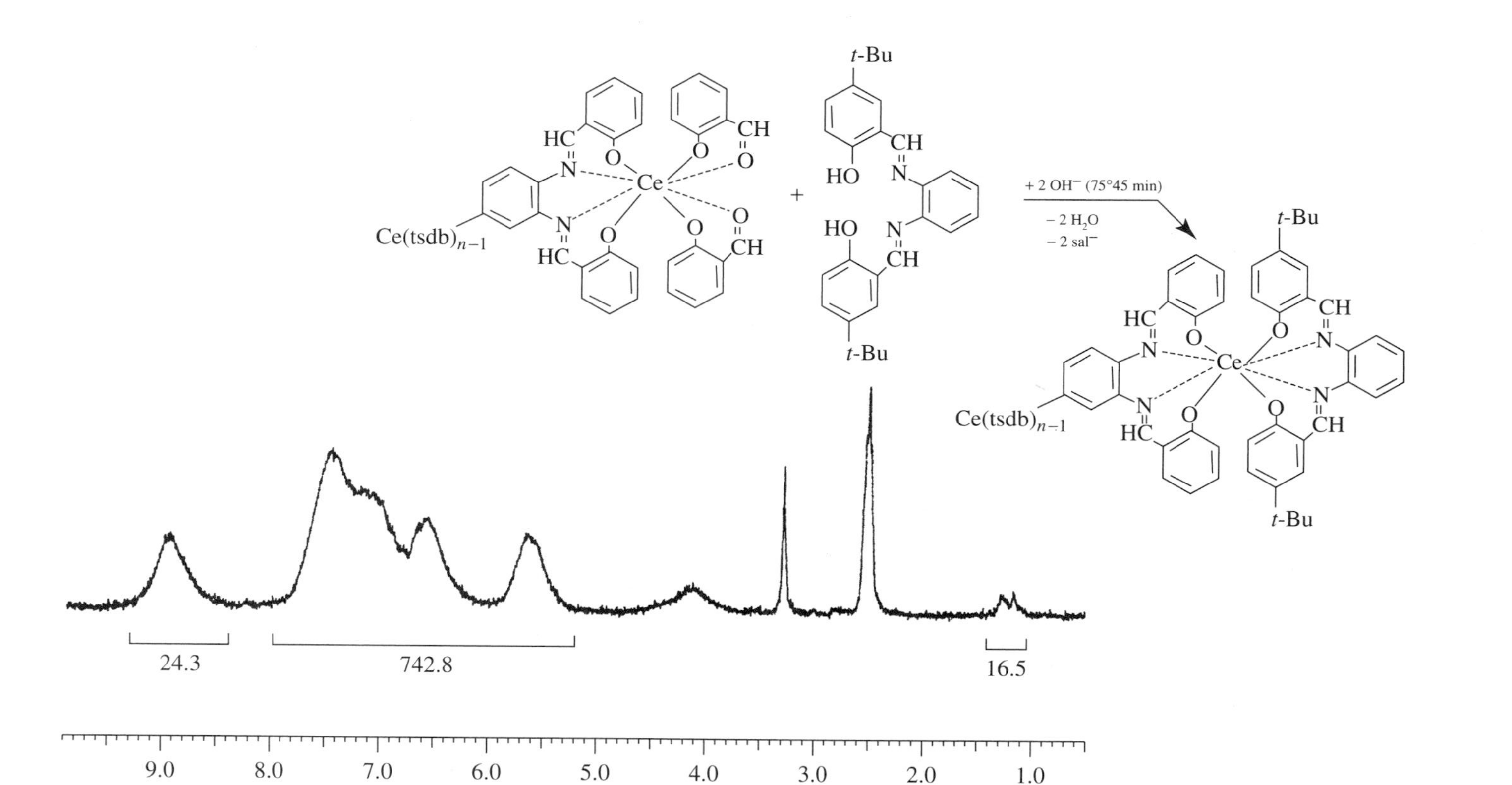

Figure 3.22 End-capping reaction and NMR spectrum for the same cerium(IV) polymer using the dianion of N,N′-bis(4-*tert*-butylsalicylidene)-1,2-diaminobenzene as an end cap. This reagent's 18 protons per end cap near 1 ppm integrate as 16.5. Using the integration of the 22 protons in the aromatic region (5.2–8 ppm) that integrate as 742.8, the number of repeating units is $(742.8/22)/(16.5/18) = 36.8$ and $\overline{M}_N = 28{,}700$.

3.2.7 Mass Spectroscopy

Mass spectroscopy is a most promising areas for molecular mass determination of inorganic polymers in the twenty-first century. Matrix-assisted laser desorption ionization mass spectrometry (MALDI-MS) provides a soft-ionization technique for mass measurements. Soft-ionization methods provide more parent ions than the usual mass spectrometry techniques that provide fragments of the molecule, with or without some of the parent ion. A polymer sample is dissolved or dispersed in an organic compound of low molecular mass that absorbs strongly at a wavelength accessible with an ultraviolet laser. Sodium or potassium metal ions are also a part of the matrix. The laser energy is transferred from the organic compound to the polymer molecules, which vaporize (desorb), taking with them one or more metal ions. Fortunately, the species with only one cation are more volatile and dominate the resulting mass spectrum, usually with a time-of-flight instrument. Individual molecular masses can be observed, even for organic polymers that have low molecular masses, if the number-average molecular mass is less than 10,000, and for inorganic polymers of high monomer molecular mass this limit should be much larger. Even when individual polymer masses cannot be discerned, the shape of the peak can be used to estimate both number-average and weight-average molecular masses (15). Other soft-ionization mass spectroscopy methods include electrospray ionization (16), field desorption (17), and laser desorption (without added matrix) (18).

3.2.8 Ultracentrifugation

An ultracentrifuge can be used to determine weight-average molecular masses of polymers using the modified Svedberg equation shown in Eq. 3.32.

$$\overline{M}_{\mathrm{W}} = \frac{s_0 RT}{D_0(1 - \overline{v}_p \rho_s)} \qquad (3.32)$$

where

s_o = sedimentation coefficient extrapolated to c = 0 and p = 0

c = concentration and p = pressure,

RT = the usual gas law constant and temperature (K),

D_o = diffusion coefficient extrapolated to c = 0,

$\overline{v}_p$ = partia specific volume (from pycnometry, and

ρ_s = fluid density.

The application of the sedimentation velocity method to calculate molecular mass distribution is extremely complex but has been successfully done in a number of instances. Details can be found elsewhere (1, 19). A schematic diagram of an ultracentrifuge and the view of an ultracentrifugation cell with monodisperse and polydisperse polymers are shown in Figs. 3.23 and 3.24, respectively.

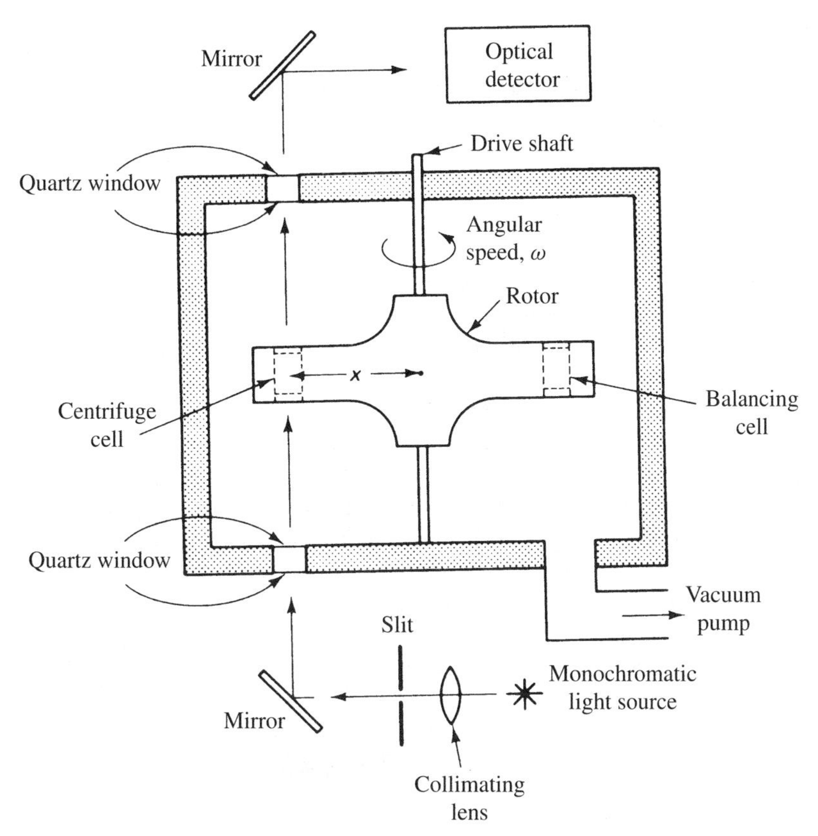

Figure 3.23 Schematic of an ultracentrifuge. From p. 368, *Contemporary Polymer Chemistry*, 2/E by Allcock/Lampe, © 1990. Adapted by permission of Prentice-Hall, Inc., Upper Saddle River, NJ.

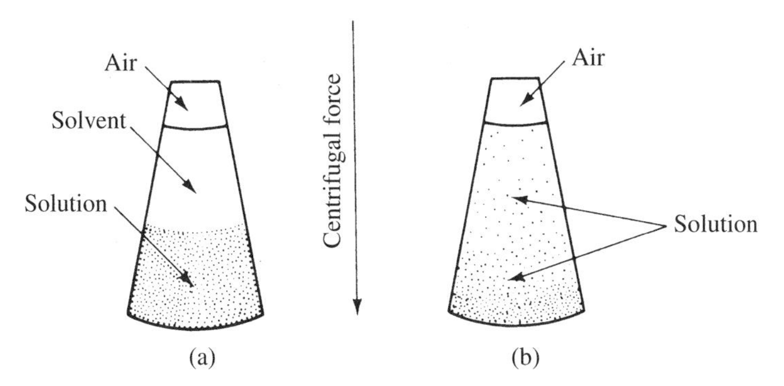

Figure 3.24 Monodisperse (a) and polydisperse (b) polymers in a cell during ultracentrifugation. From p. 369, *Contemporary Polymer Chemistry*, 2/E by Allcock/Lampe, © 1990. Adapted by permission of Prentice-Hall, Inc., Upper Saddle River, NJ.

3.3 DETERMINATIONS OF THERMAL PARAMETERS

A number of thermal parameters are important in polymer chemistry. The thermal parameters that are critical for polymer chemistry include: the glass transition temperature (T_g), the melting temperature (T_m), the ceiling temperature (T_c), and occasionally even a floor temperature (T_f), for example, $S_8 \rightarrow S_n$ has a $T_f = 432$ K (159 °C). The glassy state of a polymer, either amorphous or crystalline, becomes more flexible above its T_g. Above the T_g, an amorphous polymer becomes rubbery and a crystalline polymer becomes a flexible thermoplastic. A crystalline polymer also exhibits a T_m higher than the T_g value noted earlier and becomes a liquid. A liquid crystalline polymer has an additional T_{lc} liquid-crystalline transition between T_g and T_m. On the other hand, an amorphous polymer passes through a rubbery state and on through a gummy state and then on to a liquid state without exhibiting a definite T_m. With most polymers, as the temperature is increased a temperature is reached above which the polymer cannot exist. This temperature is called the ceiling temperature (T_c) and is related to the fact that the entropy term for a polymer relative to fragments of the polymer is almost always negative, or unfavorable relative to free energy. Because the entropy contribution to free energy is $T\Delta S$, a higher temperature eventually leads to polymer instability. A schematic representation of the temperature transitions of amorphous, crystalline, and liquid crystalline polymers is shown in Figure 3.25.

A floor temperature (T_f) exists for some polymers below which the polymer cannot be prepared. Ring systems that undergo ring-opening reactions often show this behavior. Whereas the initial ring-opening step is normally favored

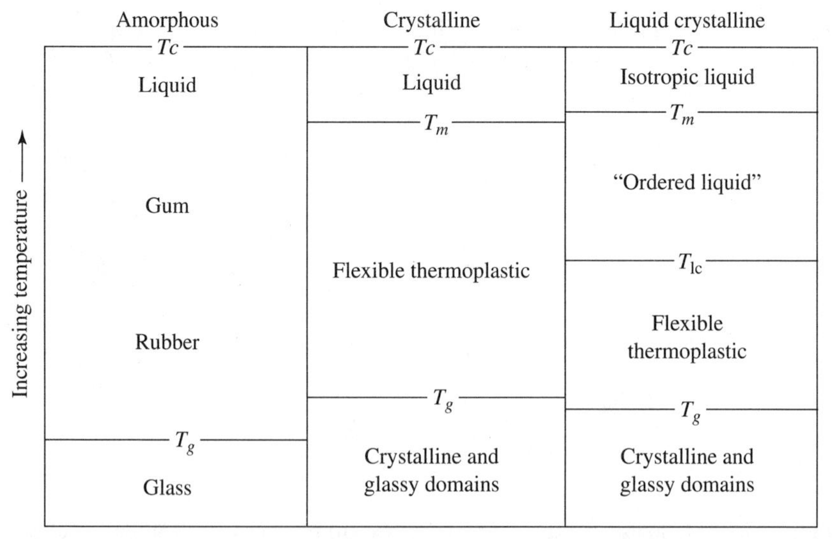

Figure 3.25 Thermal transitions of amorphous, crystalline, and liquid crystalline polymers. From p. 9, *Contemporary Polymer Chemistry*, 2/E by Allcock/Lampe, © 1990. Adapted by permission of Prentice-Hall, Inc., Upper Saddle River, NJ with T_c additions.

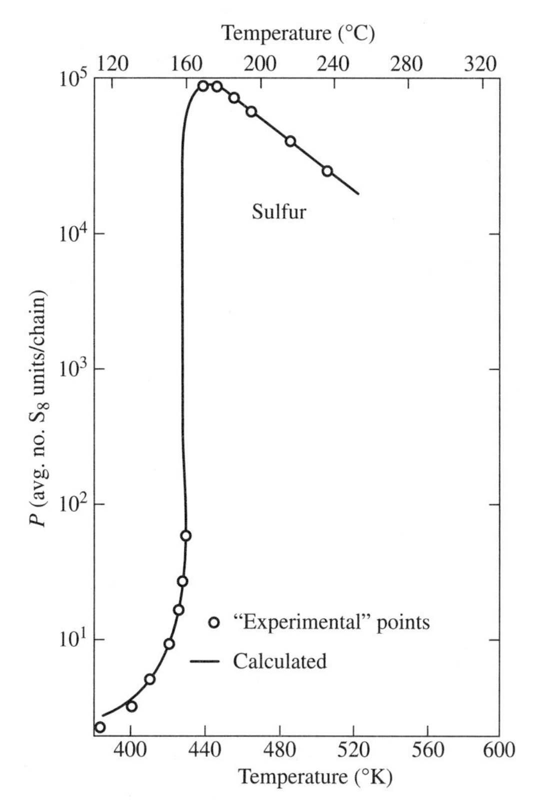

Figure 3.26 The ceiling and floor temperatures of sulfur (reprinted with permission from Tobolsky and Eisenberg, *J. Am. Chem. Soc.* 1959, **81**, 780, 2303; 1960, **82**, 289; © 1959–1960, American Chemical Society).

entropically ($\Delta S > 0$) because of the additional degrees of freedom that a chain has relative to a ring, the required bond breaking means that the enthalpy is unfavorable ($\Delta H > 0$, too). Because the entropy term is multiplied by temperature (K), a temperature should exist above which the ring can be opened because $T\Delta S$ will be greater than ΔH. Thus ΔG, which is equal to $\Delta H - T\Delta S$, will also be negative and spontaneous from a thermodynamic point of view. Naturally, systems with ring strain, typically systems with 3, 4, 7, 8, etc. membered rings, will require less energy to ring open and will have lower floor temperatures. Sulfur exhibits both floor and ceiling temperatures at about 159 °C and 170 °C, respectively (cf. Fig. 3.26).

3.3.1 Glass Transition Temperature Measurements

Whereas the glass transition temperature is the temperature at which a polymer changes from a hard glass to a flexible material, visual observation by microscopy is not the usual method of determining the glass transition temperature. A variety of methods can be used including indentation techniques (with a penetrometer), torsional rigidity determinations, broadline NMR changes, dilatometry volume changes at the transition, differential thermal analysis, and differential scanning calorimetry (1).

A penetrometer consists of a heavy needle attached to an amplification gauge. The needle can penetrate the polymer when the polymer passes through the glass transition temperature. Although this method is less accurate than some of the following methods, it is useful for the preliminary examination of polymer samples.

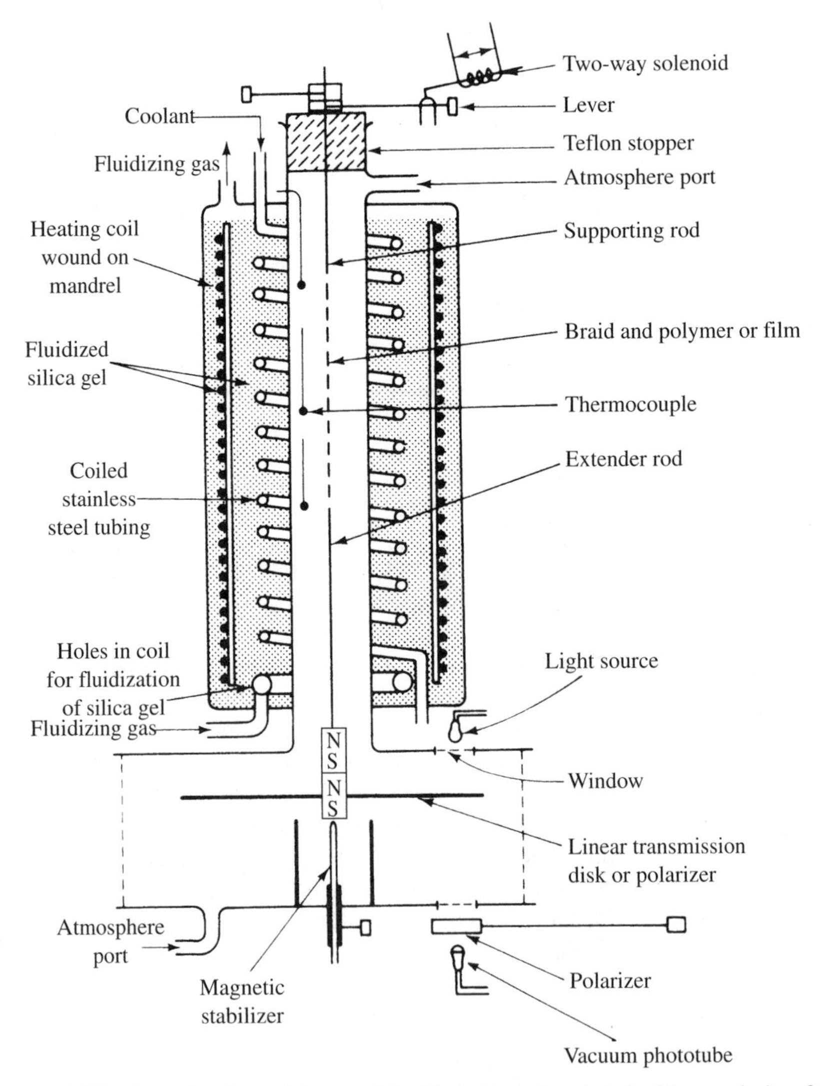

Figure 3.27 A torsional pendulum and braid analyzer (reprinted with permission from Gillham, J. K. *CRC Crit. Rev. Macromol. Sci.*, 1972–73, **1**, 83; © CRC Press, Boca Raton, Florida).

Torsional rigidity methods take advantage of the change in rigidity that occurs when a polymer passes through the glass transition temperature. One approach involves measuring the resistance to torsion and the energy loss of a bar of a polymer as the temperature is varied. A laboratory variation involves a torsional pendulum with a polymer sample impregnating a braided glass fiber. A diagram of such a torsional pendulum is shown in Figure 3.27. A solution of the polymer is administered and dried. The impregnated fiber is suspended in a constant-temperature medium, and the sample is heated. As it passes through the glass transition temperature, a large loss in rigidity occurs that can be detected by a change in the pendulum damping. It is a very sensitive method that only requires about a fourth of a gram of polymer.

Broadline NMR is another method for noting glass transition temperatures. Below the glass transition temperature the solid exhibits broad, diffuse peaks, if any can be observed at all. When the polymer passes through the glass transition temperature, the polymer becomes more like a liquid and a narrow peak or series of narrow peaks occurs. The peak width can drop by an order of magnitude over a temperature range of just a few degrees.

Dilatometry is a measurement of volume expansion versus temperature. A change in the slope of the line occurs at the glass transition temperature (and at the melting point for a crystalline polymer) as shown schematically in Figure 3.28.

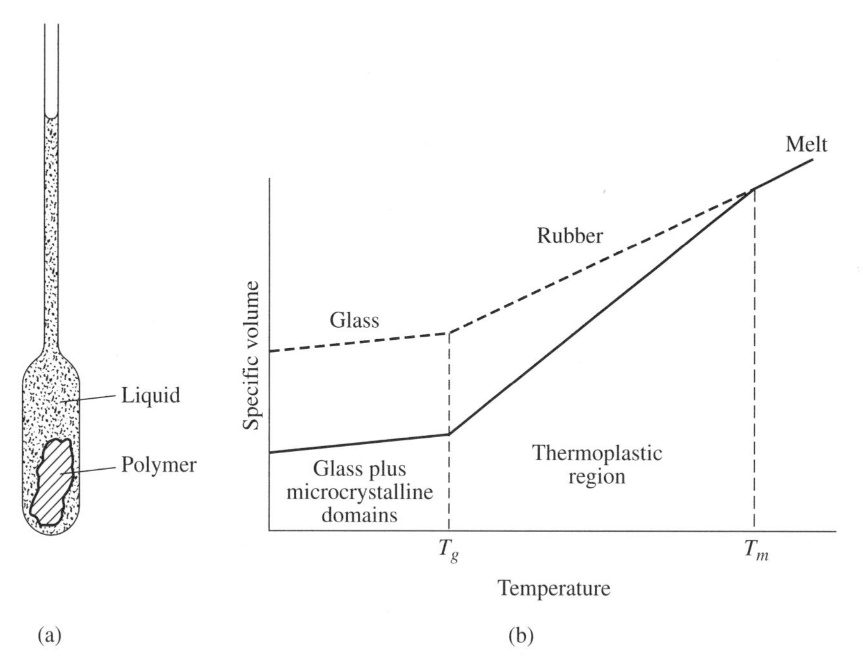

Figure 3.28 Schematic representation of a dilatometer and the variation of the volume of an amorphous polymer in a liquid as the polymer passes through its glass transition temperature. From pp. 431–432, *Contemporary Polymer Chemistry*, 2/E by Allcock/Lampe, © 1990. Adapted by permission of Prentice-Hall, Inc., Upper Saddle River, NJ.

Actually, a discontinuity can occur at the melting point temperature as the glass may have a different density than the melt.

Differential thermal analysis (DTA) is a very popular method of detecting glass transition temperatures. A small sample of the polymer is heated at a constant rate of temperature increase. Simultaneously an inert substance, such as alumina, is heated at the same rate. A temperature difference will occur

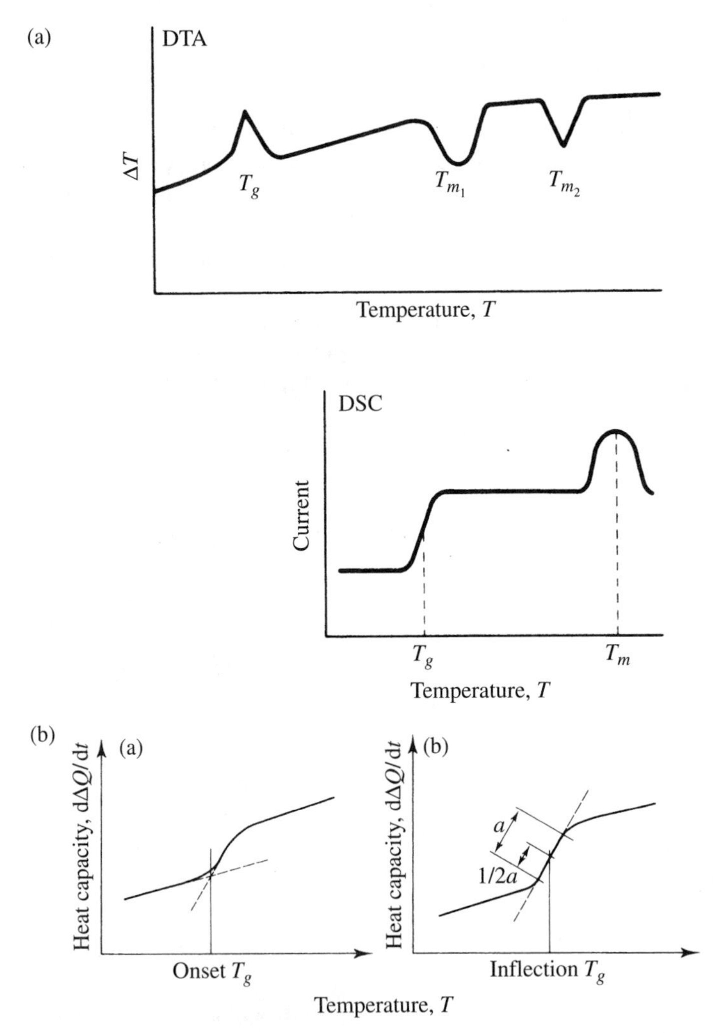

Figure 3.29 Glass transition temperature by differential thermal analysis (DTA) and by differential scanning calorimetry (DSC) at the top (based on p. 433, *Contemporary Polymer Chemistry*, 2/E by Allcock/Lampe, © 1990. Adapted by permission of Prentice-Hall, Inc., Upper Saddle River, NJ). The bottom shows how onset and inflection glass transition temperatures are obtained from DSC measurements; based on Rabek, op cit., p. 565.

between the two samples as a result of the difference in the specific heats of the two substances. When the polymer approaches its glass transition temperature, an exothermic change causes a discontinuity in the DTA curve as shown in Figure 3.29.

Differential scanning calorimetry (DSC) also uses two samples, a polymer sample and an inert reference substrate. Both the sample and the reference are heated separately by individually controlled heater elements and monitored by temperature sensors as shown in Figure 3.30. The power is adjusted continuously to the heaters in sufficient quantity as to keep the sample and the reference at identical temperatures as the heating occurs. The differential power required to keep this balance is recorded versus the programmed temperature of the system as shown in Figure 3.29. The glass transition temperature by DSC can be reported at the onset of the transition or at the inflection point as shown in Figure 3.29. Recording the change as a first derivative makes a more definitive change for the inflection point. The rate of heating is also very important, as noted below.

Some examples of glass transition temperatures and crystalline melting temperatures for a number of inorganic polymers are given in Table 3.6. Note that the siloxane and phosphazene polyelectrolytes (20) have glass transition temperatures near those of the neutral siloxane and phosphazene polymers (1). The metal Schiff-base coordination polymers (12, 21) also have glass transition temperatures near the analogous yttrium and lanthanide Schiff-base coordination polyelectrolytes (22). The values in Table 3.6 for *catena*-poly[N,N',N'',N'''-tetrasalicylidene-3,3′-diaminobenzidinatocerium(IV)] are for a heating rate of 20 °C/min. Earlier the values were reported as $T_g = 184\,^\circ\text{C}$ with an onset at 169 °C (23, 24)—these values were for a heating rate of 40 °C/min. Obviously, the species is not at equilibrium at the higher rate of heating. *Thus T_g values should be measured at heating rates as slow as the transitions can be clearly observed.*

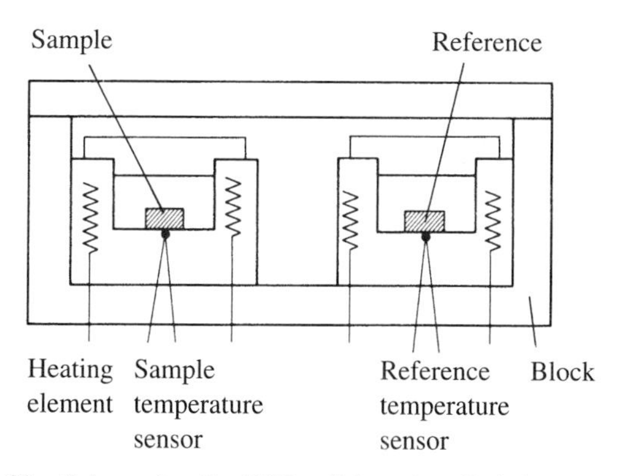

Figure 3.30 Schematic of a DSC cell based on Rabek, op cit., p. 563.

TABLE 3.6 Glass Transition Temperatures and Crystalline Melting Temperatures for Selected Inorganic Polymers.

Polymer	T_g °C		T_m °C	Ref.
Poly(dimethylsiloxane)	−123		−29	a
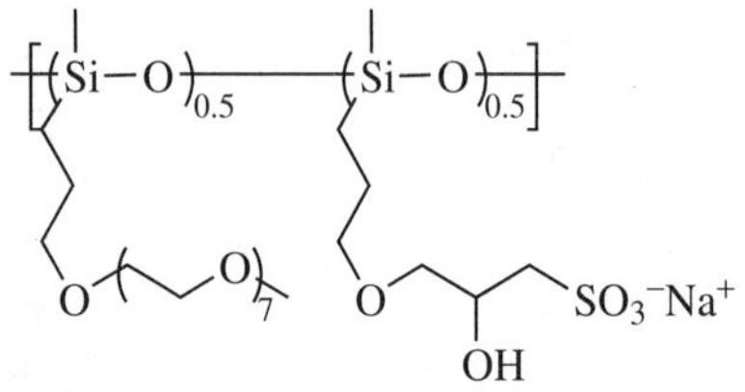 $-[(Si-O)_{0.5}-(Si-O)_{0.5}]-$ with side chains $O(CH_2CH_2O)_7$ and $OCH_2CH(OH)CH_2SO_3^-Na^+$	−65		—	b
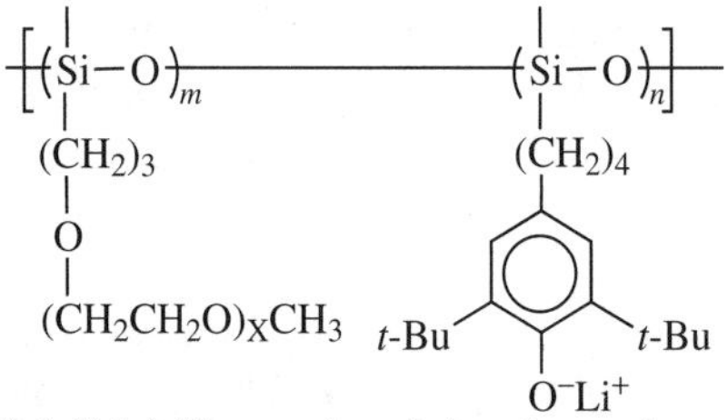 $-[(Si-O)_m-(Si-O)_n]-$ with side chains $(CH_2)_3-O-(CH_2CH_2O)_XCH_3$ and $(CH_2)_4$–(di-*t*-Bu phenyl, O^-Li^+)	−55		—	b
Poly[bis(trifluoroethoxy)phosphazene]	−66		+242	a
Poly(dichlorophosphazene)	−63		−10	a
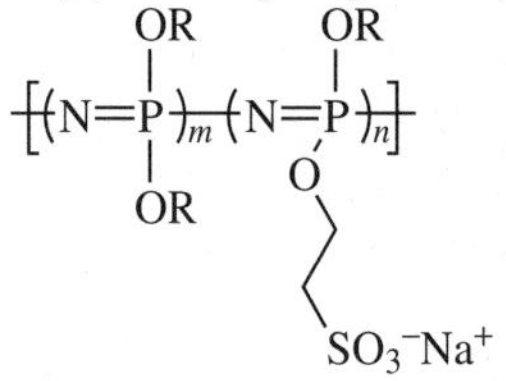 $-[(N=P(OR)_2)_m-(N=P(OR)(O-CH_2CH_2CH_2SO_3^-Na^+))_n]-$	−67		—	b
$[Zr(tsdb)]_n$	+74	(66)	—	c
$[Ce(tsdb)]_n$	+146	(140)	—	d
$[NaY(tsdb)]_n$	+162	(154)	—	e
$[NaGd(tsdb)]_n$	+143	(125)	—	e
$[NaYb(tsdb)]_n$	+179	(169)	—	e

[a]Allcock & Lampe; [b]Rawsky & Shriver; [c]Archer & Wang; $tsdb^{4-}$ = N, N′, N″, N‴-tetrasalicylidene-3,3′diaminobenzidinato(4−) ligand; onset temperature in parentheses; [d]Chen; [e]Chen & Archer

3.3.2 Other Thermal Parameters

Although crystalline and liquid crystal polymers show additional thermal parameters as noted in Figure 3.25 above, detailed discussions of these are beyond the scope of this text. Melting temperatures of crystalline polymers can be ascertained through dilatometry (Fig. 3.28); however, DTA and DSC are more commonly used (Fig. 3.29).

The thermal stability (ceiling temperature or decomposition temperature) can normally be determined by thermal gravimetric analysis (TGA). One caution in comparing results for various inorganic (or organic) polymers is the gaseous background for the measurements. TGA measurements conducted under nitrogen or argon (or *in vacuo*) give higher thermal stability results because oxidative decomposition is eliminated. On the other hand, from a practical point of view, TGA measurements conducted in air are more realistic. Early attention toward inorganic polymers was often centered on thermal stability. For example, polysiloxanes (silicones) are more thermally stable than their hydrocarbon counterparts (as oils, elastomers or cellular polymers) (25–28) and the thermal stability of some of the polymeric metal phosphonates (29, 30) sent synthetic chemists scurrying about for greater thermal stability using inorganic polymers. The end result has been some inorganic polymers with thermal stability and many others that lack good thermal stability (31). Results for Schiff-base polymers of the transition and lanthanide metals are typical of this dichotomy (12, 32–40).

3.4 SPECTROSCOPIC CHARACTERIZATIONS SPECIFIC TO INORGANIC POLYMERS

Although the spectroscopic methods used for small inorganic and organic molecules are also appropriate for inorganic polymers and need not be elaborated herein, some specific spectroscopic characterizations of inorganic polymers, especially metal coordination or organometallic polymers, are essential.

3.4.1 Nuclear Magnetic Resonance Spectroscopy

The use of nuclear magnetic resonance (NMR) spectroscopy for end-group analysis has already been noted in Section 3.2.6. Figure 3.21 provides an example of such a use.

The relative intensities of proton NMR peaks can also be used to determine whether the condensation polymerization expected has actually occurred. An example of an incomplete reaction was observed in the reaction between bis(3,4-diaminobenzene)sulfone and tetrakis(salicylaldehydato)zirconium(IV). Under mild reaction conditions the ratios of the salicylaldehydato aldimine proton NMR peaks relative to the aromatic peaks of the diaminobenzene groups (corrected for those of the salicylaldehydato aromatic protons) were only about one-half as intense as anticipated. Increased time and temperature provided peaks with appropriate ratios.

Other nuclei are typically less sensitive and/or less naturally abundant than the proton so that larger quantities of sample are necessary. The relative sensitivities for identical numbers of atoms at a constant field, the percent natural abundances, and the spin numbers (I) of several isotopes that have been used in NMR studies of inorganic polymers follow:

^{1}H	1.000	99.98%	1/2	^{2}H	0.096	0.016%	1
^{11}B	0.165	81.17%	3/2	^{13}C	0.702	1.108%	1/2
^{15}N	0.001	0.365%	1/2	^{19}F	0.834	100%	1/2
^{29}Si	0.078	4.7%	1/2	^{31}P	0.066	100%	1/2
^{117}Sn	0.045	7.67%	1/2	^{119}Sn	0.052	8.68%	1/2

Note that only fluorine-19 is close to the proton in both sensitivity and abundance (cf. Exercise 3.6).

Silicon-29 NMR peak shifts can be helpful in determining the nature of silicon-containing polymers (41). In carbosilanes (polymeric or ring) each silicon atom is surrounded by four carbon atoms and the ^{29}Si NMR peaks are close to zero, where tetramethylsilane, $Si(CH_3)_4$, is the standard. The same would be true for the proton and carbon-13 NMR peaks using the same $Si(CH_3)_4$ standard. Linear or cyclic polysilanes $(SiR_2)_n$ with an Si_n backbone (each silicon atom is attached to 2 other silicon atoms and 2 carbon atoms) have silicon-29 NMR peaks between −10 and −40 parts per million (ppm) relative to $Si(CH_3)_4$. The polysilynes $(SiR)_n$ have each silicon attached to three other silicon atoms and only one-carbon atoms are at −50 to −70 ppm. Polysiloxanes have NMR peak shifts ranging from −120 to +40 ppm. The more oxygen atoms surrounding each silicon the more negative the shift—silicates with four oxygen atoms around each silicon atom have peak shifts of −80 to −120 ppm. Silicon atoms with hydride bonds have silicon-29 NMR peak shifts ranging from 0 to −70 ppm—three SiH_3 groups provide the largest negative shifts.

Caution must also be used with silicon-29 NMR. A polycarbosilane with a cyano substituent $[Si(CH_3)(CN)CH_2]_n$ has its ^{29}Si peak at about −15 ppm (42) even though each silicon is surrounded by four carbon atoms. The ^{13}C NMR has nonseparable CH_3 and CH_2 peaks at about 1.1 ppm and the SiCN peak at about 112 ppm, whereas the proton NMR is overlapping CH_3 and CH_2 peaks at about 0.4 ppm (42).

Phosphorus-31 NMR peaks in polyphosphazenes are normally measured relative to H_3PO_4. The ^{31}P NMR shifts range from −18.6 for polyphosphazenes with two phenoxide groups to +8.3 for polyphosphazenes with two methyl groups. The ^{31}P NMR shift values for polyphosphazenes with other groups are as follows:

−3.5 to −7.5 ppm for two alkoxide groups,
+5.3 to −0.7 ppm for two R_2N groups,
−14 to −26.6 for two RNH groups,
+1.5 to +1.9 for one methyl group and one phenyl group, and
+6.8 to +7.9 for one ethyl group and one phenyl group.

Phosphorus-29 resonances can be used to follow polymerization reactions. To illustrate, the telechelic polymer syntheses of block copolymers of phosphazenes and siloxanes in Figure 2.48 that have reactants that contain PCl_3^+ groups (6 and 7) have a ^{31}P resonance at +8 ppm that is replaced by a resonance at −2 ppm as the reaction (with 5) proceeds. The replacement of the chloro groups at the end of the reaction can also be monitored by ^{31}P NMR (43) (cf. Exercise 3.7).

Examples from metal coordination chemistry related to polymers also exist. For example, carboxylate carbon-13 NMR peaks vary with monodentate and bidentate coordination to the carboxylate group (44). Uranyl dicarboxylate polymers with only bidentate carboxylato coordination exhibit a carbon-13 peak at about 185 ppm, whereas polymers with monodentate and bidentate carboxylato coordination exhibit two carbon-13 peaks at about 175 and 185 ppm, analogous to known monomeric coordination compounds. The first group has one mole of DMSO solvated to the uranyl ion and the second group has two moles. Both groups of uranyl polymers apparently have seven-coordinate uranium atoms in their chains (cf. Exercise 3.8).

The proton NMR spectra of cobalt(III) polymers with bis-β-diketonato ligands and S-leucine show the presence of diastereomers due to the leucine moiety (44). Furthermore, the proton NMR of these polymers confirms the normal O,O′-chelation for bis-β-diketonato ligands. NMR spectra are different when the central carbon of the β-diketone is coordinated to the metal ion-something that sometimes happens with "soft" metal ions. {Soft metal ions are usually of oxidation state II or lower and are found in the second and third transition series or heavy posttransition main group metal elements.}

NMR is often the best way to determine the ratio of two components in a copolymer—this is somewhat analogous to its use for determining end group populations to obtain molecular masses as noted earlier. In Figure 3.31, the proton NMR peak intensities for the aromatic peaks between 5.6 and 8.5 ppm from the

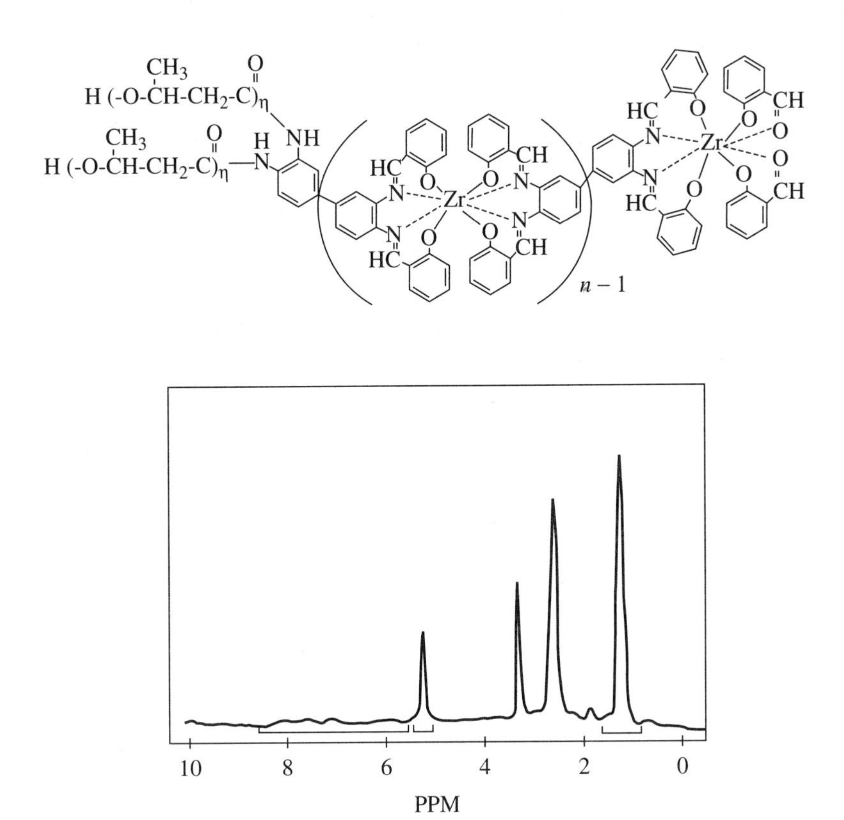

Figure 3.31 The copolymerization reaction between Zr(tsdb)$_n$ and $(\pm)\beta$-butyrolactone and the proton NMR spectrum of the product. (Spectrum taken by B. Wang.)

$Zr(tsdb)_n$ oligomer are compared with the butyrolactone peaks near 1 and 5 ppm to determine the degree of polymerization of the butyrolactone on the $Zr(tsdb)_n$ oligomer. The average number of $tsdb^{4-}$ ligands (with 22 aromatic protons per $tsdb^{4-}$ ligand) per zirconium was determined separately.

The preceding is only indicative of the many uses that NMR spectroscopy has in inorganic polymer characterization. Many more examples could be cited.

3.4.2 Electron Paramagnetic Resonance Spectroscopy

Polymers with unpaired electrons (either stoichiometric or not) can be characterized by with electron paramagnetic resonance (EPR) spectroscopy. This technique, also called electron spin resonance (ESR) spectroscopy, measures transitions between spin levels of unpaired electrons in an external magnetic field. Paramagnetic transition metal ions and organic radicals are most often characterized by this technique (45, 46), although solid-state nonstoichiometric materials are also characterized by EPR. The details of EPR use for polymer characterization have been described previously (47). Details of the theory and technique can be found in this reference. Tabulations of EPR results for coordination and organometallic transition metal compounds can be found in several Landolt-Börnstein volumes (48, 49).

Whereas NMR looks at the energies of nuclear spin state changes in an external magnetic field, EPR is based on the energies of unpaired electron spin state changes in an external magnetic field. Just as NMR spectroscopy is simplest when a nuclear spin state of $\pm 1/2$ is involved (^{1}H, ^{13}C, ^{19}F, ^{29}Si, ^{31}P, etc.), EPR spectroscopy is simplest when a single electron spin ($\pm 1/2$) is involved. The energies of the two states get further apart as the magnetic field is increased, as shown by Eq. 3.33, with the $-1/2$ spin state considered as the ground state. Transition metal complexes with d^1 or d^9 electronic configurations and one-electron free radicals of both organic and inorganic species are most often studied, although a wide range of transition metal species with other d^n electronic configurations have been studied (46, 48, 49). Samples of amorphous polymers can be studied in either the solid state as powders, in solution, or in low-temperature glasses. Aqueous and alcoholic solutions and glasses give poorer results than solvents and glasses with lower dielectric constants. Better resolution is obtained at lower temperatures, and most EPR studies are conducted at 77 K or lower. However, room-temperature EPR is often quite useful for kinetics and mechanism studies in solution.

The measurements are normally made at a constant frequency, and the external magnetic field is varied through the resonance region. X-band EPR, the most common, uses frequencies near $\nu = 9.5 \times 10^9$ Hz (9.5 GHz) and magnetic field strengths of about 3400 gauss (for $g = 2$; see below). Although the sensitivity is expected to increase with the square of the frequency, the higher-frequency Q-band (35 GHz) instruments require magnetic field strengths of about 12,500 gauss, for which homogeneity of field strength is more difficult to obtain. This drives up the cost of Q-band instrumentation. Thus many experts recommend

X-band EPR for g value measurements. However, the Q band allows a better evaluation of hyperfine coupling constants because of the greater separation that can be obtained.

Line position, shape, splitting, and intensity all provide useful information for polymers with unpaired electron density. The resonance peak positions are characterized by their g values, where the resonance energy, ΔE, is

$$\Delta E = h\nu = g\beta H_0 \tag{3.33}$$

where

h = Planck's constant ($6.626 \cdot 10^{-27}$ erg s)

ν = the frequency of the rf (radio frequency) signal used to detect the resonance

g = the g-factor relating the frequency to the magnetic field; i.e., $g = h\nu/\beta H_0$

β = the Bohr magneton ($9.274 \cdot 10^{-21}$ erg gauss^{-1})

H_0 = the strength of the applied external magnetic field

The g value for a free electron is 2.002319, and although organic radicals are typically quite close to this value, transition metal ions often have g values quite removed from the free electron value.

In a low-temperature glass or in the powdered state, the symmetry of the EPR signals can provide information on the symmetry of the site containing the free electron (50). The absorption and first-derivative signals anticipated for isotropic ($x = y = z$), axial ($x = y \neq z$), and asymmetric ($x \neq y \neq z$) sites are shown in Figure 3.32, where the axial g values are represented as $g_{\parallel}$ and $g_{\perp}$ and the asymmetric g values as g_1, g_2, and g_3 (cf. Exercise 3.9). {Note that $g_{\perp}$ for axial symmetry is at the negative peak, *not* where the derivative spectrum goes across the baseline. Some well-known investigators have incorrectly used the latter, but the difference is usually quite small.}

Whereas a single peak (possibly split by symmetry as just noted) is expected for a simple electron spin, nuclear spins interact with the electrons and give rise to $2\mathbf{I} + 1$ peaks for isotropic systems, where **I** is the nuclear spin. Each of these peaks is separated by the hyperfine coupling constant **A** for the nucleus involved. In lower symmetry the **A** values (e.g., $\mathbf{A}_{\parallel}$ and $\mathbf{A}_{\perp}$ for axial symmetry) are also nonequivalent and provide even more peaks — when they can be observed.

Hyperfine coupling **A** values are energies. The **A** values measured experimentally are based on the change in the magnetic field (ΔH in gauss). Thus the measured $\mathbf{A}_{(\mathrm{Gauss})}$ values vary with the g value and should be converted to cm^{-1} (tabulated in 10^{-4} cm^{-1} units (48, 49) to make them similar in magnitude to the gauss values) or to a frequency unit that is g invariant. For example,

$$\mathbf{A}_{(\mathrm{cm}-1)} = (g\beta/hc)\mathbf{A}_{(\mathrm{Gauss})} = 4.67 \cdot 10^{-5} g\mathbf{A}_{(\mathrm{Gauss})} \tag{3.34}$$

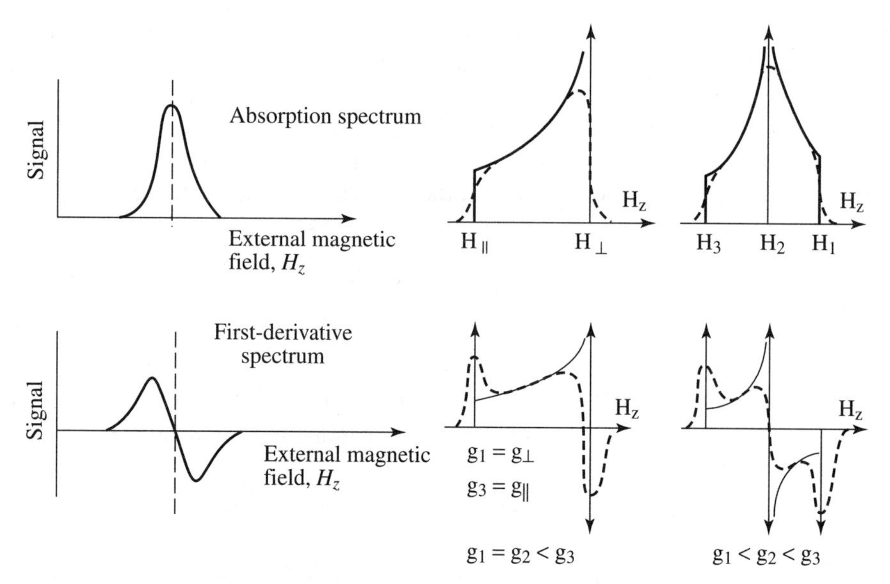

Figure 3.32 Schematic absorption and first-derivative EPR signals anticipated for isotropic (x = y = z), axial (x = y ≠ z), and asymmetric (x ≠ y ≠ z) symmetries, which give rise to an isotropic g value, to axial g values ($g_{\parallel}$ and $g_{\perp}$), and to asymmetric g values (g_1, g_2, and g_3). The theoretical spectra are given for the axial and asymmetric cases (solid lines) as well as the anticipated spectra (dashed lines) due to line broadening.

where the terms are as in Eq. 3.33, c = the speed of light ($2.998 \cdot 10^{10}$ cms^{-1}), and **A** is the measured hyperfine coupling for the units shown as subscripts. For example, a measured A value of 56 gauss at $g = 1.97$ is $51.5 \cdot 10^{-4}$ cm^{-1} whereas a 56-gauss coupling at the free electron value ($g = 2.0023$) is $52.4 \cdot 10^{-4}$ cm^{-1}. Both round to $52 \cdot 10^{-4}$ cm^{-1}. Rounding is appropriate because most results are imprecise by $\pm 3 \cdot 10^{-4}$ cm^{-1} units or more. However, at $g = 2.20$ [an average g value for many copper(II) compounds] a 56-gauss coupling is $58 \cdot 10^{-4}$ cm^{-1}, and iron(III) species often have average g values >4, where a 56-gauss coupling would have an energy of $>100 \cdot 10^{-4}$ cm^{-1}.

Irradiated organic polymers are predominantly split by the hydrogen nuclei (protons, $\mathbf{I} = 1/2$) in the radicals. The number of peaks gets quite large because nuclei in difference symmetry create separate sets. A methylene (CH_2) group on a radical has three resonances from the two equivalent protons interacting with the electron ($2n_H\mathbf{I}_H + 1 = 3$), whereas a radical with two nonequivalent CH_2 groups provides nine resonances (the product of 3 for one CH_2 times 3 for the second one). This is shown schematically in Figure 3.33.

When transition metal ions with large **I** and **A** values are involved in radicals, the details of the organic portion are often washed out. However, in simpler cases the EPR spectra show both the metal and nonmetal spin interactions — although even in monomers not all of the anticipated lines are observed. Figure 3.34 shows an EPR spectrum for bis(salicylaldiminato)copper(II), a model monomer

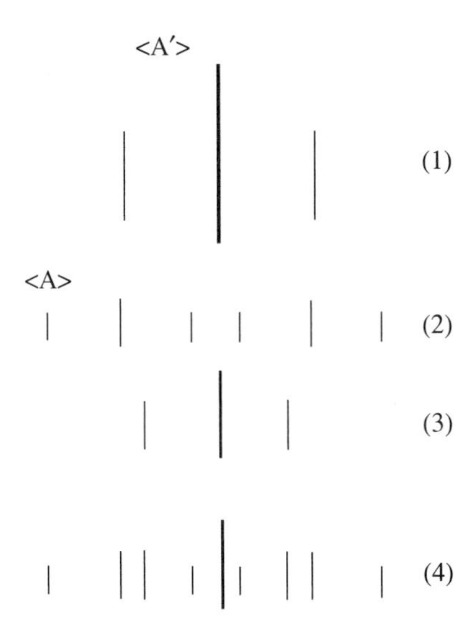

Figure 3.33 The peaks anticipated for an unpaired spin interacting with the hydrogen atoms of two nonequivalent CH_2 groups: (1) the 4:8:4 (1:2:1 increased to allow splittings that will still be whole numbers) peaks of a CH_2 with hyperfine of A; (2) the 1:2:1 splitting of two peaks of intensity 4 by the second CH_2 with hyperfine A′; (3) the 1:2:1 splitting of the peak of intensity 8 by the second CH_2 with hyperfine A′; and (4) the resultant 9-line spectrum with intensities of 1:2:2:1:4:1:2:2:1. Note how this would become the 1:4:6:4:1 spectrum expected for 4 equivalent $I = 1/2$ nuclei as A and A′ approach the same value.

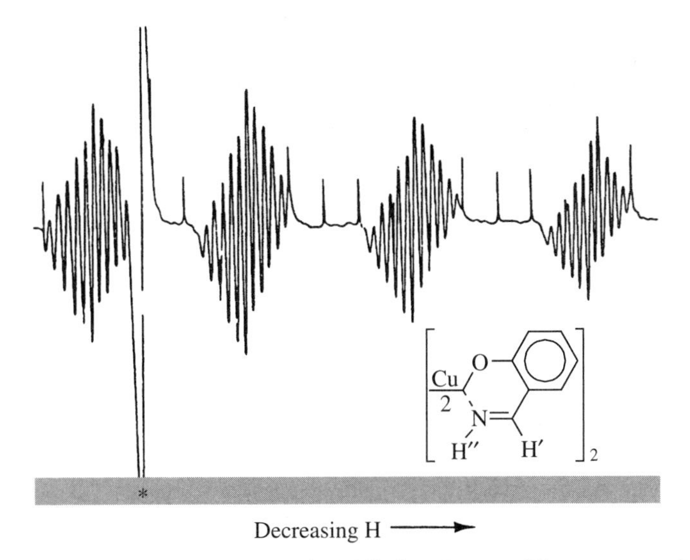

Figure 3.34 EPR spectrum of bis(salicylaldiminato)copper(II)—see text for details (reprinted with permission from Maki and McGarvey, *J. Chem. Phys.* 1958, **29**, 35; © American Institute of Physics, 1958).

for Schiff-base salicylaldimine polymers of copper(II). The four groups of lines are due to ^{63}Cu($\mathbf{I} = 3/2$). Because ^{14}N has a nuclear spin of 1, the two equivalent nitrogen atoms should exhibit five lines ($2n_{\mathrm{N}}\mathbf{I}_{\mathrm{N}} + 1$), each of which is split into three peaks by two equivalent hydrogen atoms ($\mathbf{I} = 1/2$) on the two aldimine C atoms ($2n_{\mathrm{H}}\mathbf{I}_{\mathrm{H}} + 1$). Overlap produces an 11-line pattern instead of the predicted 15 lines [$(2n_{\mathrm{N}}\mathbf{I}_{\mathrm{N}} + 1)(2n_{\mathrm{H}}\mathbf{I}_{\mathrm{H}} + 1)$]. Because the five peaks of the two-nitrogen spectrum would be in a 1:2:3:2:1 pattern and the two equivalent hydrogen nuclei would cause three peaks with a 1:2:1 pattern, appropriate overlap considerations yield the 1:2:3:4:5:6:5:4:3:2:1 pattern that is observed. Proof that the aldimine hydrogen atoms and not the hydrogen atoms or the nitrogen atoms cause the added splitting was shown when the latter were substituted with deuterium and no change in the spectral pattern was observed. The interaction with the aldimine protons is undoubtedly facilitated by the conjugated unsaturation of the chelate rings.

A detailed discussion of the use of EPR in polynuclear molybdenum coordination compounds (51) provides a background both for the use and for the complications that arise as the number of interacting metal ions increases. This article only considers monomers through tetramers. Even at the tetramer stage, the complications are quite apparent.

Detailed characterization of the unpaired electrons that arise from the irradiation of diamagnetic metal coordination polymers is helped by EPR spectroscopy. An example of such a spectrum is shown in Figure 3.35. EPR spectra are normally shown as first-derivative spectra (cf. Fig. 3.35). First-derivative spectra allow separate components of broad peaks to be observed. In the disulfide polymer illustrated as part of Figure 3.35, [acacCo^{III}(leu)acac–S–S–]$_n$, the irradiation is expected to break the sulfur-sulfur bond and produce sulfur radicals with g values near 2.00. The average g value observed for this polymer is 2.006, typical for sulfur radicals. But why is an eight-line pattern observed?

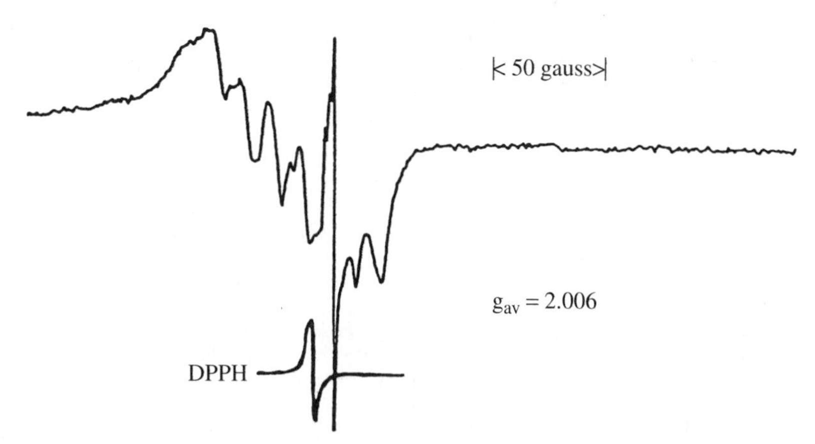

Figure 3.35 EPR spectrum of an irradiated [Co(acac-S-S-acac)(S – leu)]$_n$ polymer (Results obtained by V. Tramontano).

As noted above, each atomic nuclear spin (**I**) with which the unpaired electront spin interacts produces 2I + 1 peaks separated by the hyperfine-coupling constant for that atom. However, ^{32}S nuclei (>95% abundance) have $\mathbf{I} = 0$, ^{33}S nuclei (<1%) have $\mathbf{I} = 3/2$, and ^{34}S nuclei (4%) have $\mathbf{I} = 0$. None of these should produce an eight-line pattern. On the other hand, the ^{59}Co (100%) nucleus has $\mathbf{I} = 7/2$, which should give an eight-line spectrum if the spin density is on the cobalt nuclei. For the radical spin to be on cobalt, an electron would need to be transferred from the cobalt(III) to the sulfur radical site, which is short an electron after the S–S bond cleavage. This would result in a cobalt(IV) center. *Each* component of the eight-line spectrum should be separated by the hyperfine splitting constant observed for cobalt in cobalt(IV) species in which an unpaired electron is centered on the cobalt. The best case, $Co(1 - C_7H_{11})_4$, has $A_{Co} = 56(\pm 2) \times 10^{-4}\ cm^{-1}$. [Other cobalt(IV) species have lower hyperfine coupling constants (49), but they all contain sulfur and/or phosphorus donors that take some of the unpaired density.] However, the *total* separation for the eight peaks in the cobalt polymer is only 47.6 gauss (6.8 gauss for each of the seven spacings, so $A_{Co} = 6.8 \times 2.006 \times 4.67 \times 10^{-5} = 6.3 \times 10^{-4}\ cm^{-1}$), which is less than 12% of the $56 \times 10^{-4}\ cm^{-1}$ anticipated for the unpaired electron density on cobalt. Thus, even though the intensity of the eight line components is about 90% of the total intensity of the derivative EPR spectrum (Fig. 3.35), less than 12% of the spin density is on the cobalt centers. The g value of 2.006 is also more consistent with a sulfur-centered radical — the cobalt(IV) organometallic species noted above has a g value of 2.13. The suggestion that sulfur radicals are the primary center for the unpaired spins in the irradiated polymer seems okay, but the unsaturated β-diketone chelate ring apparently allows some of the spin density to be delocalized from the sulfur radical atom.

The preceding example illustrates that EPR intensity measurements on derivative spectra can be misleading. Even measuring the area under an EPR absorption curve, which should be proportional to the number of unpaired spins in a sample, often leads to an error of 30–50% unless

1. the line-width and shape are similar to a standard sample,
2. the number of spins of the unknown and standard are similar,
3. the sample characteristics, including shape and dielectric loss, are similar,
4. a short spin-lattice relaxation time, T_1, for both avoids signal saturation, and
5. both are time and temperature stable over the period of measurement.

If all of these conditions are met, the errors can be reduced to 5–10% (47).

Other examples of EPR spectroscopy with inorganic polymers include:

1. the evaluation of the thermal oxidative stabilization obtained from grafting HALS groups, NH_2–R–NH-(4-aza-3-dimethyl-5-dimethyl-cyclohexane), where $R = (CH_2)_{2 or 6}$, onto anhydride functionalized poly(organophosphazene) molecules (52),

2. both the characterization and stability of γ-irradiated polysilane radicals (53), and
3. the autooxidation of poly(hydrosilane)s after uptake using fusinite as a paramagnetic probe (54).

EPR studies are quite abundant in solid-state materials containing transition metal ions, too. A few representative examples include:

1. the nature of vanadium in vanado-silicate molecular sieves as a function of synthetic methodology (55),
2. the nature of copper in a polymeric copper(II) compound with a bridging dicyanamide anion (56), and
3. the magnetic ordering of a manganese(II) carboxylato polymer that becomes EPR silent below 4 K (57).

3.4.3 Electronic Spectroscopies

Ultraviolet, visible, and near-infrared spectra of metal coordination and organometallic polymers can be helpful in substantiating the arrangement of ligands about metal ions with partially filled d orbitals, the oxidation state of the metal ions, etc. The **d-d**, **charge-transfer** (also called electron-transfer), and **intervalence charge-transfer** transitions (cf. Fig. 3.36) typically occur in one or more of these spectral regions as detailed in almost all inorganic chemistry textbooks above the general chemistry level. A recent two-volume work (58) has a great deal of information on the electronic spectroscopy of metal coordination compounds including chapters on polarized absorption spectroscopy, luminescence spectroscopy, and laser spectroscopy that are not covered here. More detailed information on electronic spectra through ligand field theory was published recently by Figgis and Hitchman (59) and earlier by Lever (60, 61), Figgis (62), Jørgensen (63, 64), Ballhausen (65), and others.

Metal-containing inorganic polymers are often characterized by their electronic spectra and by their d-d spectra in particular. Examples of d-d spectra for nickel(II) coordination species in solution with various geometrical arrangements are shown in Figures 3.37 and 3.38. Note that the molar extinction coefficients (ε) for the d-d transitions of octahedral nickel(II) complexes (Fig. 3.37a) are normally less than 10 M^{-1} cm^{-1} whereas the ε values for the other geometries (Figs. 3.37b and 3.38) are typically quite a bit higher. All true d-d transitions are **symmetry forbidden** with low intensities — and except for their vibronic coupling, they would have no intensity at all. {**Vibronic coupling** implies a vibration of the same symmetry as the electronic transition coupling with the electronic transition.} These d-d transitions can be further separated into the more intense **spin-allowed** transitions, in which no change in spin occurs, and the less intense **spin-forbidden** transitions, in which a change in spin occurs during the transition. If no change in spin occurs, the ground state and the excited state have

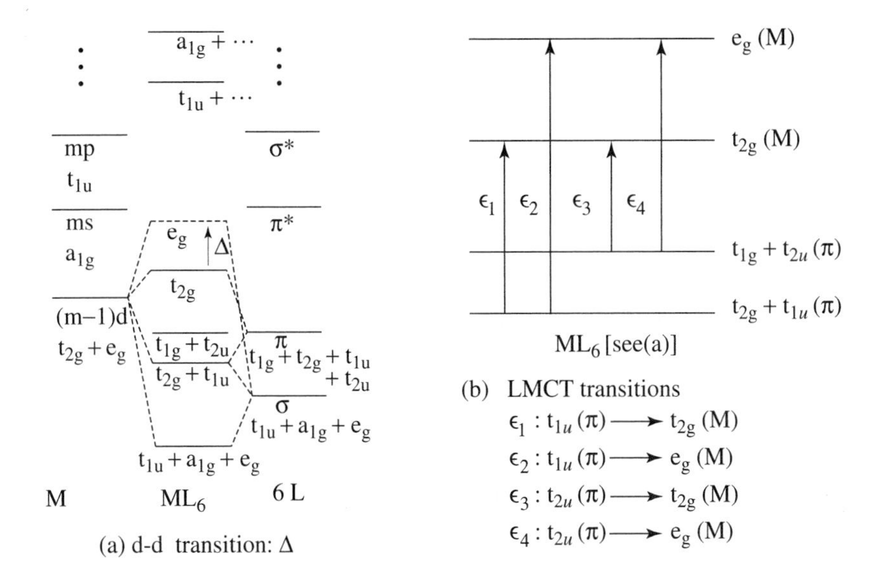

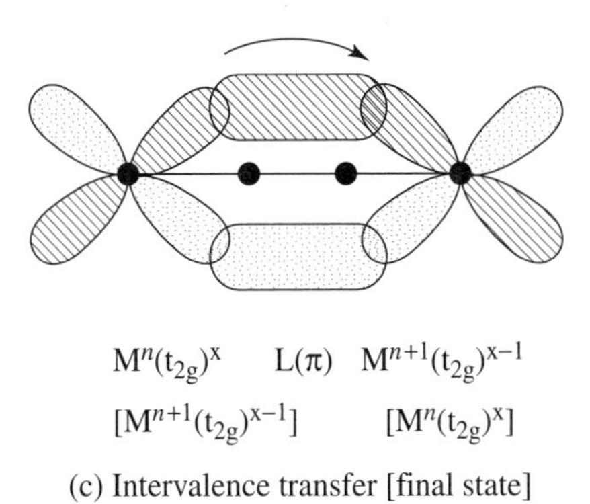

Figure 3.36 Types of electronic spectral transitions. (d-d, ligand-to-metal charge-transfer, and intervalence).

the same multiplicity. When the spin changes, a change in multiplicity occurs. The multiplicity of a state is normally shown as a left superscript. Spin-forbidden electronic transitions are typically an order of magnitude less intense than the spin-allowed transitions; however, mixing of states can make them almost as strong in a few cases. An example is one of the two bands at about 15,000 cm^{-1} in the hexaaquanickel(II) ion shown in Figure 3.37a.

Mixing of states (e.g., p states) can occur in most nonoctahedral geometrical configurations, and this also increases the intensity of the transitions. Even so, the intensities are typically of the order of 10^{1-3} M^{-1} cm^{-1}. Transitions that have molar extinction coefficients $>10^3$ are normally charge-transfer or intervalence-transfer transitions. Exceptions for molecules containing heavy transition metals

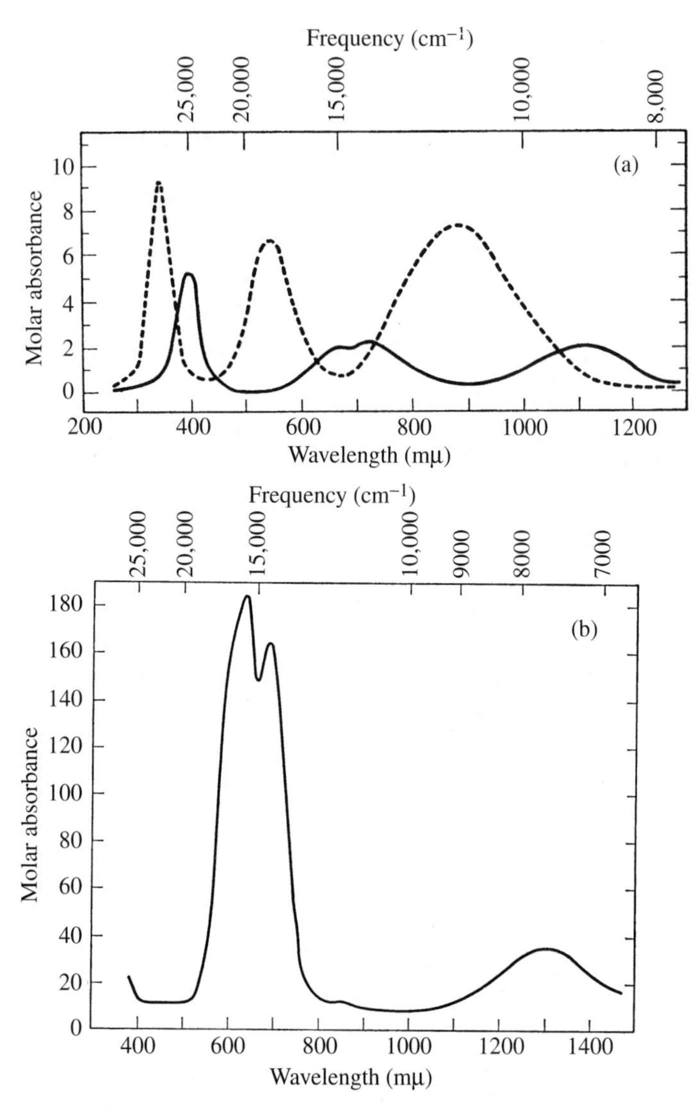

Figure 3.37 Typical visible and near-infrared absorption spectra for nickel(II) ions surrounded by ligands with different geometrical arrangements: (a) octahedral aqua $[Ni(H_2O)_6]^{2+}$ (———) and ethylenediamine $[Ni(NH_2CH_2CH_2NH_2)_3]^{2+}$ (- - - -) ions; (b) tetrahedral $Ni(Ph_3AsO)_2Br_2$, where Ph = C_6H_5. Spectra based on F. A. Cotton and G. Wilkinson, *Advanced Inorganic Chemistry*, 1st Ed., Wiley, NY, 1962, pp 735–736.

can occur; for example, $IrBr_6^{2-}$ has a band at about 700 nm (about 14, 500 cm^{-1}) with an molar extinction coefficient of 3000 M^{-1} cm^{-1} at room temperature that appears to be a symmetry-forbidden but vibronically allowed d-d transition (66). A simple tabulation of the preceding discussion is given in Table 3.7.

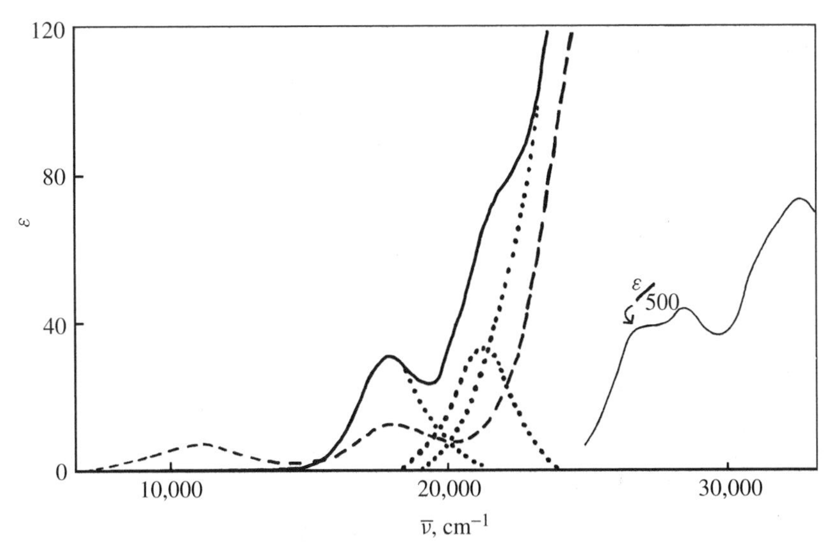

Figure 3.38 Absorption spectra for bis(1,1,1-trifluoro-4-pentene-2-ono)nickel(II): planar in dry benzene or chloroform (———) and octahedral in pyridine (- - - -). A spectral band analysis of the visible bands for the planar complex is also shown (• • • •).

TABLE 3.7 Absorption Intensities of Electronic Spectral Transitions.

Transition	Approximate ε
Spin allowed, symmetry (Laporte) allowed (charge transfer)	10^4
Spin allowed, symmetry forbidden, d-d with d-p mixing (e.g., tetrahedral coordination compounds and polymers)	10^2
Spin allowed, symmetry forbidden, d-d without d-p mixing (e.g., octahedral coordination compounds and polymers)	10
Spin forbidden, symmetry forbidden d-d	10^{-1}

The energies of the d-d transitions in octahedral complexes are related to both the metal and the ligand or ligands. For a given metal ion the energies of the d-d bands increase in the order $I^- < Br^- < Cl^- < F^- \approx$ O donors < N donors < CN^-. A more detailed list is shown in Table 3.8, where water (the aqua ligand) is the standard. Jørgensen catalogued ***f*** ligand parameter values that can be multiplied by a ***g*** metal ion parameter [$\boldsymbol{g}$ = 8700 cm^{-1} for nickel(II)] to give the Δ_o values of unknown octahedral coordination compounds; that is,

$$\Delta_o = \boldsymbol{f} \times \boldsymbol{g} \tag{3.35}$$

The order and ***f***-values of several of the ligands are based on average ligand environment results from mixed-ligand coordination compounds rather than from

TABLE 3.8 Relative Energies of d-d Transitions in Metal Coordination Species as a Function of the Ligands.

Ligand	*Relative* Δ_o *Energy*[a,b]	*Relative* d^6 Δ_o *Energy*[a,c]
Iodo	$<Br^{-d}$	0.55
Bromo	0.72	0.71
Thiocyanato-*S* = SCN^-	0.75	0.79
Chloro	0.80	0.76
Azido (N_3^-)	0.83	0.84
Diethyldithiophosphate-*S,S′*	0.83	0.82
Fluoro	~0.9	0.90
N, N′-Diethyldithiocarbamato-*S,S′*	0.90	0.95
Dimethyl sulfoxide = $(CH_3)_2S\mathbf{O}$	0.91	0.85
Urea = $(NH_2)_2C\mathbf{O}$	0.91	0.92
Hydroxo—	0.93	
Ethanol[b]/Methanol[c] = ROH	0.97	0.98
Oxalato $(C_2O_4)^{2-}$	0.99	1.00
Aqua (water) (standard)	**1**.00	1.00
Thiocyanato-*N* = $S C\mathbf{N}^-$	1.02	0.99
Nitrito (nitro-*O*) = $ON\mathbf{O}^-$	—	1.10
Glycinato = $\mathbf{N}H_2CH_2C(O)\mathbf{O}^-$	1.18	1.15[e]
Pyridine = C_5H_5N	1.23	1.26
Ammine (ammonia)	1.25	1.27
Ethylenediamine = $NH_2CH_2CH_2NH_2$	1.28	1.30
Sulfite = $\mathbf{S}O_3^{2-}$	~1.3	1.55
2,2′-Bipyridine (bpy)	1.33	1.33
1,10-Phenanthroline (phen)	1.34	1.33
Triphenylphosphine	—	1.41
Nitro = $\mathbf{N}O_2^-$	~1.4	1.53
Methyl (CH_3^-) or phenyl ($C_6H_5^-$)	$<CN^{-d}$	—
Cyano = $\mathbf{C}N^-$	~1.7	1.95
Carbonyl = $\mathbf{C}O^-$	—	2.12

[a]Jørgensen *f* parameter for highlighted donor atoms; [b]A.B.P.Lever, 2nd Ed., *op cit.*, Table 9.2, p. 748 — based on C. K. Jørgensen, *op cit.*; [c]Y. Shimura, *op cit.*, Table II $\mathbf{d}_{Co}$ ÷ 16.5 to get *f* parameter; [d]A.B.P.Lever, 2nd Ed., *op cit.*, p 739; [e]Calcd. from average of two donors

pure ML_n species (cf. Exercise 3.10). **The average ligand environment** concept is based on the fact that the Δ_o energy of a mixed ligand $ML_nL'_{6-n}$ species relative to the Δ_o values for ML_6 and ML'_6 species is as follows:

$$\Delta_o(ML_nL'_{6-n}) = [n\Delta_o(ML_6) + (6-n)\Delta_o(ML'_6)]/6 \qquad (3.36)$$

It is important to realize that the absorption maxima and extinction coefficients obtained through the use of such tabulations are approximate values. Changes in bond lengths and geometrical distortions can alter both the energies and intensities.

The energies of the spin-allowed d-d bands for octahedral d^2–d^8 ions can be obtained from Tanabe-Sugano diagrams. The Tanabe-Sugano diagrams are plots of the energy of electronic transitions for octahedral coordination species in E/B units (where B is a Racah electronic repulsion term and E is in units of 1000 cm^{-1}) versus Δ_o/B (or Dq/B, where 10 Dq $= \Delta_o$) (cf. Fig. 3.39). The spin-allowed terms (which are normally more intense) have the same degeneracy as the ground state, where the left superscript is the degeneracy. For the d^8 octahedral coordination species, the $^3A_{2g}$ state is the ground state and the excited states of the same degeneracy are the $^3T_{2g}$ and two $^3T_{1g}$ states. For such d^8 octahedral coordination species the $^3T_{2g}$ term is directly related to Δ_o. In fact, the energy of the first excited state equals Δ_o. That is,

$$\nu_1 = A_2 \rightarrow T_2 = \Delta_o \tag{3.37}$$

where the multiplicity of 3 and g for gerade have been omitted to make Eq. 3.37 more generally applicable.

The energies of the other two spin-allowed states, $^3T_{1g}(F)$ and $^3T_{1g}(P)$, are functions of both B and Δ_o. Knowing the energy of either of these $^3T_{1g}$ states relative to the energy of the first excited state makes it possible to graphically determine B. Another method of determining B comes from substituting Δ_o from Eq. 3.37 into either Eq. 3.38 or Eq. 3.39 or by solving Eqs. 3.38 and 3.39 simultaneously:

$$\nu_2 = A_2 \rightarrow T_1(F) = 1.50\Delta_o + 7.5B - 0.50(225B^2 + \Delta_o^2 - 18\Delta_o B)^{1/2} \tag{3.38}$$

$$\nu_3 = A_2 \rightarrow T_1(P) = 1.50\Delta_o + 7.5B + 0.50(225B^2 + \Delta_o^2 - 18\Delta_o B)^{1/2} \tag{3.39}$$

Again, the multiplicity of 3 and the g have been omitted, because these same Eqs. 3.37–3.39 are suitable for d^3 octahedral species [except that the terms are fourfold degenerate (^4X terms)] and for tetrahedral d^2 and d^7 species. The gerade (g) subscripts are not appropriate for tetrahedral species because no center of symmetry exists. Although unlikely with tetrahedral species, another limitation is the change of spin that occurs at high ligand fields for d^7 species.

For d^2 and d^7 octahedral species, the ground state is $T_{1g}(F)$ and no simple Δ_o transition occurs, as is evident from Eqs. 3.40–3.42:

$$\nu_1 = T_1(F) \rightarrow T_2 = 0.50\Delta_o - 7.5B + 0.50(225B^2 + \Delta_o^2 + 18\Delta_o B)^{1/2} \tag{3.40}$$

$$\nu_2 = T_1(F) \rightarrow A_2 = 1.50\Delta_o - 7.5B + 0.50(225B^2 + \Delta_o^2 + 18\Delta_o B)^{1/2} \tag{3.41}$$

$$\nu_3 = T_1(F) \rightarrow T_1(P) = (225B^2 + \Delta_o^2 + 18\Delta_o B)^{1/2} \tag{3.42}$$

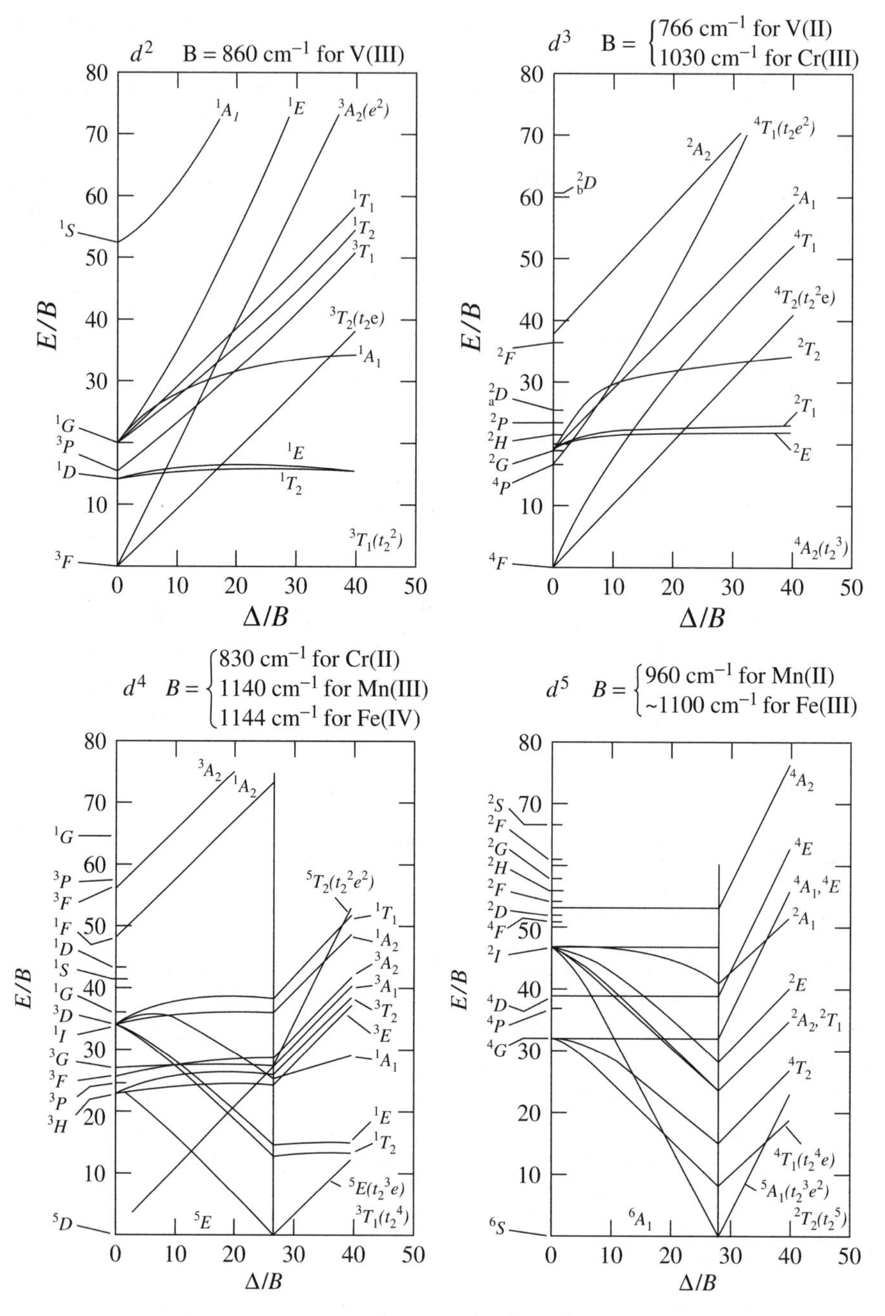

Figure 3.39 Tanabe-Sugano diagrams for d^2 to d^8 octahedral complexes.

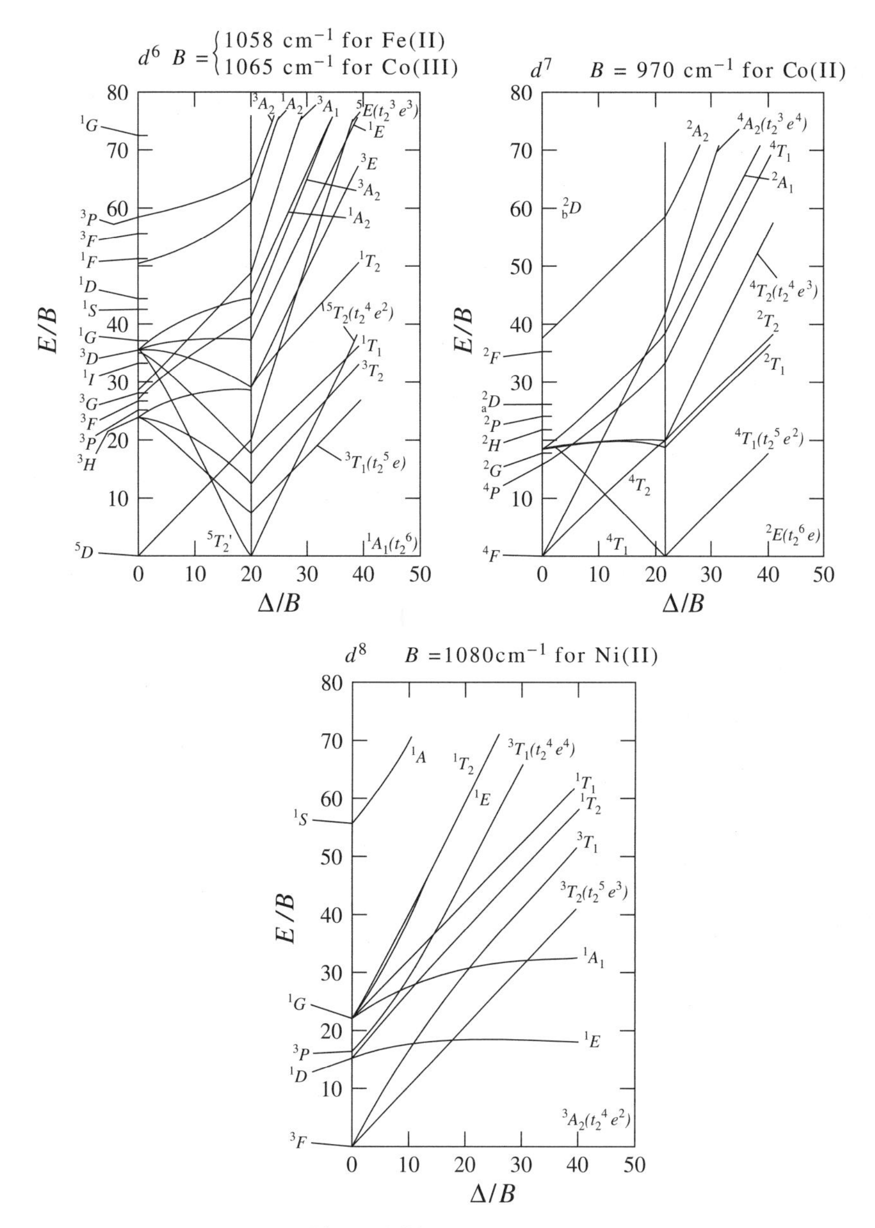

Figure 3.39 *(Continued)*.

All three states are a function of both Δ_o and B; however, ν_1 and ν_2 are separated by Δ_o, but as noted above, the spin-change possibilities for d^7 species precludes using these equations for low-spin octahedral d^7 coordination species. Equations 3.40-3.42 are suitable for tetrahedral d^3 and d^8 species. For other ions, the reader is referred to equations or matrices in the references noted above.

Note that the ground state changes in the octahedral d^4–d^7 species when Δ_o increases sufficiently for spin pairing. The relationship between spin pairing and the Racah electron repulsion parameters is given in Eqs. 3.43–3.46, where the energy separation for spin pairing is given:

$$\Delta > 6B + 5C \text{ for spin paired } d^4 \text{ octahedral species} \tag{3.43}$$

$$2\Delta > 15B + 10C \text{ for spin-paired } d^5 \text{ octahedral species} \tag{3.44}$$

$$2\Delta > 5B + 8C \text{ for spin-paired } d^6 \text{ octahedral species} \tag{3.45}$$

$$\Delta > 4B + 4C \text{ for spin-paired } d^7 \text{ octahedral species} \tag{3.46}$$

Tetrahedral coordination compounds have energies about 4/9 those of the corresponding octahedral species. In fact, the low-energy band of $Ni(Ph_3AsO)_2Br_2$ at about 7600 cm^{-1}, Figure 3.37b, is probably the second d-d band for this species — corresponding to the 15, 000 cm^{-1} band of the octahedral aqua ion in Figure 3.37a. On the other hand, the first d-d transition for tetragonal planar species (Fig. 3.38) should be (and is) at an energy much higher than that of analogous octahedral species.

As an example of how d-d spectra can be used to ascertain geometry, note the dotted absorption band analysis (dotted line) in Figure 3.38. Planar nickel(II) species with two nitrogen and two oxygen donors show two bands in the visible region (around 20,000 cm^{-1}) if the species has *trans* nitrogen (and *trans* oxygen) donors but only one if they are *cis*. Thus this nickel(II) ketoimine complex is thought to have *trans* nitrogen donors. Also note that the bands with molar extinction coefficients of about 30 M^{-1} cm^{-1} in noncoordinating solvents decrease to less than 10 M^{-1} cm^{-1} in pyridine consistent with a shift to octahedral geometry. Also note that the bands shift to energies comparable to the octahedral tris(ethylenediamine)nickel(II) ion in Figure 3.37a.

Caution: Figure 3.38 increases linearly in energy, whereas Figure 3.37,a and b increase linearly in wavelength. Although these are monomers, nickel(II) polymers would have similar absorption spectra.

The large intensity of electron-transfer transitions of polymeric iron(III) hydrous oxide (rust in the solid state) should be well known to anyone studying chemistry. Nonhydrolyzed aqueous iron(III) is almost colorless in solution. Like aqueous manganese(II), iron(III) is a spin-free d^5 ion that has d-d transitions that are both symmetry forbidden and spin forbidden. However, excited-state charge transfer occurs when an electron from an energy level primarily on a hydroxo or an oxo ligand moves to an energy level that is primarily a d level of the iron(III). This is effectively a p-d transition that is not symmetry forbidden and typically of much greater spectral intensity. This and similar charge-transfer reactions are termed **ligand-to-metal** charge-transfer transitions. These transitions are accessible (found at energies below the vacuum UV) for metal ions in higher oxidation states.

Metal-ions in low oxidation states often have charge-transfer transitions that are **metal-to-ligand** in nature. Common examples are the deep-red low-spin iron(II) and ruthenium(II) diimine complexes. An example of an inorganic coordination polymer with a metal-to-ligand transition that shows a decreasing energy with increasing polymer size is shown in Figure 3.40. This tungsten(IV) polymer is analogous to monomeric tungsten(IV) coordination compounds where the metal-to-ligand nature of the transitions has been shown. The chloroform-soluble oligomer has its charge-transfer transition at higher energy than the dimethyl sulfoxide-soluble polymer ($[\eta] = 25\ cm^3/g$) consistent with some electronic delocalization in the polymer. The insoluble portion of the polymer has its charge-transfer transition at even lower energy, which is suggestive of even longer polymer chains — consistent with the insolubility.

Metal coordination dimers, oligomers, and polymers in which the metal ions are not all in the same oxidation state often exhibit **intervalence charge-transfer**

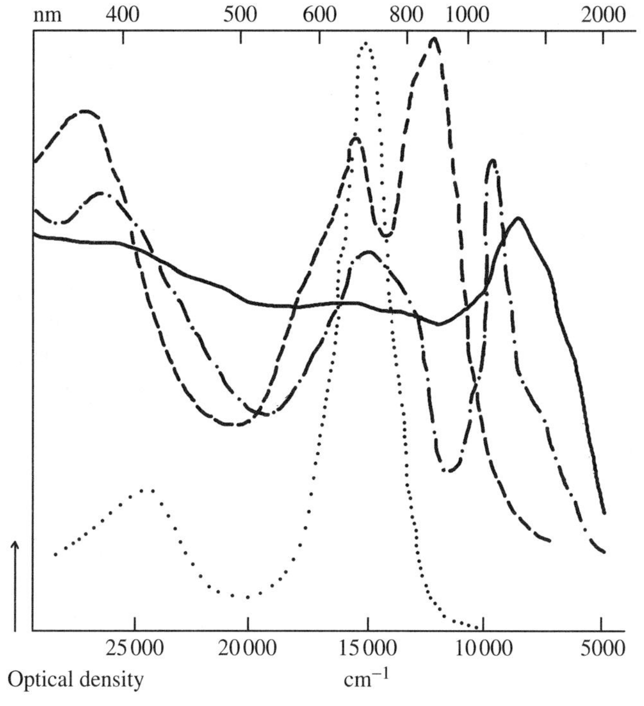

Figure 3.40 The metal-to-ligand charge-transfer bands of (a) the model tetrakis(5,7-dichloro-8-quinolinolato) tungsten(IV) chelate · · · · ; (b) the fractions of poly{bis(5,7-dichloro-8-quinolinolato)tungsten(IV)-μ-5,8-quinoxalinediolato}: the dichloromethane-soluble oligomer - - - - ; the DMSO-soluble polymer -·-·-·; and the DMSO-insoluble polymer ——.Cf. Figure 2.15 for the structure of this polymer.

transitions. As noted above, some iron(II) species have a tendency to donate electrons to ligands at low energy and iron(III) species accept electrons from ligands at low energy. Thus, if both iron(II) and iron(III) coexist in the same chain or cluster, the transfer of electrons from iron(II) to iron(III), via ligands or directly in metal-metal bonded species, is a logical possibility. The spectra resulting from such transfers are termed intervalence charge-transfer spectra or simply intervalence bands. Intervalence spectra are typically both intense and broad. Prussian blues, both soluble $KFe_2(CN)_6$ and insoluble $Fe_4[Fe(CN)_6]_3 \cdot 15H_2O$, exhibit intervalence charge-transfer bands that produce a very deep blue color when pale-colored aqueous Fe^{II} and $[Fe^{III}(CN)_6]^{3-}$ (or Fe^{III} and $[Fe^{II}(CN)_6]^{4-}$) are mixed. The three-dimensional structure of these polymeric species was shown in Figure 1.16. More will be noted about these species when Mössbauer spectroscopy is discussed in Section 3.4.5.

The interpretation of solid-state inorganic polymer spectra is not always as straightforward as the interpretation of solution spectra because of geometrical distortions that often occur and the difficulty in ascertaining ε values for the polymers (or other solids as well). Therefore, it is important to use electronic spectroscopy in conjunction with other methods to ensure unambiguous results (cf. Exercise 3.11).

3.4.4 Vibrational Spectroscopies

Infrared and Raman spectroscopies provide the same vibrational information for inorganic polymers that they do for monomeric molecules. Because the details of infrared spectroscopy are normally covered before attaining the level assumed by this book, no details of infrared spectroscopy methodology will be provided here. A few examples will be noted, however.

On the other hand, Raman spectroscopy has a special advantage in molecules with homonuclear bonds because Raman spectra are more intense for vibrations with minimal dipole change during the vibration, for example, the symmetrical Si-Si stretch would be much stronger in Raman spectra than in infrared spectroscopy, where intensity is directly related to the change in dipole during the vibration. See Tables 3.9 and 3.10 and Figure 3.41. Its limited use with inorganic polymers is based on the low intensity of Raman scattering relative to the incident light and the complications that result from absorption in resonance Raman spectroscopy (see below). Other background emission, scattering, and/or fluorescence further obscure definitive results. However, the ability to use Raman spectroscopy easily with aqueous and other polar molecule solutions should make it suitable for many investigations.

3.4.4.1 Infrared Examples

Infrared spectroscopy can see impurities or incomplete reactions, if at the one to five percent level. For example, the aromatic carbonyl stretching frequencies noted in Table 3.10 can be used to signal reaction completeness in metal-coordination Schiff-base polymerization reactions with phenolic coordination (see

TABLE 3.9 Basic Characteristics of Infrared (IR) and Raman Spectroscopy[a].

Paramater	Infrared Spectroscopy	Raman Spectroscopy
Spectroscopic phenomenon	Absorption of light: $h\nu_{IR} = \Delta E_{vibr}$	Inelastic scattering of light: $h\nu_0 - h\nu_{sc} = \Delta E_{vibr}$
Allowed transition	$\Delta\nu = +1, +2, +3, \ldots$	$\Delta\nu = \pm 1, \pm 2, \pm 3, \ldots$ (transitions for $\Delta\nu = +2, +3, \ldots$ i.e., overtones are considerably less conspicuous than in IR)
Excitation	Polychromatic IR radiation	Monochromatic radiation (ν_0) in the UV, visible, or near IR
Molecular origin	Dipole moment: $\mu = qr$	Induced dipole moment: $\boldsymbol{P} = \alpha \boldsymbol{E}$
Requirement for vibrational activity	Change in dipole moment during vibration: $(\partial\mu/\partial Q_k)_0 \neq 0$	Change in polarizability during vibration: $(\partial\alpha/\partial Q_k)_0 \neq 0$
Band intensity	$I_{IR}^{1/2} \propto (\partial\mu/\partial Q_k)_0$	$I_R^{1/2} \propto (\partial\alpha/\partial Q_k)_0$
Frequency measurement	Absolute: $\nu_{vibr} = \nu_{IR}$	Relative to the excitation frequency: $\nu_{vibr} = \nu_0 - \nu_{sc}$
Readout signal	Comparative: transmittance ($T = \Phi_s/\Phi_r$) or absorbance ($A = -\log T$)	Absolute: radiant power or intensity of scattered radiation
Spectral plot	Linear in %T or logarithmic in A vs. wavenumber (cm^{-1})	Linear: Raman intensity vs. wavenumber shift (cm^{-1})
Dominant spectral feature	Vibrations destroying molecular symmetry: antisymmetric stretching and deformation modes	Vibrations preserving molecular symmetry: symmetric stretching modes
Inactive molecule	Homonuclear diatomics	None
Centrosymmetric molecule	Only "*u*"-symmetry modes active	Only "*g*"-symmetry modes active
Medium	Water is a strong absorber and is a poor solvent for IR studies	Water is a weak scatter and is a good solvent for Raman studies

[a] h, Planck's constant; ΔE_{vibr}, energy difference of vibrational levels; ν, photon frequency; $\Delta\nu$, change in vibrational quantum number; q, charge; $\boldsymbol{r}$, charge spacing; α, molecular polarizability; $\boldsymbol{E}$, electric field; Φ_s and Φ_r, radiant powers transmitted by the sample and reference cells, respectively; Q_k, vibrational normal coordinate ($k \leq 3N - 6$); "*g*" and "*u*", normal modes of vibration symmetric (*gerade*) and antisymmetric (*ungerade*) with respect to the molecular center of inversion.
From: Czernszewicz and Spiro in Solomon and Lever, Vol 1, 353ff Table 1, p 359.

Fig. 3.42). An incomplete reaction with a metal ion shows either a shoulder at about 1280 cm^{-1} if the acid form of the ligand is present or a shoulder at about 1340 cm^{-1} if excess sodium hydroxide is used to ensure complete deprotonation of the ligand. A band at intermediate energies is indicative of the chelated phenolic group (Table 3.10).

TABLE 3.10 Intensity Differences Between Infrared and Raman Spectra.

Vibration	Wave Number (cm^{-1}) Range	Reference
Strongly Infrared Active (Asymmetric Group Vibrations, Bending Modes, and Stretching of Polar Bonds)[a]		
Si–O–Si antisymmetric stretch	1000–1100	a
C–O stretch	900–1300	a
aromatic C–O–H	1270–1290	b
aromatic C–O–M^{IV}	1295–1310	b
aromatic C–O–M^{III}	1310–1330	b
aromatic C–O–M^{+}	1335–1345	b
aromatic C–O–$(M^{III})_2^c$	1280–1290	b
C=O stretch	1600–1800	a
O–H (hydrogen-bonded)	3000–3400	a
N–H stretch (hydrogen-bonded)	3100–3300	a
Si–F stretch	550–1000	d
Strongly Raman Active (Vibrations of Symmetrical Groups and Bond Stretching of Nonpolar or Slightly Polar Bonds)[a]		
Si–O–Si symmetrical stretch	450–550	a
Si–Si stretch	460–510	e
Si–C stretch	680–710	e
R–**S–S**–R symmetrical stretch	400–500	a
alkyl–**S–S**–alkyl	420–490	d
aromatic–**S–S**–aromatic	490–520	d
N=N stretch	1575–1630	a
C–S stretch	600–700	a
C–H aromatic stretch	3000–3100	a
Strong in Both Infrared and Raman Spectra		
Si–H stretch	2100–2300	a
C–H aliphatic stretch	2800–3000	a
C–Cl stretch	400–800	a,d

[a]Rabek (19), p. 286 quoting Brown & Harvey in Brame and Gresselli, eds, IR and Raman Spectroscopy, vol 1, Part C, Dekker, 1977, Ch 12; [b]H. Chen, Ph.D. Dissertation, University of Massachusetts, Amherst, 1995; [c]M–O(R)–M bridge; [d]Drago (45); [e]Shilling, Bovey et al. (69)

Infrared spectra can also be used to follow the course of a reaction in main group polymers. For example, in the telechelic synthesis of phosphazene-siloxane block copolymers discussed in the synthesis chapter (Section 2.6.2), the hydride functionality of the polysiloxane (**1** in Fig. 2.48) has an infrared Si–H stretch at 2100 cm^{-1} that slowly disappears during the 12-h reaction.

The infrared spectra of a sizable number of phosphazene, siloxane, carbosilane, carbosiloxy, phthalocyaninato siloxy and fluoroaluminum, metal coordination, and copolymers containing inorganic segments have been compiled in the *Atlas*

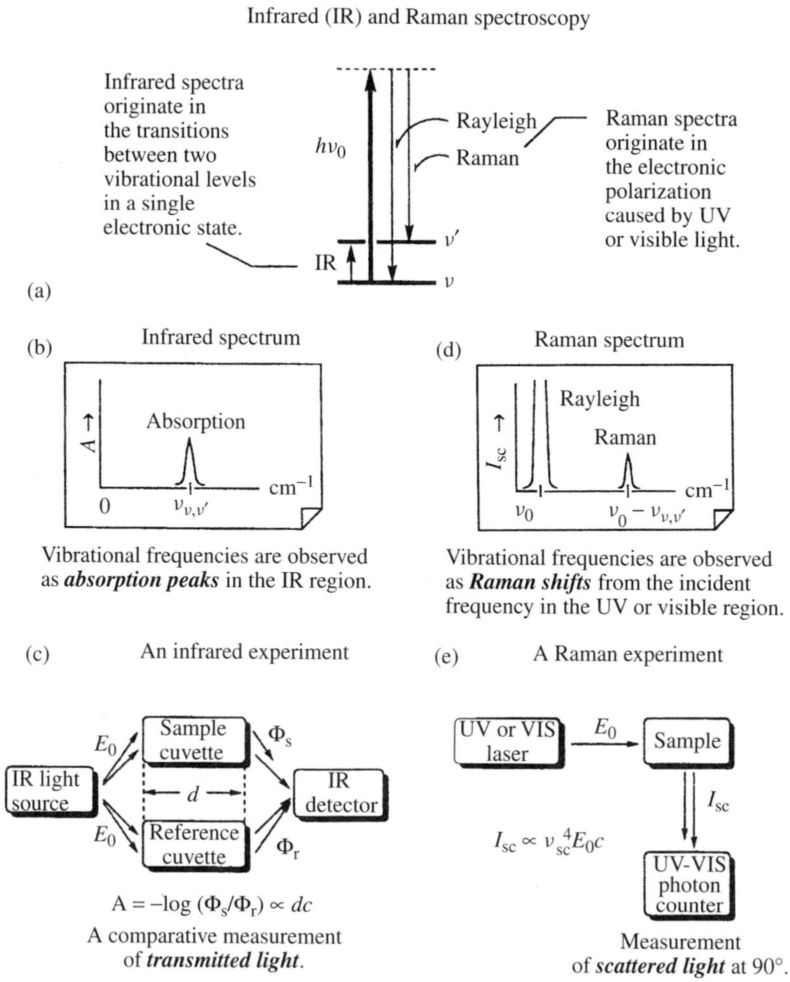

Figure 3.41 A comparison of infrared absorption and Raman scattering spectroscopies. The infrared absorption spectrum (b) is a direct measure of the difference in energy between the ground state ν and the excited state ν'; although most instruments show the percent transmission (%T). The Raman spectrum is superimposed on the Rayleigh scattering spectrum (c) and differs in energy from the Rayleigh peak by the difference in energy between the ground and excited state. From Czernuszewicz, R. S. and Spiro, T. G. in *Inorganic Electronic Structure* and Spectroscopy; Solomon, E. I. and Lever, A. B. P., Ed., John Wiley & Sons: New York, 1999; Vol. I, p. 356.

of Polymer and Plastics Analysis (67). Although almost 150 spectra are shown, a number of them have inorganic components only as side groups. This collection of infrared spectra provides a good starting point for comparing new polymers, although it predates the ring-opening silaferrocene polymers and it contains no polysilanes.

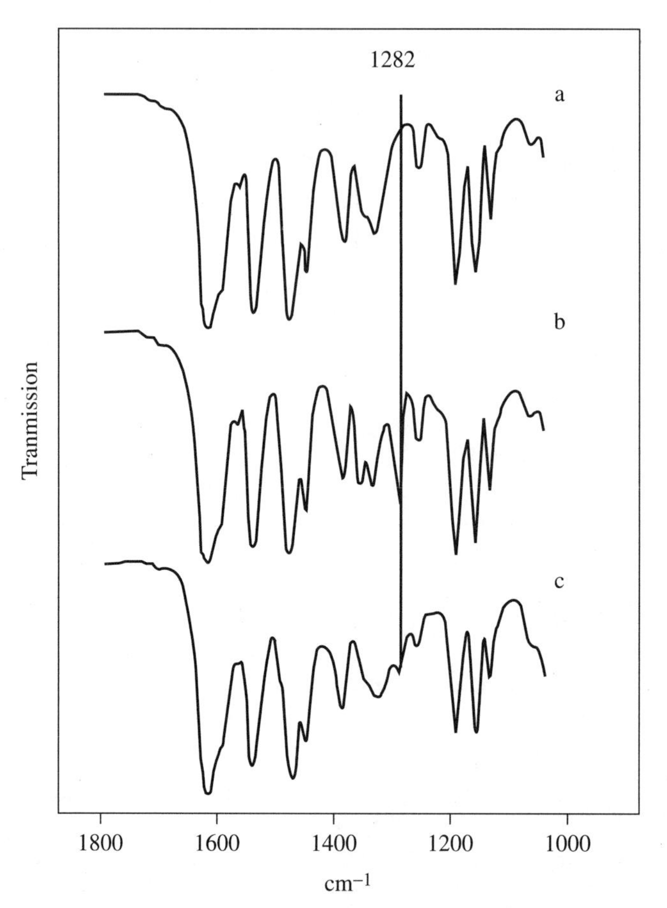

Figure 3.42 The infrared spectra of (a) the $[NaY(tsdb)]_n$ polyelectrolyte in pure form, (b) the product of $Y(acac)_3$ and H_4tsdb without added base, and (c) the attempted preparation of $[LiY(tsdb)]_n$ using the method that worked for $[NaY(tsdb)]_n$. N,N′,N″,N‴-tetrasalicylidine-3, 3′-diaminobenzidinato = $tsdb^{4-}$ and acetylacetonato (or 2, 4 − pentanedionato) = $acac^-$.

3.4.4.2 Raman Spectroscopy

Raman spectra result from inelastic scattering of light as shown schematically in Figure 3.41 (68). A Raman shift from the energy of the incident radiation less the energy of a vibrational mode as shown in the diagram is called a Stokes' shift. Stokes' shifts are more intense than the shifts that result from scattering by excited-state vibrational levels to the nonexcited ground-state level, so-called anti-Stokes' shifts. Because typically less than 1 photon in a million exhibits Raman scattering in transparent regions of the visible or ultraviolet, strong laser excitation is used in Raman studies.

However, when the frequency of the excitation laser approaches an allowed electronic transition of the molecule under study, but not so close as to have

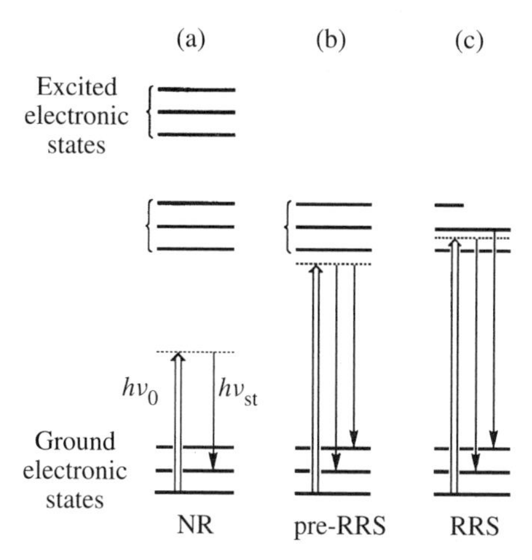

Figure 3.43 Diagram of (a) normal Raman scattering that occurs in a transparent region of a molecule; (b) preresonance Raman scattering as the energy approaches the energy of an electronic transition (about 10- to 100-fold intensity increase); and (c) resonance Raman scattering in which the interacting state is dominated by a few vibronic levels in the interior of the excited state (and may be as high as 10^3 to 10^6 fold increase from normal Raman scattering). From Czernuszewicz, R. S. and Spiro, T. G. in *Inorganic Electronic Structure and Spectroscopy*; Solomon, E. I. and Lever, A. B. P., Ed., John Wiley & Sons: New York, 1999; Vol. I, p. 380.

significant absorption that might lead to decomposition, the vibrational modes that are vibronically coupled with the transition exhibit enhanced preresonance Raman intensities (Fig. 3.43) (68). An example of how a change in the energy of such an allowed-electronic transition can change the observed Raman spectrum for an inorganic polymer follows.

The Raman spectra of a poly(di-*n*-hexylsilane) film have intense Si–Si and Si–C bands in the lower-temperature spectra at the top and bottom of Figure 3.44. The apparent disappearance of the Raman spectrum on warming is related to a confirmational change that the polymer undergoes and is related to the ultraviolet absorption band shift to higher energy (Fig. 3.45). The Raman measurements were made with irradiation above 458 nm. At the lower temperatures, an absorption band at about 370-nm dominates the near-ultraviolet spectrum of the film. The low-energy tail is sufficiently strong to provide enhanced Raman spectra. At higher temperatures the band at 317 nm dominates and insufficient absorption occurs at 458 nm to provide for more than a very weak Raman spectrum. Cooling the sample back to 17 °C and then re-recording the Raman spectrum shows the reversibility of the confirmational change. In hexamethyldisiloxane solution the analogous confirmational change occurs at about −30 °C. Details of the confirmational change have been discussed elsewhere (69).

Other references on the Raman spectroscopy of polymers or inorganic species can also be found in Jan Rabek's book on polymer characterization (47), Kazoo

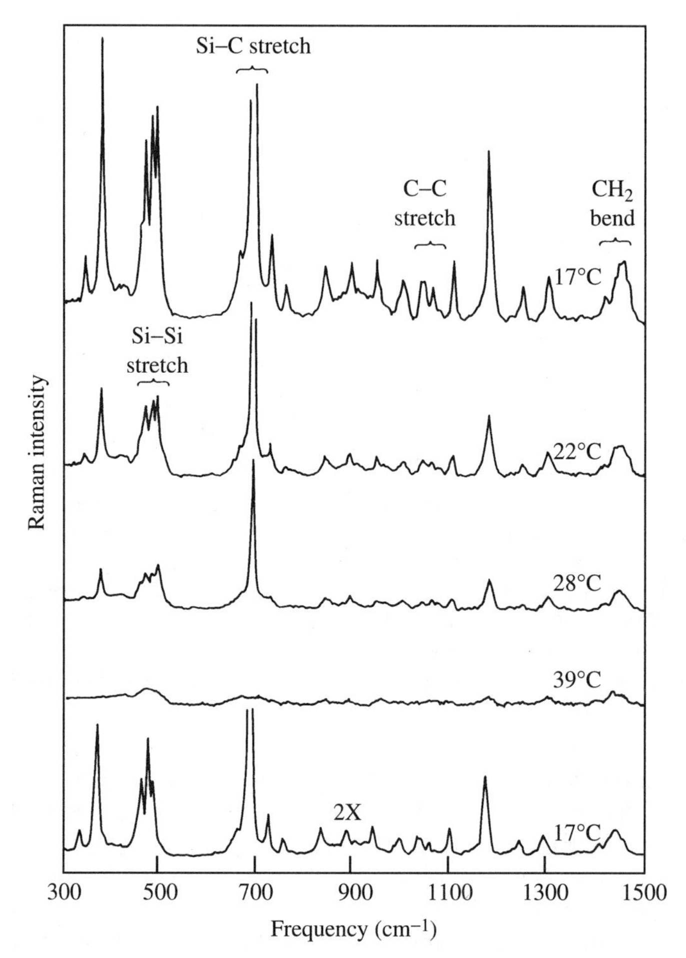

Figure 3.44 The temperature dependence of the Raman spectra of a poly(di-*n*-hexylsilane) film. Reprinted with permission from Kumzany et al., *J. Chem. Phys.* 1986, **85**, 7413; © American Institute of Physics, 1986.

Nakamoto's books on infrared and Raman spectra of inorganic and coordination compounds (the 5th edition has two volumes) (70), Laserna's book on modern techniques in Raman spectroscopy (71), Russell Drago's book on physical methods of chemistry (45), or Sidney Kettle's physical inorganic chemistry book (72).

3.4.5 Mössbauer Spectroscopy

Mössbauer spectroscopy is the study of the chemical environment as measured by the resonance absorption of gamma radiation by nuclear energy levels of nuclei that have accessible excited states(73). The most widely studied nucleus is ^{57}Fe, for which a 14.4-keV excited state is obtained by the decay of ^{57}Co that can be

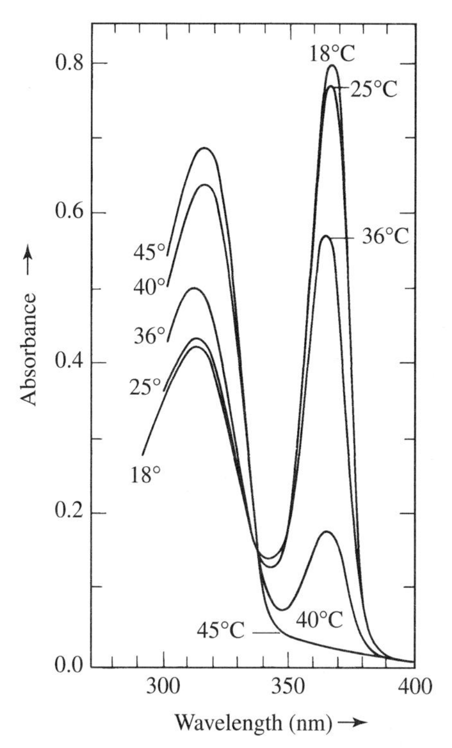

Figure 3.45 The temperature dependence of the absorption spectra of a poly(di-*n*-hexylsilane) film. Reprinted with permission from Kumzany et al, op cit., © American Institute of Physics, 1986.

prepared by the ^{56}Fe(d,n)^{57}Co nuclear reaction using deuterons from a cyclotron source. At least 99.8% of the ^{57}Co decays by electron capture (270-day half-life) and produces ^{57}Fe in an excited state.

The exact energy of the ~14.4-keV gamma radiation is dependent on the chemical environment of the iron and is measured using the Doppler effect as illustrated in Figure 3.46. The figure shows the situation in which the source S and the absorber A are in equivalent environments so that the maximum overlap occurs when the velocity $\nu = 0$. If the absorber has a different chemical environment the maximum will normally occur when $\nu \neq 0$. This shift in energy is called an isomer shift and is normally designated as δ. The exact position depends on the oxidation state of the iron (Fe^0, Fe^{II}, Fe^{III}, Fe^{IV}, etc.), its spin state (high spin, intermediate spin, low spin), the covalency of the bonding, and the electronegativity of the ligands. Obviously the last two are interconnected; that is, if the ligands have low electronegativities the bonding will be more covalent than if the ligands have high electronegativities. The ranges of isomer shifts relative to metallic iron are shown in Figure 3.47. Similar diagrams are also available for ^{99}Ru, ^{119}Sn, and ^{151}Eu isotope environments (74), three other elements found in inorganic polymers that exhibit Mössbauer resonance absorption. See Figure 3.48

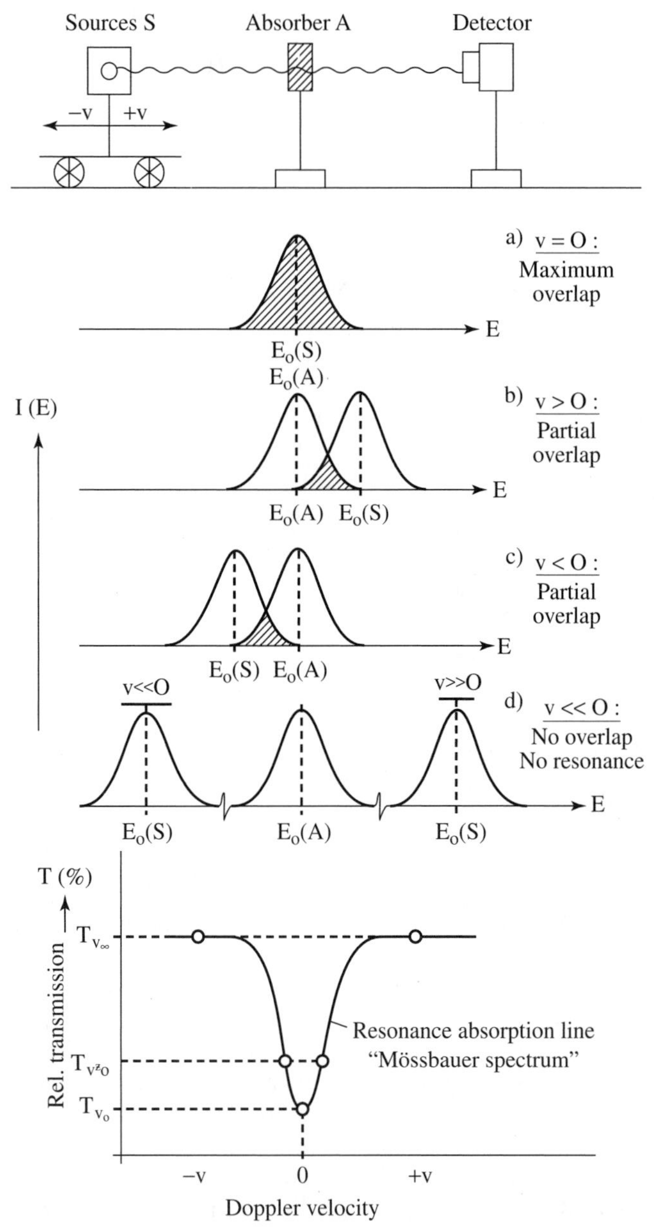

Figure 3.46 The schematic representation of an experimental setup for a Mössbauer experiment for a recoilless situation with the source S and the absorber A in identical chemical environments and identical sizes for the ground and excited states. Normally, the maximum resonance absorption will be at $v \neq 0$ if the absorber and source are chemically nonequivalent or due to other factors cited above. From Gütlich and Ensling, op cit., p 172, © 1999, John Wiley and Sons.

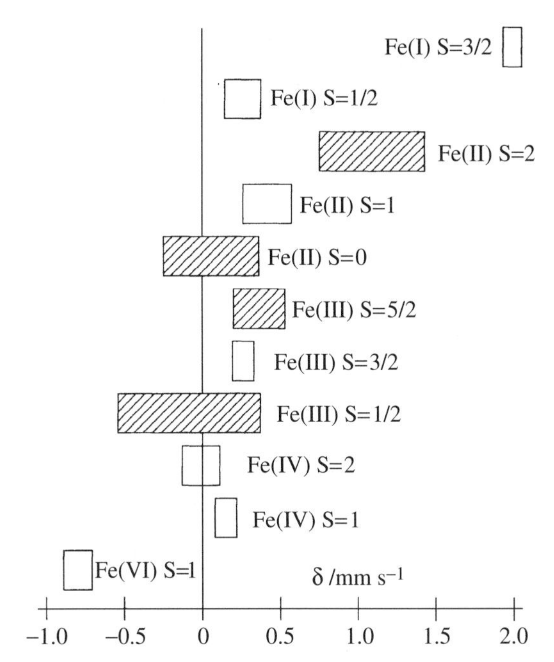

Figure 3.47 Ranges of isomer shift (δ) values observed for iron compounds, coordination species, and polymers relative to metallic iron. S = spin quantum number. From Gütlich and Ensling, op cit., p 181, © 1999, John Wiley and Sons.

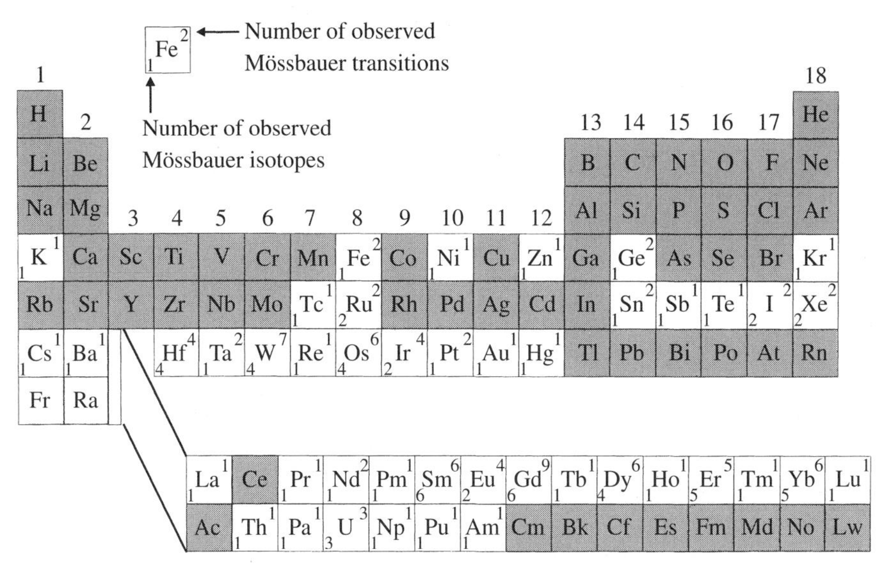

Figure 3.48 Periodic table of the elements for which the Mössbauer effect has been observed (in white) and those for which such an effect has not been observed (in gray). From Gütlich and Ensling, op cit., p 163, © 1999, John Wiley and Sons.

for a complete list of other nuclei with Mössbauer possibilities. Note that this is not a suitable technique for normal siloxanes or phosphazenes.

In addition to isomer shifts, two other effects complicate the observed Mössbauer spectra. For Mössbauer nuclei with nuclear spins of greater than 1/2, quadrupole splitting is observed unless a cubic (nongradient) field exists about the nucleus being studied. To illustrate with a monomer, salts of hexacyanoferrate(II) show no quadrupole splittings, but salts of hexacyanoferrate(III) have a small quadrupole splitting (0.28 $mm\,s^{-1}$) as a result of the hole in the t_{2g} energy level (t_{2g}^5) that is filled in the low-spin Fe^{II} species (t_{2g}^6). Low-symmetry complexes can have much higher quadrupole splittings. For example, low-spin Fe^{III} species with quadrupole splittings of up to 1.8 $mm\,s^{-1}$ are known. The quadrupole splitting, usually designated ΔE_Q, is also related to oxidation state, spin state, and bonding properties, as well as to the molecular symmetry of the species.

Magnetic dipole interactions between the magnetic dipole moment of the nucleus and the magnetic field at the nucleus lead to a magnetic splitting parameter, ΔE_M. This is observed in ferromagnetic, antiferromagnetic, and ferrimagnetic species as well as a few paramagnetic Fe^{III} species with slow magnetic state relaxation times, even at room temperature.

A sizable number of iron and tin polymers have been studied by Mössbauer spectroscopy.* These include phthalocyanine "shish kebab" polymers (75,76), ferrocene-containing polymers (77), mixed spin coordination systems (78), etc. Earlier examples include the characterization of solid-state polymeric amorphous Fe^{II}(bpy)(NCS) by Mössbauer spectroscopy along with absorption and vibrational spectroscopies and magnetic susceptibility measurements (79). The zero-field Mössbauer results show that a phase change takes place between 130 and 200 K with quite different quadrupole splittings (Fig. 3.49); however, other results indicate that no change in coordination number occurs, and the authors conclude that both phases contain a zigzag chain structure like that shown in Figure 3.50.

The coordination polymers $\{[(CH_3)_3Sn^{IV}]_3Fe^{III}(CN)_6\}_n$ and $\{A[(CH_3)_3Sn^{IV}]_3Fe^{II}(CN)_6\}_n$ where A = $(CH_3)_3Sn^+$, $(C_2H_5)_4N^+$, $(C_5H_5)_2Fe^+$, etc., have been studied by both iron-57 and tin-119 Mössbauer spectroscopy (80). The tin-119 Mössbauer spectral parameters for the polymers indicate that all of the tin atoms in the polymers are coordinated to three methyl groups (as expected) and two cyano groups consistent with the tin(IV) centers having trigonal bipyramidal structures. The iron-57 Mössbauer parameters of the $\{A[(CH_3)_3Sn^{IV}]_3Fe^{II}(CN)_6\}_n$ polymers show that the iron(II) is in a low-spin state. This latter conclusion could have been as easily determined by magnetic susceptibility measurements

3.4.5.1 Mixed-Valence Species

Although Mössbauer spectroscopy is often used to complement other spectroscopic techniques, in special cases unique information is obtained. A classic

* A search from 1993 through early 2000 yielded 96 references to Mössbauer studies on polymers, virtually all of them inorganic or organometallic polymers.

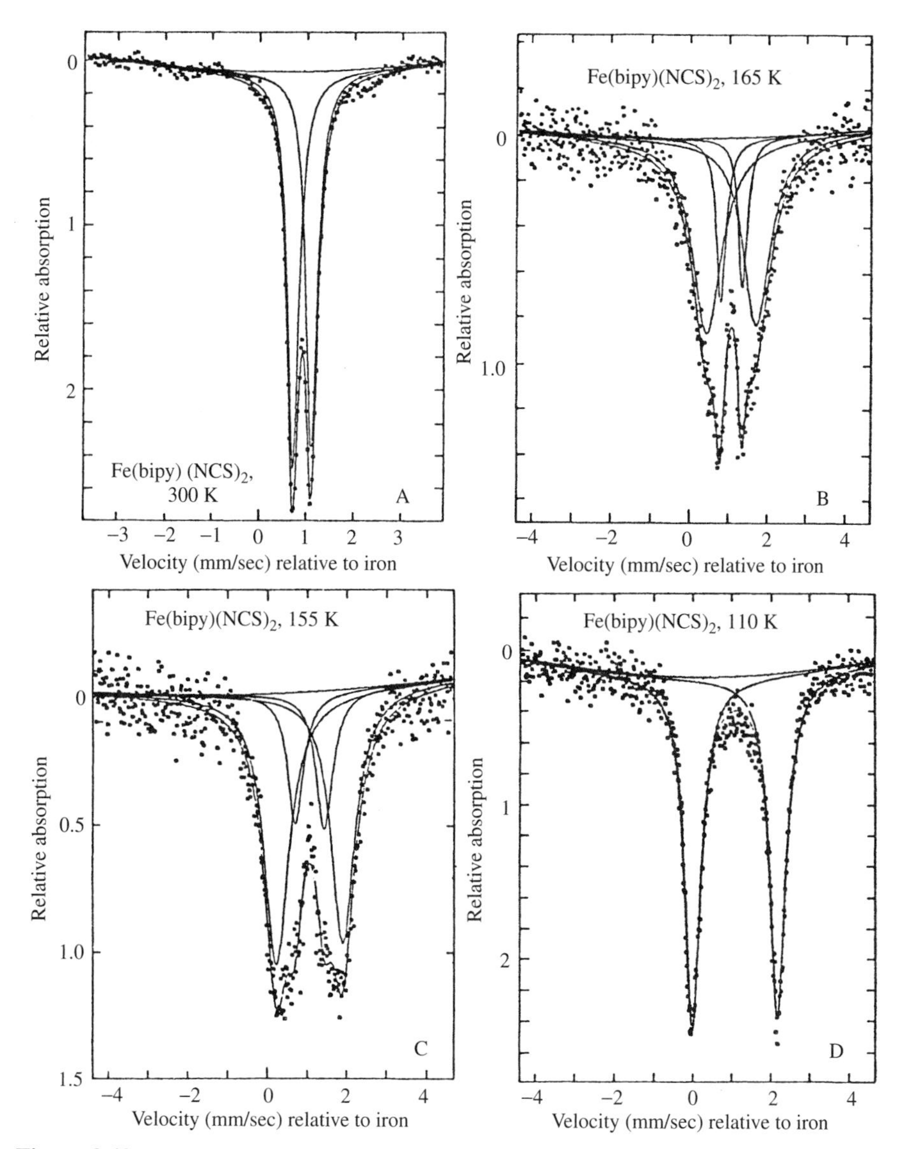

Figure 3.49 Temperature-dependent zero-field Mössbauer spectra of polymeric $Fe(bpy)(NCS)_2$. Reprinted with permission from Ref. 79; © 1982 American Chemical Society.

example is the Mössbauer spectroscopy of Prussian blue or Turnbull's blue, the mixed cyano polymer of iron(II) and iron(III). Whereas classically Prussian blue was synthesized from aqueous iron(III) and hexacyanoferrate(II) salts and Turnbull's blue was synthesized from aqueous iron(II) and hexacyanoferrate(III) salts, Fluck and coworkers (81) showed by Mössbauer spectroscopy that the product of both reactions is iron(III)[hexacyanoferrate(II)]. Although the electron transfer

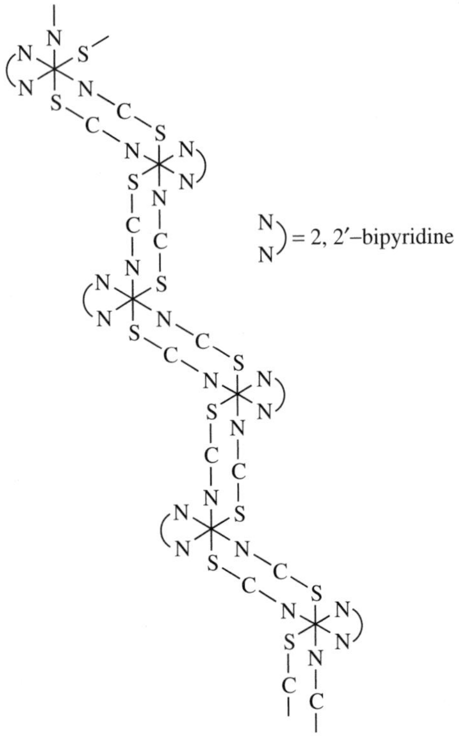

Figure 3.50 Zigzag chain structure of polymeric $Fe(bpy)(NCS)_2$. Reprinted with permission from Ref. 79; © 1982 American Chemical Society.

is rapid, both iron(II) and iron(III) centers are observed by Mössbauer spectroscopy (Fig. 3.51). These results, together with the intervalence charge-transfer absorption spectrum (Section 3.4.3) and crystallographic investigations on the solid crystalline polymer (Fig. 1.16), show that Prussian blue is a class II mixed-valence compound using the Robin and Day classification scheme for two centers labeled A and B (61):

Class I: A and B are very different with ions in different ligand fields, etc.
Trapped valences
Intervalence transitions at high energy
Optical spectra those of constituent centers

Class II: A and B in similar but not totally equivalent surroundings
Valences distinguishable but nonzero delocalization
Intervalence transitions observable—often in visible or near-infrared region
Optical spectra (other than intervalence bands) those of constituent centers
(if not obscured by intervalence bands)

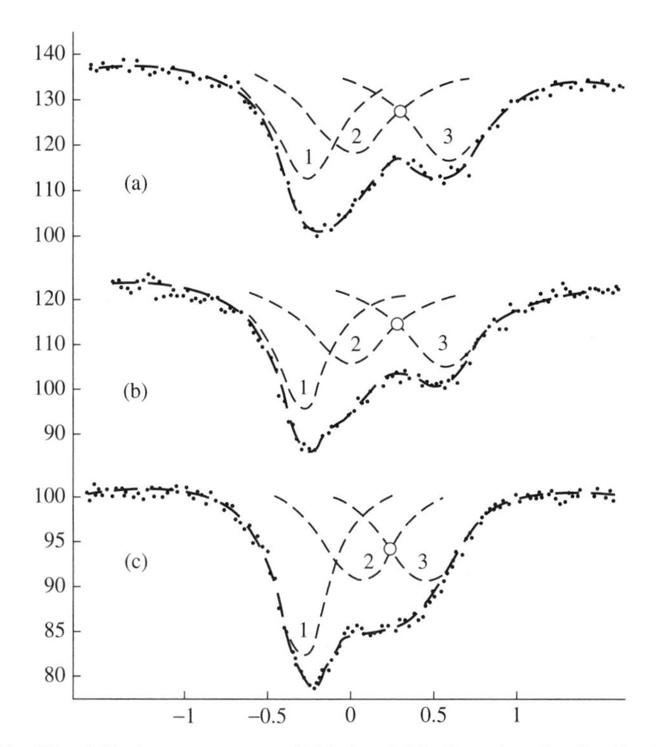

Figure 3.51 The Mössbauer spectra of (a) insoluble Prussian (or Berliner) blue, (b) Turnbull's blue, and (c) soluble Prussian blue at −130 °C. From Fluck et al. *Angew. Chem.* 1963, **75**, 461.

Class III: A and B are not distinguishable
Valences delocalized in molecular units
Transitions often observable in visible or near-infrared region
Optical spectra of constituent centers not distinguishable

Prussian blue is made up of iron(II) centers surrounded octahedrally by six C-bonded cyano ligands and iron(III) centers surrounded octahedrally by six N-bonded cyano ligands. Thus they are different, but they do show broad low-energy intervalence charge-transfer bands and belong in the Class II category of Robin and Day.

A number of other mixed-valence inorganic polymers have also been studied by Mössbauer spectroscopy, for example, mixed-valence ferrocene polymers (82, 83) and polyphenylferrisiloxane photolysis with oligoorganosilanes (84). Also, a useful and detailed quantitative treatment of mixed-valence species has recently been provided by Schatz (85).

3.4.6 Other Spectroscopic Methods

Other spectroscopic methods, such as **photoelectron spectroscopy**, have been used in selected instances, but at the present time relatively little use has been made of this technique with inorganic or organometallic polymers that provides information not available by other methods. Examples of its use include the **X-ray photoelectron spectra** (XPS) of seven silicon-based polymers including polysilanes, polycarbosilanes, and polysiloxanes (86) that have been analyzed by density-function calculations by comparison with model compounds and **UV photoelectron spectra** of ten polymers with main chains composed of silicon and germanium atoms (87). The polysilanes provide spectra that appear to be the overlap of spectra of their constituent parts plus sigma-pi mixing in the polymers with aryl side groups. The polygermanes and Si-Ge block copolymers are analogous to the polysilanes, whereas random Si-Ge copolymers show the Si-Ge bond effects.

Other electronic excitation spectroscopic techniques, such as **circular dichroism** and **luminescence** spectroscopies, are useful for characterizing coordination polymers (and other inorganic polymers as well) that have optically active centers and ligands or metal centers that emit radiation subsequent to irradiation, respectively. For example, circular dichroism proved useful in the stereoselective condensation syntheses of Δ- and Λ-bis(1,10-phenanthroline-5,6-dione)bipyridineruthenium(II) ions with Δ- and Λ-bis(1,10-phenanthroline-5,6-diamine)bipyridineruthenium(II) ions that produced the homochiral Δ- and Λ-coordination polyelectrolytes and a meso Δ, Λ-coordination polyelectrolyte (Figure 3.52) (88).

Among the large number of papers that involve the luminescence of europium in polymeric systems, two stand out because the polymeric nature is retained in

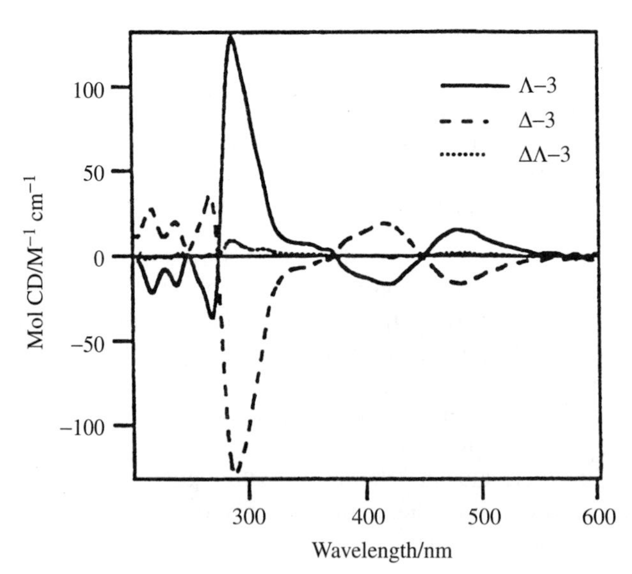

Figure 3.52 The circular dichroism (CD) spectra of Λ-, Δ-, and Δ, Λ(*meso*)-*catena*-poly(2,2′-bipyridineruthenium(II)-μ-tetrapyrido[3,2-a:2′,3′-c:3″2″-h:2‴,3‴j]phenazine hexafluorophosphate in acetonitrile. From Chen and MacDonnell, *Chem. Commun.* 1999, 2529; reproduced by permission of The Royal Society of Chemistry.

solution. One investigation uses bis(4-pyridine-2,6-dicarboxylic acids) as bridging ligands (89) and the other uses Schiff bases N, N', N'', N'''-tetrakis(salicylidene) 3,3′,4,4′-tetraaminodiphenylmethane and the analogous diphenyl derivative as bridging ligands (cf. Fig. 2.13) (90). Both papers note enhanced luminescence for mixed yttrium/europium polymers relative to pure europium polymers. The enhancement may be due to a transfer of absorbed energy from the yttrium centers along the polymer chain to the europium centers and/or a decreasing of europium-europium adjacent site self-quenching.

Examples of luminescence investigations also exist for ruthenium polymers. One recent paper (8) indicates that the ruthenium(II) centers appear electronically independent of one another in polymers in which interaction might be expected [cf. Fig. 3.11, especially based on earlier studies with ruthenium(II) dendrimers by Balzani, his coworkers, and others (91)]. Whereas large dendrimers with as many as 54 dendritic branches have absorption and emission spectra very similar to those of the parent tris(bipyridine)ruthenium(II) monomers; the dendrimers exhibit a more intense emission and a longer excited-state lifetime than the monomers in aerated solutions (92).

For discussions of other spectroscopic methods and further details on those methods noted above, the reader is referred to the references cited at the end of this chapter.

3.5 VISCOELASTICITY MEASUREMENTS

Polymers (chain or lightly crosslinked) are often said to be viscoelastic materials possessing some characteristics of both liquids and solids. Although the methods of study are analogous to organic polymers and will not be dealt with here in any detail, a number of concepts and definitions are important for anyone studying polymeric molecules.

Polymers do not behave in exactly the same way that normal solids and liquids do. They can be stretched (i.e., change shape like a liquid) when a force is applied. When the force is released, they return to their original shape like a solid.

The "stiffness" of a polymer can be evaluated quantitatively. The **modulus of elasticity** (E), also called the **tensile modulus, Young's modulus**, or just **modulus**, is based on the magnitude of the applied force per cross-sectional area or **tensile stress** (σ) that produces an induced stretch or **tensile strain** (ε). Tensile strain (ε) is the change in length (Δl) divided by the original length (l):

$$\varepsilon = \Delta l / l \tag{3.47}$$

Together,

$$E = \sigma / \varepsilon \tag{3.48}$$

However, these are not linear properties for typical polymers. See the schematic in Figure 3.53. The initial maximum is where the polymer "yields" or deforms.

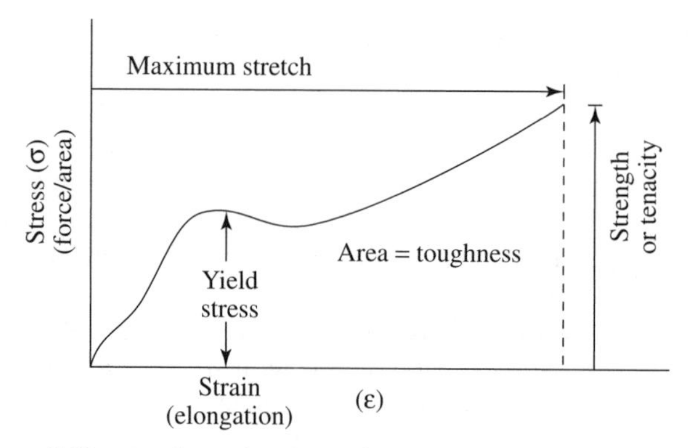

Figure 3.53 A schematic of a typical polymer stress-strain curve.

Below the **yield point**, the polymer's elongation is reversible. At the yield point the stress causes some disentanglement of the molecules. Therefore, the elongation above the yield point is not reversible. For a given stress, the strain (or elongation) is greater for a thermoplastic or a fiber than for an elastomer. Eventually, the stress and strain reach a point where the polymer breaks or fractures.

Other definitions:

Tensile strength is the load a material can bear without breaking.

Tenacity is the stress at the breaking point.

Toughness is the total energy input to the breaking point, that is, the area under the stress-strain curve.

Impact strength is the resistance to breakage with a sharp blow and is related to the toughness. Impact tests are normally used for determining the toughness because of the relative ease of testing.

All of these factors are temperature dependent. The temperature dependence of the modulus is also quite important. Below the glass transition temperature a polymer is a glassy solid with a high modulus. A major drop in the modulus occurs at the glass transition temperature because the polymer becomes more elastic. The deformations include easier bond distortion as well as chain disentanglement and slippage. Another drop in modulus occurs at higher temperatures when the polymer becomes more like a liquid with more rapid chain flow. This latter drop in modulus can be minimized with crosslinking. A large drop also occurs at the melting temperature of crystalline polymers. All of these effects are shown schematically in Figure 3.54.

Shear is the deformation of a polymer without a volume change by a force applied parallel to the resting surface of polymer (Fig. 3.55). The resistance to the shear deformation is called the **shear modulus**. The shear modulus (G) is the ratio of the **shear stress** (τ) (force F per unit area) to the **shear strain** (γ):

$$G = \tau/\gamma \tag{3.49}$$

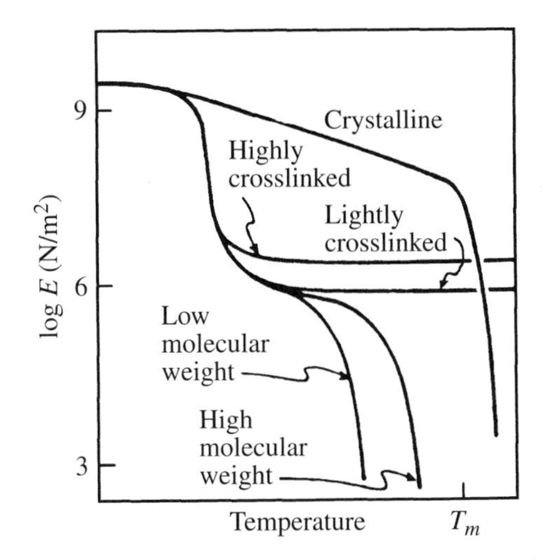

Figure 3.54 The effect of temperature on the modulus (log E scale) for various polymers shown schematically. T_m = crystalline melting temperature. Reproduced with permission from Aklonis, *Journal of Chemical Education*, 1981, **58(11)** 892–897; copyright © 1981, Division of Chemical Education, Inc.

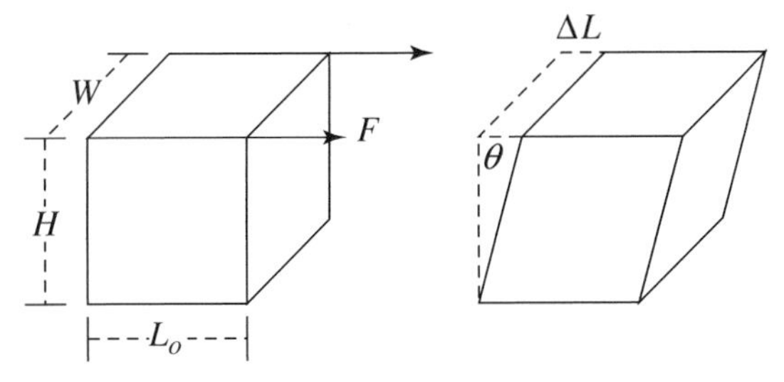

Figure 3.55 Schematic representation of shear (93). © 1997 VCH.

With reference to Figure 3.55, the shear strain (γ) is the skewness of top plane relative to the base plane. That is,

$$\gamma = \Delta L/H = \tan\theta \tag{3.50}$$

The shear modulus is very low for graphite or molybdenum disulfide, both of which have layers that can slide by each other very easily.

Polymers normally exhibit a decrease in volume when a strong hydrostatic pressure is applied. The resistance to the volume decrease is the **bulk modulus**, B or K.

Several time-dependent parameters are also commonly encountered. **Creep** or **cold flow** is the slow change in shape under constant force, for example, gravity, and is typically measured by attaching a weight at one end of a sample and measuring the long-term elongation or by suspending a sample between two supports and measuring the long-term sag that gravity causes. **Stress relaxation** is a measure of the decrease in stress that occurs when a sample is elongated rapidly to constant strain. **Shear rate** ($\dot{\gamma}$) or **velocity gradient** (93) is the rate at which the molecules flow relative to one another or

$$\dot{\gamma} = d\gamma/dt \tag{3.51}$$

where γ is the shear strain defined in Eq. 3.50 and t is time. Ideal or Newtonian liquids follow Newton's law of viscosity; that is, the shear stress τ is proportional to $\dot{\gamma}$

$$\tau = \eta\dot{\gamma} \tag{3.52}$$

where viscosity η is the proportionality constant. Because of entanglements, polymers often deviate from such Newtonian behavior, but a full discussion of this topic is beyond the scope of this book. Shear thinning, in which the increase in stress is less than anticipated as the shear rate increases, is one common deviation that is thought to occur because as the shear rate increases the molecules become disentangled and the effective viscosity decreases at higher shear rates.

The brittleness of a polymer is related to crystallinity and/or short chains (especially if $M_n \approx M_w$). Brittleness can be minimized with crosslinks. Alternatively, fillers can be added to interrupt crack propagation in potentially brittle polymers,. or an impact-absorbing second phase can be included in the polymer to avoid rupture of the polymer.

Polymers are often categorized as to elasticity (rubbery state) vs. the flexible thermoplastics of crystalline polymers. Elastomers have **resilience**; that is, they bounce back. Elastomers also have a tendency to swell and to absorb solvents or plasticizers. Furthermore, elastomers need crosslinks for shape retention, whereas crystallinity provides thermoplastics with shape stability through ordered intermolecular attractions.

3.6 CRYSTALLIZATION CHARACTERIZATION

As noted above, the crystallinity of polymers provides different characteristics from those of amorphous polymers. Often only a small percentage of the polymer is crystalline, with the bulk of the polymer being amorphous. Even then, appreciable property modification is possible. A number of techniques are used to determine the crystallinity of polymers. We will only briefly consider birefringent microscopy together with X ray (wide and small angle), small-angle polarized light, electron, and neutron scattering. These measurements are most often useful in the region between the glass transition temperature (T_g) and the melting temperature (T_m) of the polymer where crystallites and spherulites form.

3.6.1 Birefringent Microscopy

Optical birefringence is studied with a polarizing microscope. A temperature-controlled hot stage allows the polymer to be carefully heated, and the polarizing microscope allows the investigator to see developing crystallites and then see them disappear at the T_m. Sometimes a pseudo T_m is observed when one phase goes to another. This is also a first-order process, and simple thermal curves cannot distinguish between them. Studies of the rates of crystal formation vs. temperature show the maximum rate of crystallite formation is about halfway between the T_g and the T_m of the polymer.

3.6.2 Wide-Angle X-Ray Scattering

X rays from cathode-ray tubes are widely used to determine the structures of crystalline molecules. **Wide-angle X-ray scattering** (WAXS) from crystalline materials is the **normal X-ray diffraction** with which most inorganic chemists have some familiarity. The scattering from crystalline materials is **coherent**. Amorphous polymer samples only provide a hazy, not well-defined background under wide-angle scattering conditions. When a microcrystalline polymer sample is placed in an X-ray beam, a wide-angle pattern of concentric circles is obtained (Fig. 3.56) that becomes small arcs if the sample is oriented (e.g., by stretching). The diffuseness and arcs (as opposed to spots for single crystals; cf. Fig. 3.57) can be used to interpret the degree of orientation or crystallinity of the sample. The rates vs. temperature noted above are also applicable.

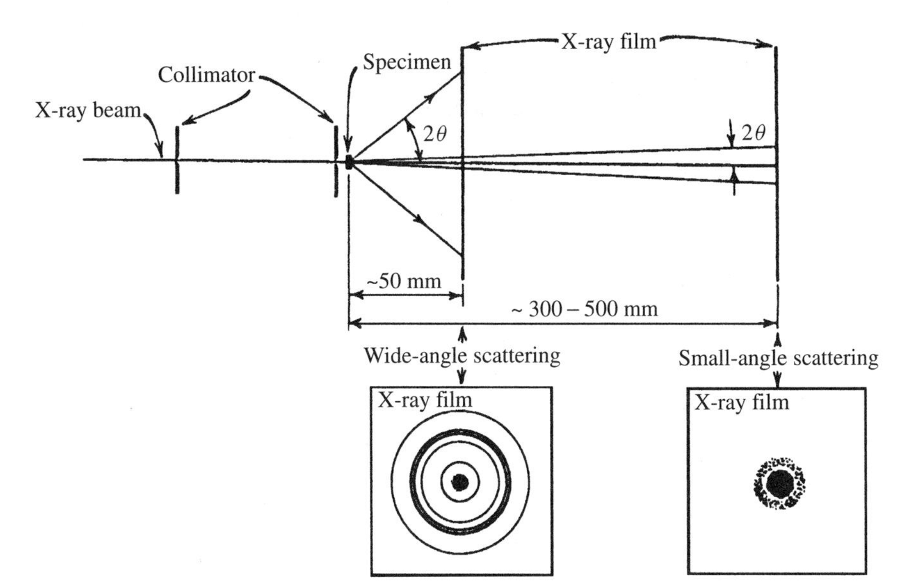

Figure 3.56 Wide-angle and small-angle X-ray diffraction of polymer samples. From Collins, E. A., Bares, J., and Billmeyer, F., *Experiments in Polymer Science*. NY: Wiley-Interscience, 1973.

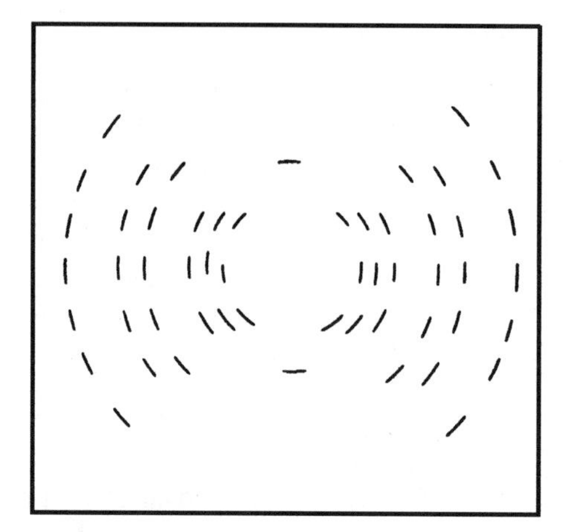

Figure 3.57 Schematic wide-angle X-ray pattern of an oriented microcrystalline polymer or an oriented helical polymer. From p. 482, *Contemporary Polymer Chemistry*, 2/E by Allcock/Lampe, © 1990. Adapted by permission of Prentice-Hall, Inc., Upper Saddle River, NJ.

3.6.3 Small-Angle X-Ray Scattering

Information about the macrostructure of a polymer such as the dimensions and packing of crystallites, spherulites, lamellae, separated phases, and voids; particle size and shape in solution or colloids; and information on branched polymers and the deformation and annealing of polymers can be obtained from **small-angle X-ray scattering** (SAXS). The reflections that lie very close to the beam stop are necessary for this information. To obtain sufficient clarity synchrotron radiation is often used. If the sample is reactive under high intensity X-radiation, this method is unsuitable.

3.6.4 Small-Angle Polarized Light Scattering

Laser radiation can be used to detect the size of spherulites and obtain rate data as well. The ready availability of laser sources and polarizing media make this an attractive method for obtaining size information. Once again, if the polymer were sensitive to decomposition at the wavelength of the laser, the results would be in error.

3.6.5 Electron Scattering

An electron microscope set in the diffraction mode (rather than in the usual imaging mode) allows a diffraction pattern to be projected on the electron microscope's screen. Crystallite dimensions, degree of crystallinity, and other morphological information can be obtained by analyzing the observed pattern. Advantages include higher intensities, small sample size, and the fact that both diffraction and transmission measurements are possible with the same sample. A major disadvantage is that electrons may cause free radical reactions in the sample that lead to crosslinking or chain scission. Thus the final conclusions from the diffraction results would be different from those of the polymer under investigation.

3.6.6 Neutron Scattering

Neutron scattering has the advantage that the inelastic scattering by hydrogen atoms is significantly higher than for other atoms. Thus, for polymers with hydrogen atoms, it provides a means of studying the movement of hydrogen atoms in the polymer. Neutrons are also more sensitive to short-chain segments, as in crystalline lamellae, and can be studied in solution or the solid state. The biggest disadvantage is the nuclear reactor needed as a neutron source and the costly neutron counters required to obtain the data.

3.7 CONCLUDING STATEMENT

Whereas entire books have been written on polymer characterization, this chapter has provided the reader with an opportunity to learn some of the fundamentals of characterizing inorganic polymers that remain as polymers through phase changes and in solution. However, laboratory practice characterizing actual polymers is the only way to really appreciate these methods and their limitations.

REFERENCES

1. Allcock, H. R., Lampe, F. W., *Contemporary Polymer Chemistry*; 2nd ed., Prentice-Hall: Engelwood Cliffs, NJ, 1990.
2. Flory, P. J. *Principles of Polymer Chemistry*; Cornell University Press: Ithaca, NY, 1953.
3. Flory, P. J. *Principles of Polymer Chemistry, op cit.*, Ch. 8.
4. Ambler, M. R., Mate, R. D. *J. Polym. Sci. Part A-1* 1972, **10**, 2677.
5. Trathnigg, B. *Prog. Polym. Sci.* 1995, **20**, 615.
6. Kraemer, E. O. *Ind. Engr. Chem.* 1938, **30**, 1200; quoted by Flory in Principles of Polymer Chemistry book.
7. Carraher, C. E., Jr. *Macromolecules* 1971, **4**, 263.
8. Kelch, S., Rehahn, M. *Macromolecules* 1999, **32**, 5818.
9. Zimm, B. H. *J. Chem. Phys.* 1948, **16**, 1099.
10. Brice, B. A., Nutting, G. C., Halwer, M. *J. Am. Chem. Soc.* 1953, **75**, 824.
11. Cooper, A. R. in *Encyclopedia of Polymer Science and Engineering*; Kroschwitz, J. I., Mark, H. F., Bikales, N. M., Overberger, C. G. and Menges, G., Ed., John Wiley and Sons: New York, 1987; Vol. 10, pp 1–19.
12. Archer, R. D., Wang, B. *Inorg. Chem.* 1990, **29**, 39.

13. Archer, R. D., Chen, H., Cronin, J. A., Palmer, S. M. in *Metal-Containing Polymeric Materials*; Pittman, C. U., Jr., Carraher, C. E., Jr., Zeldin, M., Sheats, J. E. and Culbertson, B. M., Ed., Plenum Press: New York, 1996, pp 81–91.
14. Berry, K. L., Peterson, J. H. *J. Am. Chem. Soc.* 1951, **73**, 5195.
15. Bahr, U., Deppe, A., Karas, M., Hillenkamp, F. *Anal. Chem.* 1992, **64**, 2866.
16. Hohmi, T., Fenn, J. B. *J. Am. Chem. Soc.* 1992, **114**, 3241.
17. Prókai, L. *Field Desorption Mass Spectrometry*; Marcel Dekker: New York, 1989.
18. O'Malley, R. M., Randazzo, M. E., Weinzierl, J. E., Fernandez, j. E., Nuwaysir, L. N., Castro, J. A., Wilkins, C. L. *Macromolecules* 1994, **27**, 5107.
19. Rabek, J. F. *Experimental Methods in Polymer Chemistry*; John Wiley & Sons: Chichester, 1980.
20. Rawsky, G. C., Schriver, D. F. in *Metal-containing polymeric materials*; Pittman, C. U., Jr., Carraher, C. E., Jr., Zeldin, M., Sheats, J. E. and Culbertson, B. M., Ed., Plenum Press: New York, 1996, pp 383–93.
21. Chen, H., Unpublished glass transition temperatures vs. rate of measurement results.
22. Chen, H., Archer, R. D. *Macromolecules* 1995, **28**, 1609.
23. Chen, H., Cronin, J. A., Archer, R. D. *Macromolecules* 1994, **27**, 2174.
24. Chen, H., Cronin, J. A., Archer, R. D. *Inorg. Chem.* 1995, **34**, 2306.
25. Rochow, E. G. *An Introduction to the Chemistry of the Silicones*; John Wiley & Sons, Inc.: New York, 1951.
26. Seymour, R. B. *Introduction to Polymer Chemistry*; McGraw-Hill Book Company: New York, 1971.
27. Zeigler, J. M., Fearon, F. W. G. *Silicon-based polymer science: a comprehensive resource*; Zeigler, J. M., Fearon, F. W. G., Ed., American Chemical Society: Washington, DC, 1990.
28. Ohshita, J., Shinpo, A., Kunai, A. *Macromolecules* 1999, **32**, 5998.
29. Block, B. P. in *Coordination Chemistry: Papers presented in honor of Prof. J. C. Bailar, Jr.*; Kirschner, S., Ed., Plenum Press: New York, 1969, pp 241–247.
30. Bailar, J. C., Jr. in *Organometallic polymers*; Carraher, C. E., Jr., Sheats, J. E. and Pittman, C. U., Jr., Ed., Plenum Press: New York, 1978, pp 313–321.
31. Carraher, C. E., Jr. *J. Macromol. Sci., Chem.* 1982, **A17**, 1293 and references cited therein.
32. Aswar, A. S., Bahad, P. J., Pardhi, A. V., S., B. M. *J. Polym. Matl.* 1988, **5**, 233.
33. Dutt, N. K., Nag, K. J. *Inorg. Nucl. Chem.* 1968, **30**, 2493.
34. Pandya, M. L., Machwe, M. K. *Curr. Sci.* 1978, **47**, 946.
35. Patel, M. N., Patil, S. H. *J. Macromol. Sci.-Chem.* 1981, **A16**, 1429.
36. Patel, M. N., Sutaria, D. H., Patel, J. R. *Synth. React. Inorg. Met.-org. Chem.* 1994, **24**, 401.
37. Seminara, A., Giuffrida, S., Musumeci, A., Fragala, I. *Inorg. Chim. Acta* 1984, **95**, 201.
38. Wang, B., Archer, R. D. *Polym. Mater. Sci. Engr.* 1988, **59**, 120; 1988, **60**, 710.
39. Xiong, W., Wu, S., Li, Z. *Thermchim. Acta* 1988, **133**, 377.
40. Zhong, N., Ru, W. *Synth. React. Inorg. Met.-org. Chem.* 1991, **21**, 965.

41. Ouyang, G., Simons, R., Tessier, C. in *Metal-containing polymeric materials*; Pittman, C. U., Jr., Carraher, C. E., Jr., Zeldin, M., Sheats, J. E. and Culbertson, B. M., Ed., Plenum Press: New York, 1996, pp 189–197.
42. Lienhard, M., Wiegand, C., Apple, T., Interrante, L. V. *Polymer Preprints* 2000, **41**, 570.
43. Prange, R., Allcock, H. R. *Macromolecules* 1999, **32**, 6390, and references therein.
44. Archer, R. D., Wang, B., Tramontano, V. J., Lee, A. Y., Ochaya, V. O. in *Inorganic and organometallic polymers*; Zeldin, M., Wynne, K. J. and Allcock, H. R., Ed., American Chemical Society: Washington, D.C., 1988; Vol. 360, pp 463–68.
45. Drago, R. S. *Physical methods for chemists*; 2nd ed., Saunders: Ft. Worth, 1992.
46. McGarvey, B. R. *Transition Metal Chem.* 1966, **3**, 89–201 a classic review.
47. Rabek, J. F. *Experimental Methods in Polymer Chemistry*; John Wiley & Sons: Chichester, 1980.
48. König, E. *Magnetic properties of coordination and organometallic transition metal compounds*; Springer-Verlag: Berlin, 1966; Vol. 2.
49. König, E., König, G. *Magnetic properties of coordination and organometallic transition metal compounds (Supplements.*; Springer-Verlag: Berlin, 1976; Vol. 8, 10, 11, 12.
50. Kneubühl, F. K. *J. Chem. Phys.* 1960, **33**, 1074.
51. McCleverty, J. A., Ward, M. D. *Acc. Chem. Res.* 1998, **31**, 842.
52. Minto, F., Borzatta, V., Gleria, M. J. *Inorg. Organometall. Polym.* 1996, **6**, 171; based on Web of Science abstract.
53. Seki, S., Cromack, K. R., Trifunac, A. D., Yoshida, Y., Tagawa, S., Asai, K., Ishigure, K. *J. Phys. Chem. B* 1998, **102**, 8367.
54. Chatgilialoglu, C., Guerrini, A., Lucarini, M., Pedulli, G. F., Carrozza, P., Da Roit, G., Borzatta, V., Lucchini, V. *Organometallics* 1998, **17**, 2169.
55. Sen, T., Rajamohanan, P. R., Ganapathy, S., Sivasanker, S. *J. Catalysis* 1996, **163**, 354.
56. van Albada, G. A., Quiroz-Castro, M. E., Mutikainen, I., Turpeinen, U., Reedijk, J. *Inorg. Chim. Acta*, 2000, **298**, 221.
57. Tangoulis, V., Psomas, G., Dendrinou-Samara, C., Raptopoulou, C. P., Terzis, A., Kessissoglou, D. P. *Inorg. Chem.* 1996, **35**, 7655.
58. Solomon, E. I., Lever, A. B. P. *Inorganic Electronic Structure and Spectroscopy*; Solomon, E. I., Lever, A. B. P., Ed., John Wiley and Sons, Inc.: New York, 1999; Vol. 1 Methodology; 2 Applications and case studies.
59. Figgis, B. N., Hitchman, M. A. *Ligand Field Theory and Its Applications*; Wiley-VCH: New York, 2000.
60. Lever, A. B. P. *Inorganic electronic spectroscopy*; Elsevier: Amsterdam, 1968.
61. Lever, A. B. P. *Inorganic electronic spectroscopy*; 2nd ed., Elsevier: Amsterdam, 1984.
62. Figgis, B. N. *Introduction to Ligand Fields*; Interscience Publishers: New York, 1966.
63. Jørgensen, C. K. *Absorption Spectra and Chemical Bonding in Complexes*; Pergamon Press: Oxford, 1962.
64. Jørgensen, C. K. *Modern Aspects of Ligand Field Theory*; North Holland: Amsterdam, 1971.

65. Ballhausen, C. J. *Introduction to Ligand Field Theory*; McGraw-Hill Book Co., Inc.: New York, 1962.
66. McCaffery, A. J., Dickinson, J. R., Schatz, P. N. *Inorg. Chem.* 1970, **9**, 1563.
67. Hummel, D. O. *Atlas of Polymer and Plastics Analysis*; VCH and Hanser Publishers: Weinheim & Munich, 1991; Vol. 1 Defined Polymers, part a.
68. Czernuszewicz, R. S., Spiro, T. G. in *Inorganic Electronic Structure and Spectroscopy*; Solomon, E. I. and Lever, A. B. P., Ed., John Wiley & Sons, Inc.: New York, 1999; Vol. I, pp 353–441.
69. Schilling, F. C., Bovey, F. A., Lovinger, A. J., Zeigler, J. M. in *Silicon-Based Polymer Science*; Ziegler, J. M. and Fearon, F. W. G., Ed., Amer. Chem. Soc.: Washington, DC, 1990; Vol. 224 Advances in Chemistry Series, pp 341–378.
70. Nakamoto, K. *Infrared and Raman Spectra of Inorganic and Coordination Compounds*; John Wiley & Sons, Inc.: New York, 1997.
71. Laserna, J. J. *Modern Techniques in Raman Spectroscopy*; Laserna, J. J., Ed., John Wiley & Sons, Inc.: New York, 1996.
72. Kettle, S. F. A. *Physical Inorganic Chemistry*; Spektrum Academic Publishers: Oxford, 1996.
73. Gütlich, P., Ensling, J. in *Inorganic Electronic Structure and Spectroscopy*; Solomon, E. I. and Lever, A. B. P., Ed., John Wiley & Sons, Inc.: New York, 1999; Vol. I, pp 161–211.
74. Shenoy, G. K., Wagner, F. E. *Mössbauer Isomer Shifts*; North Holland Publishing Co.: Amsterdam, 1978.
75. Hanack, M. *Synthetic Metals* 1995, **71**, 2275; also *New Matls: Conj. Dbl. Bond Syst.* 1995, **191**, 13.
76. Hanack, M. in *Metal-Containing Polymeric Materials*; Pittman, C. U. J., Carraher, C. E., Jr., Zeldin, M., Sheats, J. E. and Culbertson, B. M., Ed., Plenum Press: New York, 1996, pp 331–336.
77. Foucher, D. A., Edwards, M., Burrow, R. A., Lough, A. J., Manners, I. *Organometallics* 1994, **13**, 4959.
78. Kitazawa, T., Takahashi, M., Enomoto, M., Miyazaki, A., Enoki, T., Takeda, M. *J. Radioanal. Nucl. Chem.* 1999, **239**, 285.
79. Dockum, B. W., Reiff, W. M. *Inorg. Chem.* 1982, **21**, 391.
80. Katada, M., Fujita, M., Yamada, H., Kawata, S., Sano, H. *Nucl. Instr. & Meth. in Phys. Res. Sect. B-Beam Interact. with Mater. & Atoms* 1993, **76**, 310.
81. Fluck, E., Kerler, W., Neuwirth, W. *Angew. Chem.* 1963, **75**, 461.
82. Wagener, W., Hillberg, M., Feyerherm, R., Stieler, W., Litterst, F. J., Pohlmann, T., O., N. *J. Physics-Condensed Matter* 1994, **6**, L391.
83. Rosenblum, M., Nugent, H. M., Jang, K. S., Labes, M. M., Cahalane, W., Klemarczyk, P., Reiff, W. M. *Macromolecules* 1995, **28**, 6330.
84. Klapshina, L. G., Semenov, V. V., Kornev, A. N., Rusakov, V. S., Shchegolikhina, O. I., Zhdanov, A. A., Domrachev, G. A. 1998.
85. Schatz, P. N. in *Inorganic Electronic Structure and Spectroscopy*; Solomon, E. I. and Lever, A. B. P., Ed., John Wiley and Sons, Inc.: New York, 1999; Vol. 2, pp 175–226.
86. Kuroki, S., Endo, K., Maeda, S., Chong, D. P., Duffy, P. *Polymer J.* 1998, **30**, 142.

87. Seki, K., Yuyama, A., Narioka, S., Ishii, H., Hasegawa, S., Isaka, H., Fujino, M., Fukiki, M., Matsumoto, N. in *Polymeric Materials for Microelectronic Applications*; American Chemical Society: Washington, DC, 1994; Vol. 594 ACS Symp. Series, pp 398–407.
88. Tor, Y., Glazer, E. C. *Polym. Prepr.* 1999, **40(1)**, 513.
89. Latva, M., Takalo, H., Simberg, K., Kankare, J. *J. Chem. Soc.-Perkin Trans. II* 1995, 995.
90. Chen, H., Archer, R. D. *Macromolecules* 1996, **29**, 1957.
91. Venturi, M., Serroni, S., Juris, A., Campagna, S., Balzani, V. in *Dendrimers*; Vögtle, F., Ed., Springer: Berlin, 1998; Vol. 197, pp 193 and references cited therein.
92. Issberner, J., Vogtle, F., DeCola, L., Balzani, V. *Chem.-Eur. J.* 1997, **3**, 706.
93. Elias, H.-G. *An Introduction to Polymer Science*; VCH: Weinheim, 1997, Chapter 7.

EXERCISES

3.1. Integrate Eq. 3.6 as suggested and make the suggested substitutions and compare your results with Eq. 3.11.

3.2. Calculate $\overline{M}_N$ values for polymers with a $\overline{M}_W$ value of 20,000 and ρ values of (a) 0.80, (b) 0.95, (c) 0.99, and (d) 0.999 for a step polymerization reaction (cf. Eq. 3.17).

3.3. **a.** Assume the y-axis values in the graph of Figure 3.7 are proportional to the number of molecules for each point along the x-axis. Integrate and calculate the $\overline{M}_N$ value of the polymer. The integration can be done by a commercial computer program or by hand. In the latter case, separate the total area under the curve into a large number of equal Δx portions for which approximate areas can be determined using the average y value for each portion. Add all the Δx areas together to get the total area and then determine the x value where 1/2 of the area is larger and 1/2 of the area is smaller—for this you are on your own. The author has used a small computer program and a digitizer to do this in his laboratory.

b. Following the procedure just noted and multiplying each Δx area by the molecular mass for the average x of that subarea, calculate the $\overline{M}_W$ value of the polymer in Figure 3.7.

3.4. In light scattering measurements of molecular mass, the intramolecular dissymmetry of the scattered light used to determine the molecular mass requires that corrections of a double extrapolation be made for polymers that exceed about 1/20 of the wavelength of the light, λ. Estimate the molecular masses for a rigid-rod inorganic polymer molecule that would have the dimensions of 0.05λ assuming $\lambda = 6000$ Å. Indicate the assumptions you use in estimating the size versus molecular mass.

3.5. Repeat Exercise 3.4 for an inorganic random coil polymer. Again, indicate all of the assumptions that you use.

3.6. The ^{3}H nucleus is more sensitive (1.21:1.00) for NMR measurements than is the proton. It also has a spin of 1/2, which is considered ideal for NMR measurements. Why is ^{3}H NMR not commonly used?

3.7. Make tables or charts of the ^{29}Si and ^{31}P NMR shifts given in Section 4.3.1. Using your tabulations and your knowledge of chemical trends, suggest reasons for the trends that you observe.

3.8. Sketch the repeating units of the uranyl carboxylate polymers discussed in Section 3.4.1 verifying the seven-coordination suggested by the text.

3.9. Note that the symmetry variation in EPR g values is logical from the symmetry of the sites. That is, an isotropic site where $x = y = z$ leads to $g_x = g_y = g_z$ or a single g value. Similarly an axial site where $x = y \neq z$ leads to $g_x = g_y \neq g_z$ or $g_\perp$ and $g_\parallel$, respectively. An asymmetric site, where $x \neq y \neq z$ leads to $g_x \neq g_y \neq g_z$ or g_1, g_2, and g_3. Using a set of character tables, determine the point groups that lead to isotropic, axial, and asymmetric g values.

3.10. Compare the relative Δ_o values in the two columns of Table 3.8 for ligands that are saturated π donors, unsaturated π donors, only σ donors, and π acceptors. Explain any differences you note for one group versus another. Note: A lack of a trend for some ligands may be to the result of Jorgensen having primarily d^6 species when he determined his general ***f*** values.

3.11. It has been shown {Archer; Bonds; Pribush, *Inorg. Chem.*, 1972, **11**, 1550} that tungsten(V) coordination compounds analogous to the polymer molecules exhibit ligand-to-metal charge-transfer transitions [whereas the tungsten(IV) species were previously shown to exhibit metal-to-ligand transitions]. **If** a local excess of the dione in the tungsten polymer synthesis caused **some** of the tungsten to be oxidized to tungsten(V), suggest an alternate interpretation for the broad low-energy spectrum of the insoluble polymer of tungsten noted in Section 3.4.3.

CHAPTER 4

PRACTICAL INORGANIC POLYMER CHEMISTRY

Inorganic and organometallic polymers have a multitude of actual and potential uses (1–6). In this chapter we will explore a number of the more exciting uses for inorganic and organometallic polymers. Although the emphasis is on tractable inorganic and organometallic polymers, the practicality of some insoluble inorganic polymeric materials will also be noted (cf. Exercise 4.1).

4.1 INORGANIC POLYMER ELASTOMERS

Polysiloxane elastomeric materials (silicone rubbers) have been a major use for polysiloxanes for over 50 years and make up approximately 50% of the applications of siloxane polymers (7). On the other hand, poly(dichlorophosphazene) was called "inorganic" rubber over a century ago, but the reactivity of the chloro groups made the dihalophosphazene polymer impractical until other more stable derivatives were synthesized. The numerous dental and biomedical uses of the siloxane and phosphazene elastomers will be discussed separately (Sections 4.3–4.4).

The relationship between elastomers (rubbers) relative to other classes of polymers was shown in Figure 3.25. In fact, gums (which are further from the T_g than the rubber or naturally elastic state) are moved into the rubber category by

Inorganic and Organometallic Polymers, by Ronald D. Archer
ISBN 0-471-24187-3

crosslinking. However, too much crosslinking provides a rigid polymer instead of an elastomer.

4.1.1 Polysiloxane Elastomers

Polysiloxane elastomers (8–15) (commonly called **silicones**) are widely employed because of their utility over a wide temperature range (−100 to +300 °C) and their chemical resistance, which also manifests itself in weathering resistance. Substitution of phenyl, vinyl, and carborane units for the methyl groups of the original poly(dimethylsiloxane) elastomers have improved the thermal stability and decreased the flammability of the elastomers. The vinyl groups are used for crosslinking purposes as noted below. Even so, virtually all silicone elastomers are either crosslinked polydimethylsiloxane or a crosslinked copolymer of polydimethylsiloxane. Longer Si–O bonds and side groups only on alternate atoms provide greater flexibility than organic polymer backbones. These bond differences also make polysiloxanes more permeable to oxygen — important for contact lenses and possibly for artificial gill devices for extracting dissolved oxygen from water for divers.

The thermal decomposition of vulcanized (crosslinked) siloxane rubber shows that decomposition starts at about 300 °C and is complete by about 400 °C. The residue of over 50% is basically silicic acid. Whereas these elastomers do ignite with high (90%) char production, flame-resistant siloxanes incorporate halogenated organic groups, silica or metal oxide fillers, dibutyltin dichloride, and/or platinum catalysts remaining in the polymers (14).

A number of crosslinking techniques have been used to produce elastomers from linear siloxanes. Two cure systems take advantage of the availability of vinyl functional polymers. Condensation reactions are also used for crosslinking siloxanes.

Siloxane polymers with vinyl functionality and siloxane polymers with hydride functionality undergo a platinum-catalyzed addition cure hydrosilylation reaction as shown in Figure 4.1. This cure system has no byproducts (other than the catalyst). This allows the fabrication of parts with dimensional stability. The system can be adjusted to give quick curing at room temperature (5–10 min — room temperature vulcanizing), cures between 50 and 130 °C (called low-temperature vulcanizing), and high-temperature vulcanizing (>130 °C). For transfer and impression molding elastomers, Gelest (8) suggests end-functionalized polymers with molecular masses of the order of 28,000 and hexamethyldisilazane treated

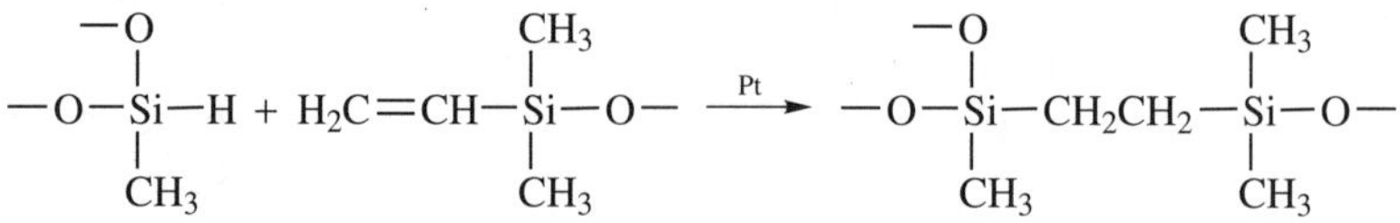

$$-O-\underset{CH_3}{\overset{-O}{\overset{|}{\underset{|}{Si}}}}-H + H_2C{=}CH-\underset{CH_3}{\overset{CH_3}{\overset{|}{\underset{|}{Si}}}}-O- \xrightarrow{Pt} -O-\underset{CH_3}{\overset{-O}{\overset{|}{\underset{|}{Si}}}}-CH_2CH_2-\underset{CH_3}{\overset{CH_3}{\overset{|}{\underset{|}{Si}}}}-O-$$

Figure 4.1 Hydrosilylation reaction for coupling siloxanes with vinyl and hydride functionality.

$$\begin{array}{c} -\mathrm{O} \\ | \\ -\mathrm{O}-\mathrm{Si}-\mathrm{CH_3} \\ | \\ \mathrm{CH_3} \end{array} + \mathrm{H_2C{=}CH}-\begin{array}{c} \mathrm{O}- \\ | \\ \mathrm{Si}-\mathrm{O}- \\ | \\ \mathrm{CH_3} \end{array} \xrightarrow[\text{-ROH}]{\mathrm{RO}\bullet} \begin{array}{c} -\mathrm{O} \\ | \\ -\mathrm{O}-\mathrm{Si}-\dot{\mathrm{C}}\mathrm{HCH_2CH_2} \\ | \\ \mathrm{CH_3} \end{array}-\begin{array}{c} \mathrm{O}- \\ | \\ \mathrm{Si}-\mathrm{O}- \\ | \\ \mathrm{CH_3} \end{array}$$

Figure 4.2 Peroxide-activated cure for siloxanes with vinyl and methyl functionality.

fumed silica* for strength and to control the consistency or flow of the elastomer. To get a quick room temperature cure, 150–200 ppm of a 3–3.5% platinum catalyst complex of tetramethyldivinyldisiloxane in tetramethyldivinyldisiloxane is used. (Platinum complexed in a cyclic vinylsiloxane is recommended for high-temperature cures.) The 150 ppm at 3–3.5% provides the typical 5-ppm platinum level used in this type of cure. Hardness is controlled by crosslink density.

Free radical coupling is also possible between vinyl functionality and siloxane methyl groups using peroxide-activated cure systems that induce free radical coupling between the methyl groups and the vinyl groups. Subsequent reactions take place between the crosslinking sites and other methyl groups until the reaction is quenched (cf. Fig. 4.2). Methylvinylsiloxane-dimethylsiloxane copolymers of extremely high molecular mass are typical base stocks for siloxane elastomers cured by this method. The base stocks or gums typically have molecular masses of 500,000–900,000. A number of peroxides are used; for example, dibenzoylperoxide in silicone oil. A peroxide loading of 0.2–1.0% is used for 140–160 °C cures. A short postcure at even higher temperatures is used to remove the volatile peroxide decomposition products and stabilize the polymer.

Terpolymer gums containing low levels of phenyl groups are used for low-temperature applications. At higher phenyl concentrations, high temperature- and radiation-resistant applications are possible when stabilized with fillers such as iron oxide. The phenyl groups reduce crosslinking efficiency and produce rubbers with lower elasticity. Solvent resistance can be obtained with fluorosilicone polymers. Although the peroxide cure method has classically been used for high-molecular-mass materials, it is being used in low-viscosity specialty systems as well.

An interesting approach to providing a siloxane network for an elastomer is shown in Eq. 4.1, where the ethoxide is replaced with an oxo-poly(dimethylsiloxane) group with hydroxyl terminal groups that can react further if desired.

$$\mathrm{Si(OCH_2CH_3)_4} + 4\mathrm{HO[Si(CH_3)_2O]}_n\mathrm{H} \rightarrow \mathrm{Si\{O[Si(CH_3)_2O]}_n\mathrm{H\}_4} + 4\mathrm{CH_3CH_2OH} \quad (4.1)$$

Other condensation cure reactions react in multiple steps, given schematically in Eqs. 4.2–4.4. Note that the first step (Eq. 4.2) involves reaction with an excess

* Fumed silica provides high surface area, great reinforcement, and excellent electrical insulation. Silicas from aqueous solution provide only moderate reinforcement and less pronounced insulation because of water adhering to the silica (4).

of the methyltriacetoxysilicon so that the product reacts as shown, leaving two acetates on each terminus. This tetrafunctional polymer (usually with molecular masses of 15,000–150,000) can be stored in the absence of moisture until needed. Exposure of the end groups to moisture causes a rapid crosslinking reaction to occur.

$$\mathrm{HO[Si(CH_3)_2O]_nH + 2CH_3Si(O_2CCH_3)_3 \rightarrow}$$
$$\mathrm{CH_3Si(O_2CCH_3)_2O[Si(CH_3)_2O]_nSi(O_2CCH_3)_2CH_3 + 2CH_3CO_2H} \quad (4.2)$$

Replacing the polymer molecules $CH_3Si(O_2CCH_3)_2O[Si(CH_3)_2O]_nSi(O_2CCH_3)_2CH_3$ with simple $\equiv Si(O_2CR)$ functionality, where $\equiv Si$ represents the three bonds to silicon not involved in the condensation reaction and R = methyl:

$$\mathrm{\equiv Si(O_2CR) + H_2O \rightarrow \equiv Si\text{–}OH + RCO_2H} \quad (4.3)$$
$$\mathrm{\equiv Si\text{–}OH + \equiv Si(O_2CR) \rightarrow \equiv Si\text{–}O\text{–}Si\equiv + RCO_2H} \quad (4.4)$$

Similar moisture cure reactions occur with enoxy, oxime, alkoxy, and amine functionalities. Tin catalysts such as di-*n*-butyldiacetoxytin at about 1 part per 2500 SiOR are required, but many formulations use ten times this amount for more rapid cures.

A number of other siloxane reactions and radiation can be used for curing polysiloxanes (cf. Fig. 4.3). Not all of the reactions suggested in Figure 4.3 produce useful elastomers, but the wide variety of reactions that have been developed is impressive (cf. Exercises 4.2 and 4.3).

Table 4.1 provides a summary of the mechanical properties of a number of polysiloxane rubber materials. Note that these properties are adjusted by varying the degree and type of crosslinking, the molecular mass of the raw polymer, and the side groups on the siloxane. However, compromises are often necessary. For example, substituting fluoroalkyl groups (e.g., $CF_3CH_2CH_2$) for alkyl or aryl groups makes the siloxane polymer more solvent resistant, especially to swelling, but the T_g increases and the elastomer cannot be used at low temperatures.

Table 4.2 provides typical electrical properties for a typical polysiloxane elastomer. These elastomers are very good insulators unless filled with metal oxides that conduct electricity.

Finally, it should be noted that polysiloxane foams are also available for fire resistant insulation on pipes and for fire stops. More details on polysiloxane elastomers can be found in the references cited at the beginning of this subsection.

4.1.2 Polyphosphazene Elastomers

Allcock and others learned how to substitute alkoxy and other substituents in place of the chloro groups, as was noted in Chapter 2, thus providing hydrolytically stable polymers that have many useful properties (16–20). Although there have been attempts to commercialize the polyphosphazenes, to

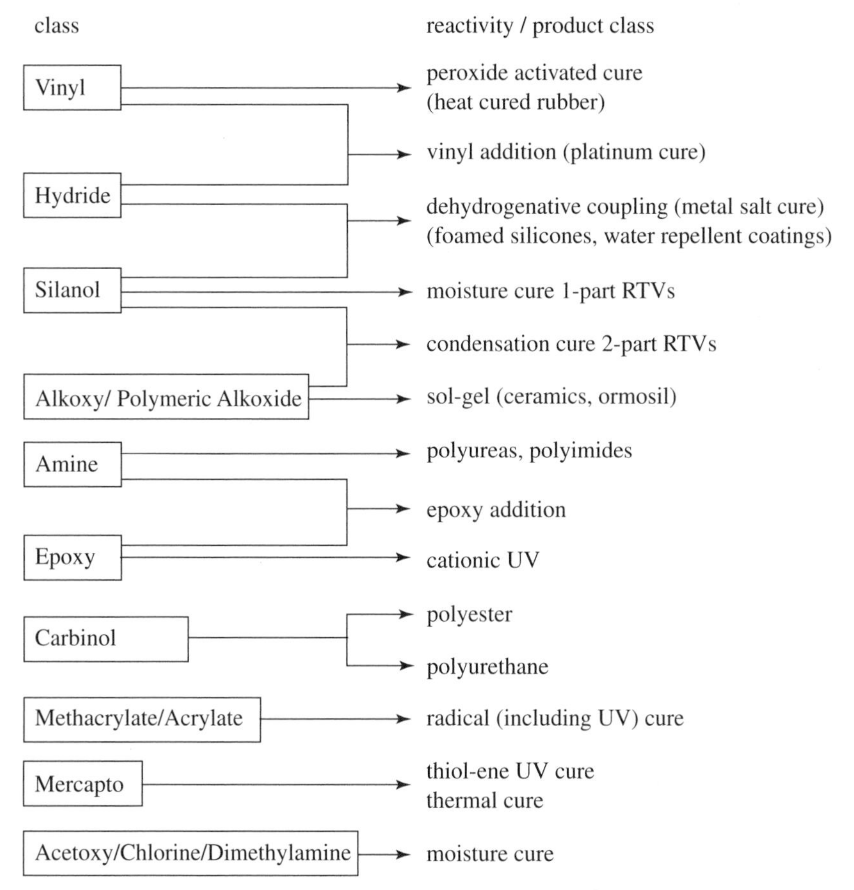

Figure 4.3 Reactivity of functional siloxanes (8).

date their cost has relegated them to small niche markets. Furthermore, the newer direct methods are even more expensive (19). However, a solvent-free method for synthesizing $(NPCl_2)_n$ (where $n \geq 1000$) from PCl_5 and $(NH_4)_2SO_4$ (21) has the potential for lowering the costs by as much as 40% (22).

4.1.2.1 Fluoroalkoxyphosphazene Elastomers

The **fluoroalkoxyphosphazene elastomers (FZ)*** (19) with trifluoroethoxy and higher fluoroalkoxy substituents (Fig. 4.4a) possess good thermal and fluid resistance. The elastomer marketed in the early 1990s has about 65% trifluoroethoxy, 35% $OCH_2(CF_2CF_2)_{1-3}H$, and 0.5% allylic substituents. The allylic site is used as a crosslinking site. The elastomer is made by substitution on

* FZ is the ASTM designation for the elastomers of this class (19).

TABLE 4.1 Properties of Different Classes of Polysiloxane Elastomers[a].

Class	Hardness, Durometer	Tensile Strength, MPa[b]	Elongation, %	Compression Set[c], %	Useful Temperature Range, °C Min	Useful Temperature Range, °C Max	Tear Strength. J/cm[2c]
general purpose	40–80	4.8–7.0	100–400	15–50	−60	260	0.9
low compression set	50–80	4.8–7.0	80–400	10–15	−60	260	0.9
extreme low temperature	25–80	5.5–10.3	150–600	20–50	−100	260	3.1
extreme high temperature	40–80	4.8–7.6	200–500	10–40	−60	315	
wire and cable	50–80	4.1–10.3	100–500	20–50	−100	260	
solvent-resistant	50–60	5.8–7.0	170–225	20–30	−68	232	1.3
high strength flame retardant	40–50	9.6–11.0	500–700				2.8–3.8

[a] Based on Table 6 of Rich et al., op cit. © John Wiley & Sons, 1997.
[b] To convert MPa to psi, multiply by 145.
[c] At 150 °C, 22 h.
[d] To convert J/cm^2 to lbf/in., multiply by 57.1.

TABLE 4.2 Electrical Properties of Typical Polysiloxane Elastomers[a].

Property	Value
volume resistivity, Ω · cm	$1 \times 10^{14} - 1 \times 10^{16}$
electric strength. V/25.4 μm(= V/mil)	400–700
dielectric constant, 60 Hz	2.95–4.00
power factor. 60 Hz	0.001–0.01
surface resistance, Ω	$3.0 \times 10^{13} - 4.5 \times 10^{14}$
dielectric loss factor, tan δ	$5 \times 10^{-4} - 4 \times 10^{-3}$

[a] Based on Table 9 of Rich et al., op cit. © John Wiley & Sons, 1997.

the dichlorophosphazene polymer. Such syntheses were discussed in Chapter 2, and the substituent pattern should be largely statistical.

The translucent pale-brown FZ elastomer has a T_g of -70 ± 2 °C with a density of 1.75 g/cm^3 and a refractive index of 1.41. This gum can be crosslinked using a peroxide; for example, cure times of 20–30 min at 160 °C are possible with dicumyl peroxide. Surface-treated silica and semi reinforcing carbon black are the preferred fillers, although clays, silicates, and nonreinforcing carbon

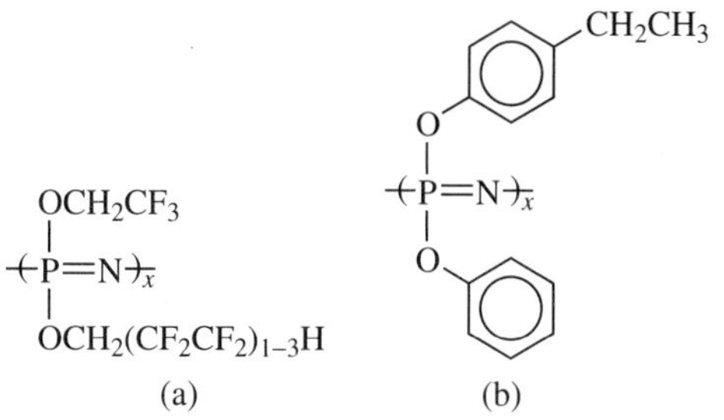

OCH_2CF_3
$-\!\!\left[P=N\right]\!\!-_x$
$OCH_2(CF_2CF_2)_{1-3}H$
(a)

CH_2CH_3
O
$-\!\!\left[P=N\right]\!\!-_x$
O
(b)

Figure 4.4 Commercially developed (a) fluoroalkoxy (FZ) and phenoxy (PZ) polyphosphazene elastomers.

blacks are suitable extending fillers. Acidic fillers such as precipitated or fumed silicas and small-particle carbon blacks inhibit the peroxide cure and lead to poorer thermal stability. The elastomer has been used to prepare small O-rings (<1 g) to large, complex seals (>20 kg). Silane-type primers provide good metal adhesion during the cure.

The cured FZ compounds have densities of 1.75–1.85 g/cm^3 and exhibit good tensile strength (better than FVMQ fluorosiloxane at all temperatures and better than FKM fluorocarbon above 55 °C). FZ is also less toxic than the fluorosilanes. The fatigue resistance of FZ is superior to FVMQ and organic elastomers at extension ratios of 2.20; however, FZ shows lower ultimate elongation than FVMQ at all temperatures (−35 to +180 °C) and has poor fatigue resistance at >2.25 when measured by the Monsanto fatigue-to-failure test. There is little loss in either tensile strength or ultimate elongation by FZ for 800 hr at 150 °C; however about a 30% decrease is noted when aged at 175 °C for the same time period. These and other properties would make FZ an ideal elastomer for many commercial uses were it not >$200/kg in cost. Military and civilian aircraft have used FZ fuel seals and diaphragms. FZ shock mounts have been used on aircraft engines, and FZ hydraulic seals have been used on military aircraft. Because of the useful temperature range as well as good fuel, fire, and fatigue resistance, FZ has been used for large fabric-reinforced boot seals in the air intake system on the M-1 tank. Small diaphragms on some autos have used FZ elastomers.

4.1.2.2 Aryloxyphosphazene Elastomer

The commercial **aryloxyphosphazene elastomer (PZ)*** has phenolate and *p*-ethylphenolate substituents on the phosphorus atom (Fig. 4.4b) with approximately 52 mol% phenolate and 43% *p*-ethylphenolate plus 5% allylic substituent for crosslinking with either peroxide or sulfur (19). This PZ elastomer has found use in applications where fire is a potential hazard. It is quite fire retardant and contains no halogens. However, T_g is −18 °C even before curing; therefore this elastomer is not useful for low-temperature applications. This polymer can accept fairly high filler loading to provide even lower flammability. The toxicity of the

* PZ is the ASTM designation for the elastomers of this class (19).

combustion products is rated about 1.5 times that of Douglas fir, whereas other common synthetic elastomers are rated from 6 to 25 times as toxic as fir. The elastomer has been used in hull and pipe insulation on naval ships and submarines. Other potential uses such as wire and cable insulation are favored by the fire resistance of the PZ elastomer (23), but its modest tensile properties (5.2–12.2 mPa vs. 6.9–13.8 mPa for FZ) and its high dielectric (4–5 at 10,000 Hz) have limited interest. Even so, the naval use of PZ (about 20 metric tons) is about double the tonnage of FZ.

4.1.3 Other Inorganic Elastomers

4.1.3.1 Siloxane-Carborane Elastomers

As noted in Chapter 1, siloxane $-[Si(R)_2O]_n-Si(R)_2-$ chains ($R = CH_3$) bridged with the m-dicarbodecaborane $-CB_{10}H_{10}C-$ provide copolymers with very high thermal stabilities (>400 °C) and softening temperatures, but with relatively low T_g (−42 to −88 °C for $n = 2$–6). These elastomers have been used as O-rings, gaskets, and cable and wire coatings.

4.1.3.2 Polysilane Elastomers

Although polydimethylsilane is an intractable solid, some of the higher homologs of the poly(methylalkylsilane) series, where alkyl = n-butyl to n-dodecyl, are elastomers even before curing. Poly(phenylhydrosilane) and copolymers with methylalkylsilane or methylarylsilane groups are low-melting solids or liquids that can be crosslinked by the hydrosilylation reactions noted above for polysiloxanes. However, the sensitivity of polysilanes to ultraviolet light precludes these and related polysilanes from achieving practicality as commercial elastomers. This light sensitivity of the polysilanes is useful for other applications as noted below.

4.2 INTERFACE COUPLING REACTIONS

4.2.1 Silicon Coupling Agents

Silicon alkoxy or acetoxy hydrolysis is probably the most popular interface coupling reaction, especially for the binding of polymers to glass and metal surfaces (24–28). With glass the coupling reactions are quite evident:

$$\equiv Si{-}X + H_2O \rightarrow \equiv Si{-}OH + HX \tag{4.5}$$

where $X = OC_nH_{2n+1}$ or CH_3CO_2 or even Cl or NH_2 and ηSi represents the three bonds attached to silicon that are not involved in the condensation reaction.

$$\equiv Si{-}OH + H{-}O{-}Si(\text{surface of glass}) \rightarrow \equiv Si{-}O{-}Si(\text{surface of glass}) + H_2O \tag{4.6}$$

Note that the water is cycled in and out (and is effectively catalytic), but the alcohol, acetic acid, HCl, and NH_3 are byproducts of the coupling reaction. The reader may be familiar with the odor of acetic acid from certain silicone caulking formulations. Although the alkoxy formulations are stable for hours in water at room temperature, the hydrolysis is both acid and base catalyzed. Thus derivatives that produce acids or bases on hydrolysis provide more rapid hydrolysis and coupling than the alkoxides. However, a catalytic amount of an acid or base is often added to enhance the hydrolysis and coupling of alkoxy formulations.

As the coupling progresses, the silicon atoms of the coupling agent are not only bonded to the surface but also bonded to each other (Fig. 4.5). That is, the typical $RSiX_3$ silicon coupling agent — which after hydrolysis is a $RSi(OH)_3$ molecule — has a functional group R that can bond to a polymeric molecule and a hydrolyzable group X such as alkoxy, acetoxy, amino, or halo. The functional groups include vinyl ($CH_2{=}CH–$), chloropropyl ($ClCH_2CH_2CH_2–$), epoxy (O-bridged $CH_2-CH-CH_2OCH_2CH_2CH_2-$), primary amine ($H_2NCH_2CH_2CH_2–$), etc.

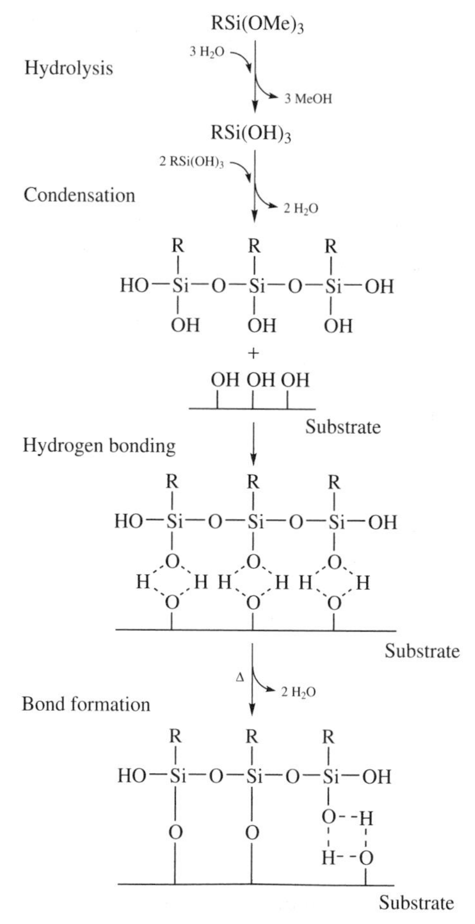

Figure 4.5 A schematic representation of a silicon coupling reaction sequence (Reprinted with permission from Ref. 28; © 1977 American Chemical Society).

Note that the functionality is normally on the γ-carbon atom. Electronegative substituents on α or β carbon atoms causes ready reactions at the Si–C bond. Both nucleophilic and electrophilic reactions can occur at the Si–C bond if the electronegative group is placed on a β carbon atom and nucleophilic reactions can occur if the group is placed on an α carbon atom.

The analogous reactions with metal surfaces are similar because almost all metals have thin hydrous oxide surfaces that possess hydroxyl groups that can react with the Si–OH group. The weakness of this coupling is that only one bond to each metal atom occurs. However, the hydrophobic nature of the siloxane surfaces and the extensive crosslinking between the silicon atoms by oxygen atoms provide very stable interfaces.

The water that is necessary to initiate the hydrolysis reaction can usually come from the substrate surface or from the atmosphere. In some cases, a small amount of water is actually added.

Note that the term interface coupling should be used to distinguish the ionic/covalent bonding that occurs in these situations (generating 200–1000 kJ/mol) from the weaker interface adhesion resulting from London dispersion, dipole-induced dipole, dipole-dipole, and hydrogen-bond interactions that typically generate 10–30 kJ/mol/interaction. However, a multitude of these latter interactions (sticky feet)* per polymeric molecule often provide very strong adhesion (29). The adhesion of paints and other coatings takes advantage of this multifunctionality and effectively inhibits corrosion and marine fouling. Although polymeric substrates for paints are usually organic, organic silicone oils are often a part of such formulations and the pigments are normally polymeric inorganic oxides; for example, titanium dioxide is a white pigment. Polymeric lead and cadmium pigments used earlier have been replaced by nontoxic materials. Note that pigments must be colorfast under severe weather conditions. Most organic dyes fail to be acceptable for that reason (cf. Exercise 4.4).

4.2.2 Metal Coupling Agents

Titanium and zirconium alkoxide coupling agents (26) have been marketed by Kenrich Petrochemicals. These reagents are of the general type $M(OR)_x(OR')_{4-x}$, where R = isopropyl and R′ = stearyl or a functional alkyl group. As in the case of the silane coupling agents, hydrolysis provides interactions suitable for coupling with a glass or metal surface. However, these are monomers to which polymers can be attached.

The present author observed that glassware used for studying a zirconium Schiff-base polymer could not be cleaned with the solvent in which the polymer is soluble, even using an ultrasonic bath. As a result, Wang and Archer developed functionalized zirconium Schiff-base polymeric coupling agents (Figs. 4.6–4.8) that adhere strongly to glass and metal (actually metal oxides on metal) surfaces (30–32). The coupling is so dominant that the polymers must be end capped

* Term often used by T. J. McCarthy to discuss these interactions.

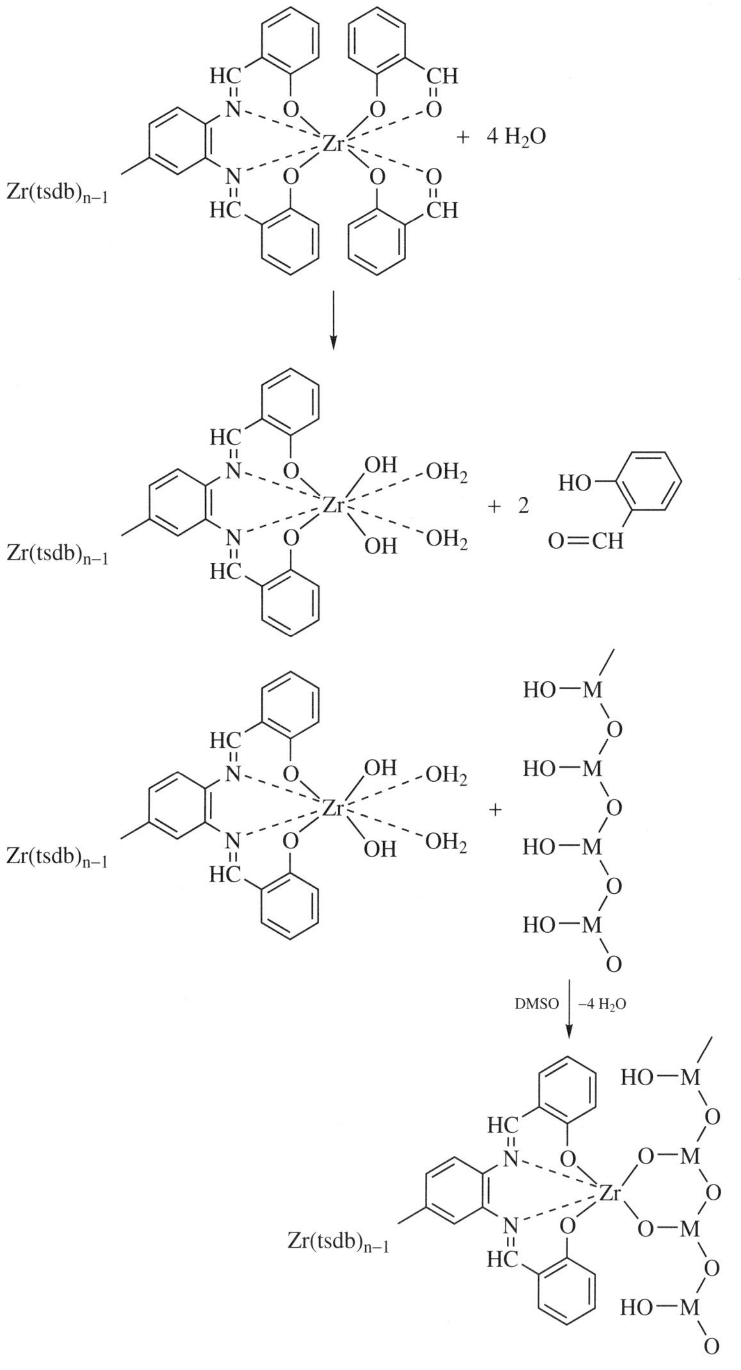

HC
N
O
O
CH
O
Zr
+ 4 H_2O
$Zr(tsdb)_{n-1}$
N
O
O
O
HC
CH

HC
N
O
OH
OH_2
HO
Zr
+ 2
$Zr(tsdb)_{n-1}$
N
O
OH
OH_2
O=CH
HC

HO—M
O
HC
N
O
OH
OH_2
HO—M
Zr
+
O
$Zr(tsdb)_{n-1}$
N
O
OH
OH_2
HO—M
HC
O
HO—M
O

DMSO
–4 H_2O

HO—M
O
HC
N
O
O—M
Zr
O
$Zr(tsdb)_{n-1}$
N
O
O—M
HC
O
HO—M
O

Figure 4.6 A schematic representation of a zirconium Schiff-base coupling reaction.

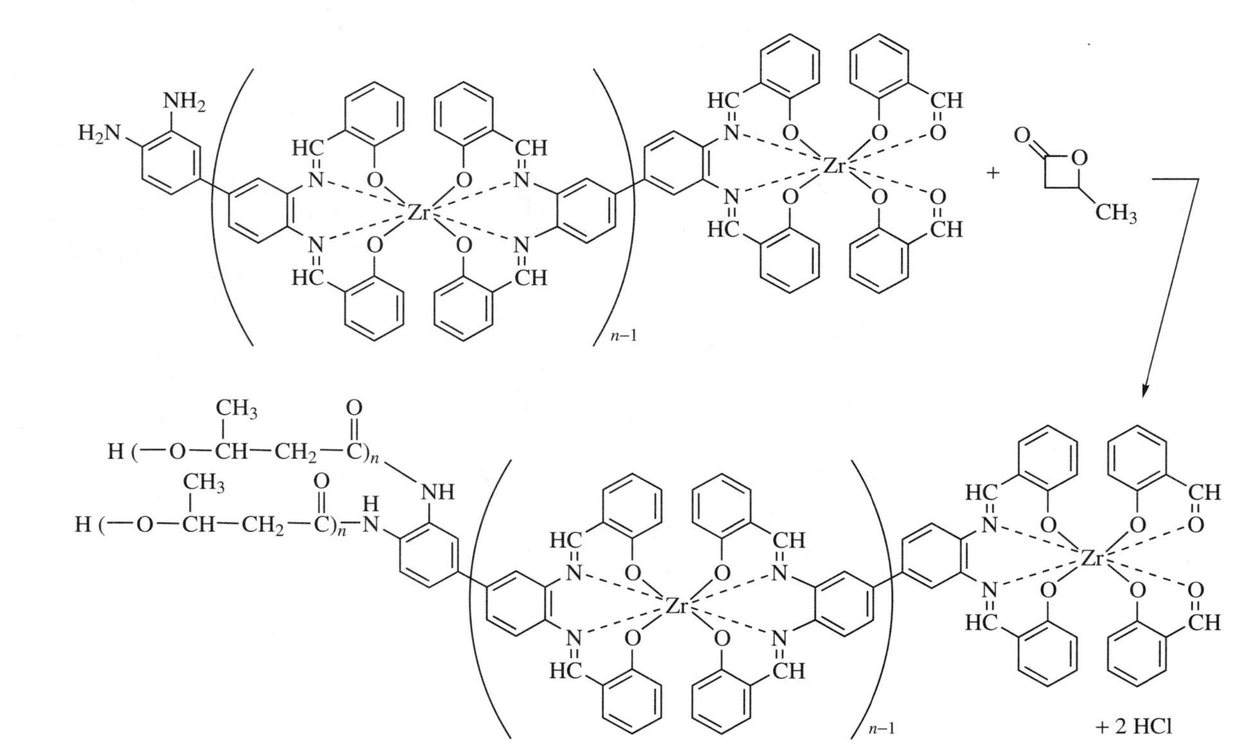

Figure 4.7 A schematic representation of a zirconium Schiff-base/butyrolactone copolymer that enhances the adhesion of polyesters and poly(methylmethacrylate) to glass (e.g., Pyrex® and silica) and metal oxide surfaces on metals.

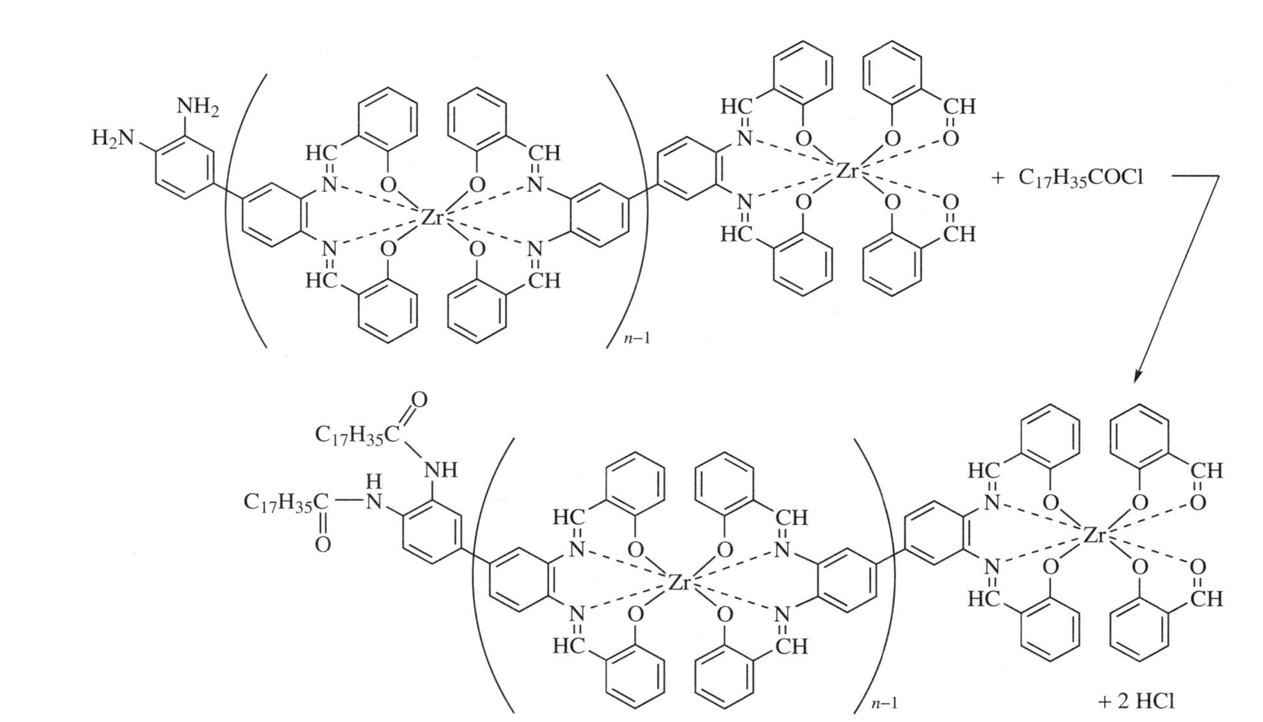

Figure 4.8 A schematic representation of a zirconium Schiff-base/stearoyl copolymer that enhances the adhesion of polyethylene and polypropylene to glass and metal oxide surfaces.

before characterization in dehydrating solvents such as DMSO—otherwise the polymers adhere to glass, metal, and other oxy substrate surfaces. When β-butyrolactone is polymerized by the basic amine end groups of a zirconium Schiff-base polymer (Fig. 4.7), polylactone groups are actually condensed with the end groups and provide increased adhesion of poly(methylmethacrylate) and other polyesters to glass and to aluminum (31, 33). Similarly, condensation of stearyl chloride with the amine end groups (Fig. 4.8) provides markedly increased adhesion of polyethylene and polypropylene to both glass and aluminum (Table 4.3) (31). Proof of the magnitude of the increases in adhesion was determined by peel, indentation-debonding, and scribe-stripping tests.

More recently, Byrd et al. (34) have used a phosphonic acid surface on quartz (35) to build up zirconium Schiff-base polymers and provide a self-assembly surface adduct. Zirconium salicylaldehyde and tetrasalicylidene-3,3′-diaminobenzidine were alternately added to the phosphonated quartz surface for up to 12 layers. The roughness of the coverage as measured by atomic force microscopy indicates a root mean square roughness of less than one monolayer depth after seven layers had been added, suggesting a very even coverage. Absorption measurements in the ultraviolet also indicate an evenness of addition for each of the 12 layers; that is, the increase is linear versus the number of assembly cycles.

Commercialized chromium coupling agents have taken advantage of the inertness of chromium(III) species. Aqueous chromium(III) coordinated to methacrylate or fulmarate provide both the olation (OH) bridging and a site for polymerization with unsaturated organic molecules. The chromium fulmarate system (Volan 82®) was developed by duPont for coupling polyethylene to aluminum.

TABLE 4.3 Adhesion Tests for the Polymeric Coatings.

Test	Polymer	Substrates	Results
Scribe-stripping	Polylactone	Glass only	edge rough
Scribe-stripping	Polylactone	Glass/ZRPL	Edge straight
Scribe-stripping	PE	Glass only	edge rough
Scribe-stripping	PE	Glass/ZRDS[a]	edge straight
Scribe-stripping	PP	Glass only	edge rough
Scribe-stripping	PP	Glass/ZRDS	edge straight
ID[b]	Polylactone	Glass only	50 NT debond
ID[b]	Polylactone	Glass/ZRPL	No[c]
ID[b]	PMMA	Glass only	13.6 NT debond
ID[b]	PMMA	Glass/ZRPL	19 NT debond
Peel	PMMA	Glass only	3.5×10^2 J/m^2
Peel	PMMA	Glass/ZRPL	1.25×10^3 J/m^2
Peel	PMMA	Al only	2.4×10^2 J/m^2
Peel	PMMA	Al/ZRPL	9.0×10^2 J/m^2

[a]ZRDS: Zr(tsdb)-distearamide ZRPL: Zr(tsdb)-polylactone; [b]Indentation debonding; [c]No debonding

4.3 INORGANIC DENTAL POLYMERS AND ADHESIVES

Although most of the dental use of polymeric inorganic species (36) involves three dimensional silica-like materials, appreciable use is made of soluble siloxane materials in forming mold impressions (37), for elastomers in making soft maxillofacial prostheses, and for soft denture linings.* Dentures can be relined to adapt the denture to the changing contours of the soft tissue. However, a resilient relining resin placed directly on the denture can absorb some of the forces produced during chewing. Polysiloxanes (as well as acrylics) are used for this purpose, although no one formulation is ideal. The siloxanes are available as either one- or two-component systems that polymerize much like any polysiloxane elastomer. However, neither adheres well to the denture base, they cannot be polished well, and some siloxanes support bacteria propagation. The soft acrylics adhere well to the dentures but have poor elasticity and harden with age as the plasticizer is lost. Derivatives that are hydrophilic tend to swell when placed in water, and this changes the contours of the denture on the tissue-bearing surfaces. Acrylic-based siloxanes seem a realistic compromise; however, a recent study has suggested that the high-temperature vulcanized polysiloxane is superior to other types of hard and soft denture base resins. A fluoroalkylphosphazene (FZ) polymer (Section 4.1.2) including methacrylate functions was used as a soft liner for dentures because of its extended life, shock isolation, and resistance to microbial attack (19). However, as this book was being written, the polyphosphazenes were not commercially available.

External maxillofacial prosthetics is the art of restoring missing and/or defective facial tissues with polymeric biocompatible materials. Such materials should have toughness, strength, and durability along with good elastomeric properties (softness and resiliency), along with biocompatibility, dimensional stability, inertness, solvent and stain resistance, a blending into adjacent tissues in terms of appearance and color, translucence, low thermal conductivity, and easy fabrication. Again, polysiloxanes are one of the polymers in the forefront of this field. Problems of color stability and colorant effects on these elastomers have also been investigated, with the suggestion that colorants should be added before crosslinking so that the color remains intact as the elastomer "weathers" (38).

Polysiloxanes are also quite important elastomer impression materials for dentistry. Both condensation and addition polysiloxanes are used. Room-temperature vulcanizing polysiloxanes with a variety of viscosities and catalysts are available. For the condensation curing, a hydroxyl-terminated poly(dimethylsiloxane), which is a viscous liquid, is mixed with colloidal silica or other finely powdered metal oxide to form a paste that is mixed with an alkyl silicate (i.e., an alkoxy silane such as tetraethoxysilane) plus a small amount (1–2%) of a catalyst, typically an organic tin compound. Ethanol is evolved during the curing. Stannous octanoate effects a cure (crosslinked polymerization)

* A Web of Science search for "dental AND (siloxane or silicone)" yielded 84 references for 1993–early 2000.

within 30 s to 2 min, whereas zinc octanoate requires 1–4 days. Dibutyltin dilaurate, another common tin catalyst, requires about 10–20 min. Heat and moisture also accelerate the rate of curing. Limited shelf life is a problem for this type of system. Both moisture and oxidation sensitivities make for a maximum of a 2-year shelf life.

The addition polysiloxanes used for dental impressions are typically two-paste systems that can be simultaneously ejected with a dual syringe for automatic mixing. One component is a polysiloxane with terminal vinyl groups and a reinforcing filler. The other component consists of a hydride-terminated siloxane oligomer, a filler, and a chloroplatinic acid catalyst. These systems have replaced earlier systems that required a more exact mixing technique on the part of the dental professional.

Silane coupling agents (Section 4.2.1) are used with porcelain or ceramic surfaces (39), but only as monomeric materials such as 3-(methacryloxy)propyltrimethoxysilane, but the polymers used in restorative resins are normally organic resins filled with inorganic fillers pretreated with silanes. The fillers are normally 1- to 10-micron glass particles containing some barium, strontium, zirconium, or zinc to obtain some opacity for X-rays.

Attempts to avoid the processing problems in such restorative resin composite systems (high viscosity and inorganic-organic separation during aging), silsesquioxane epoxide or methacrylate hybrid polymers have been studied as possible alternates (40, 41). Other inorganic-organic hybrid composites for dental restorative materials are also being developed (42, 43).

In an altogether different application, toothpastes with a poly(dimethylsiloxane) oil and triclosan (a well-known bactericide) appear to provide marked plaque reduction (44). Both the polysiloxane and the triclosan have some effect when used alone. In vitro studies have shown that the silicone oil retains the triclosan and slowly releases it — inhibiting bacterial growth over a longer time period than when applied alone.

Poly(acrylic acid) with iron(II and III) chloride gelation in water or in 30% hydrogen peroxide provides a means of blocking the microscopic channels in tooth dentin and protects against tooth decay (45). These metal-containing polymers have the metals on side chains (Type III) and are essentially organic polymers crosslinked with metal ions. Similar polymers with OTiO and OZrO type crosslinks have also been described for similar purposes (46).

4.4 INORGANIC MEDICAL POLYMERS

4.4.1 Polysiloxanes as Biopolymers

Figure 4.9 provides a pictorial sketch of the multiple medical uses for polysiloxanes (silicones) in the early 1980s (15). A few diverse examples will be elucidated here.

Polysiloxanes can be used in biodegradable artificial skin to cover third-degree burns and close deep wounds. The base membrane layer consisting of bovine hide,

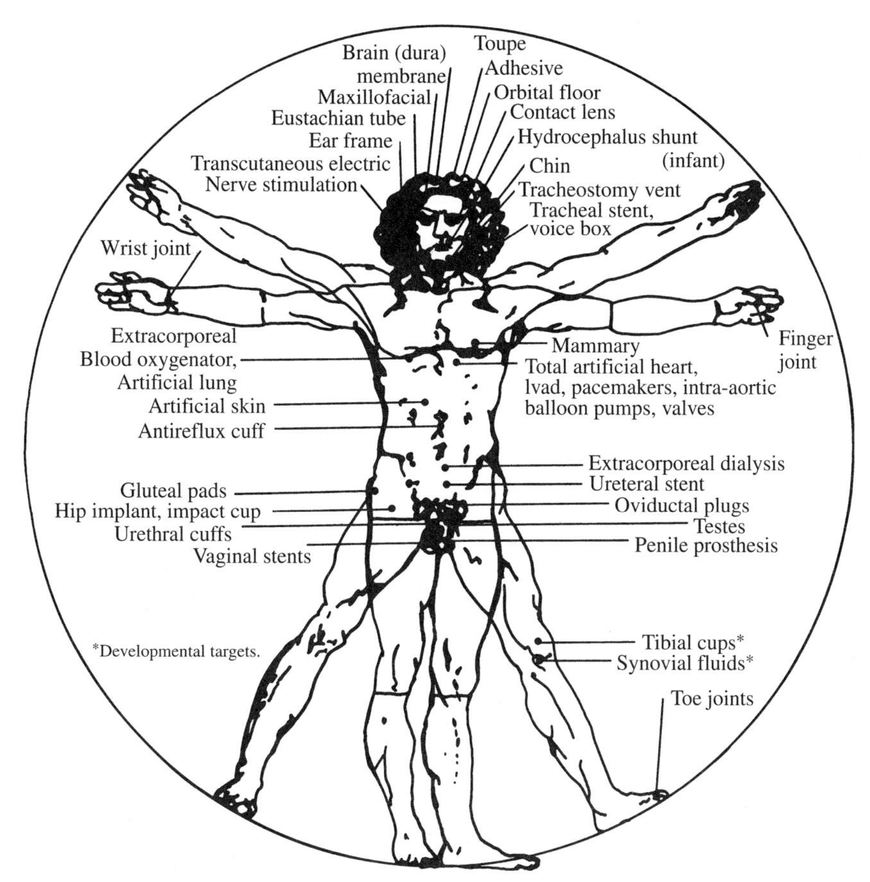

Figure 4.9 Polysiloxane medical devices from the early 1980s (Reprinted with permission from Ref. 15; © 1983 American Chemical Society). Catheter and drainage devices not shown. lvad = left ventricular assist device.

collagen, and chondroitin 6-sulfate (from shark cartilage) is covered with a 0.5- to 0.8-mm top layer of condensation-cured polysiloxane. The bilayer membrane seals the wound, preventing fluid loss and the proliferation of bacteria. The lower layer provides a template for the regeneration of tissue. The siloxane membrane is pliable but tough, and it also allows both oxygen and body moisture to diffuse during the regenerative process. Eventually the lower layer is physiologically degraded and the polysiloxane layer is slowly ejected (see Fig. 4.10).

The use of polysiloxanes for maxillofacial applications was noted in Section 4.3. Although poly(dimethylsiloxane) is also the dominant polymer for this application, phenylethyl and other side chains help with pigment compatibility.

Transcutaneous nerve stimulators have been fabricated from conductive polysiloxanes. Electrodes for administering low-level voltage to underlying nerves can be constructed by adding either carbon or silver into a polysiloxane

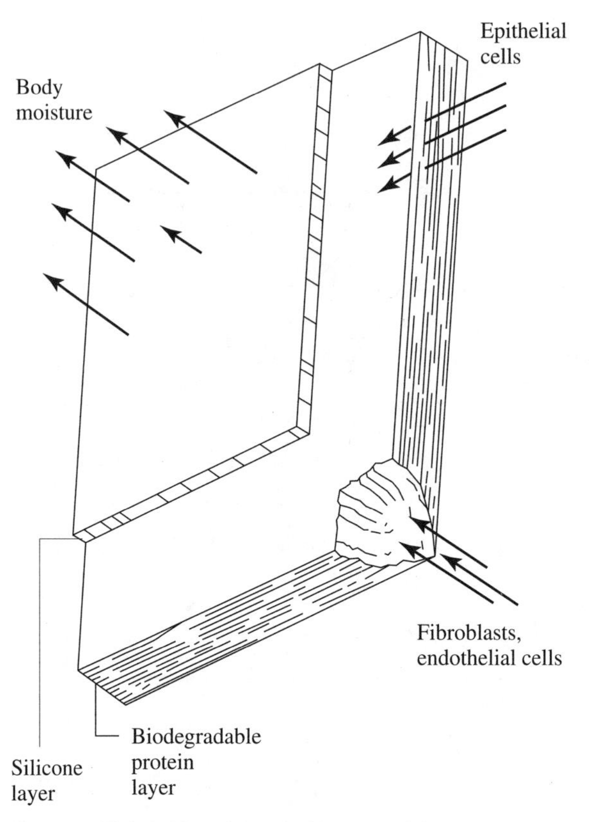

Figure 4.10 Bilayer artificial skin with polysiloxane (silicone) outer layer that provides a moisture- and oxygen-permeable mechanical support and barrier for the inner layer derived from shark cartilage that acts as a template for new skin formation (Reprinted with permission from Ref. 15; © 1983 American Chemical Society).

before curing. This prevents the transmission of pain impulses from the nerves to the brain.

Contact lens fabrication is one area in which polysiloxanes have been of extreme importance, primarily because of their oxygen permeability. Although many of the lens formulations have siloxane groups (from monomers to oligomers) as side chains, block copolymers of polysiloxane and poly(alkylene oxide) produce compositions that are highly oxygen permeable, are flexible, and have good wettability.

A variety of other compositions and siloxane substituents have been used in attempts to provide the best properties. For example, radical bulk polymerization of methacrylate end-capped fluoro polysiloxanes with side groups such as 3-(2,2,3,3-tetrafluoropropoxy)propyl or larger with hydrophilic monomers, such as dimethylacrylamide and *N*-vinyl pyrrolidinone, resulted in transparent hydrogels possessing a wide range of water content, high oxygen permeability, and a low modulus of elasticity. The fluoro side groups had been added through platinum-catalyzed hydrosilation of hydride-containing polysiloxanes made from acid-catalyzed ring-opening polymerizations of tetramethylcyclotetrasiloxane and other cyclotetrasiloxanes (47).

A review of contact lens chemistry (48) lists 12 rigid gas-permeable lens materials. The lenses listed have oxygen-permeable ratings (*Dk* values) of 14–100 barrer (where 1barrer $= 1 \times 10^{-11}$ cm^3O_2•cm/s•cm_2•mmHg), with both the highest and the lowest being "silicone acrylates." A new soft contact lens material (Lotafilcon A) has a high *Dk* value of at least 140 barrer (49). Lotafilcon A is a biphasic block-copolymer with a highly permeable siloxane-based polymeric phase coupled with a water (hydrogel) phase. An exact value is not possible because of the water phase, but the 140 barrer value is a minimum.

The role of polysiloxanes (silicones) in breast implants is well known. However, given the biocompatibility of polysiloxanes and their use in many other medical devices, the claims of health problems related to the devices appears suspect and could not be scientifically shown, although the lawsuits regarding the claims forced Dow Corning into bankruptcy.

Devices fashioned out of polysiloxanes and other polymers can be used to control incontinence. The bladder control devices have a fluid-filled cuff that surrounds the urethra with enough pressure to prevent urination. Squeezing the attached manual pump forces the liquid up into a balloon, reduces the pressure on the urethra, and allows urination. Devices are available for both men and women. As is evident from Figure 4.9, the list of uses for polysiloxanes goes on and on.

Another type of use for polysiloxanes is drug encapsulation (50). In fact, the concept of a drug delivery device with dispersion of a drug in a polysiloxane in which it is incompatible goes back to the 1970s, when Alza Corporation developed a series of applications including a contraceptive intrauterine device that releases progesterone at a constant rate, pilocarpine release for glaucoma, and scopolamine release for motion sickness. Not only does this approach provide for a more even distribution of a drug, but also the local administration of the drug reduces the amount of the drug that is required. All of these factors reduce drug-related side effects.

4.4.2 Polyphosphazenes as Biopolymers

Polyphosphazenes are also potential biomedical polymers (4). Many of the uses outlined above for the polysiloxanes can be envisioned for polyphosphazenes as well. In fact, for many of the uses where wettability is important, the hydrophilic polyphosphazenes definitely would have an edge over the polysiloxanes were it not for the greater cost of the polyphosphazenes. As noted above, hydrophilic groups must be added to the polysiloxanes to obtain compatibility with water. The ease with which substitution of various organic groups can be made with polyphosphazenes provides almost endless possibilities. By changing the side groups almost any property desired can be provided. For example, either bio-erodable or bio-inert polyphosphazenes can be synthesized as desired.

Examples of bio-erodable polyphosphazenes that can be used for controlled release of encapsulated drugs include ethyl esters of amino acids substituted on poly(dichlorophosphazene) that slowly hydrolyze to produce ethanol, ammonium dihydrogen phosphate,* the amino acid, and the free drug. Another example is for a steroid such as 2-estradiol to be attached as a side group to the polyphosphazene. Then when the hydrolysis takes place, the steroid is released along with ammonium dihydrogen phosphate. To provide bio-inert polymers, $-OCH_2CF_3$, $-OC_5H_6$, and siloxane hydrophobic units have been used as polyphosphazene substituents.

Water-soluble polyphosphazenes are possible. The bio-inert species noted above are water insoluble, whereas alkyl amino groups or certain alkoxy groups or even aryloxy groups with phenolic or carboxylic acid functionality hydrogen bond to water and can dissolve in water if not crosslinked.

Surface charge or the presence of ionic surfaces is necessary for blood or tissue compatibility. The sulfonation of aryloxy substituents on polyphosphazenes with concentrated sulfuric acid can overcome this potential problem.

Other examples could be provided such as heparinized surface or immobilized enzymes; however, the high cost of polyphosphazenes has driven them off the commercial market, at least for the time being. Synthetic strategies that can reduce the cost of the phosphazene polymers are sorely needed.

4.4.3 Metal-Containing Polymers for Medical Purposes

Metal-containing polymers have also been studied with respect to medical uses. For example, polyferrocenes (51) and polyphosphazenes with ferrocene side groups (52) have been suggested as mediators in amperometric bio-sensors. Polymeric derivatives of *cis*-diamminedichloroplatinum(II), the well-known cancer drug, have also been suggested for better drug administration and for a more even dosage (53).

Tin-containing polymers inhibit bacteria and have been suggested for a variety of uses such as paints, wood protection, and protection against ship hull biofouling.

4.5 INORGANIC HIGH-TEMPERATURE FLUIDS AND LUBRICANTS

Polysiloxane or silicone fluids are well known for their outstanding properties relative to hydrocarbon fluids. A wide range of siloxane fluids exist, but in general they have the following characteristics (54):

1. wide service temperature range
2. low viscosity changes vs. temperature
3. thermal stability
4. low flammability
5. shear stability
6. dielectric stability
7. high compressibility
8. chemical inertness
9. low surface tension
10. low toxicity

* Although some authors suggest the products are ammonia and phosphoric acid, simultaneous and stoichiometric release yields ammonium dihydrogen phosphate.

As a result of these properties, liquid polysiloxanes (silicones) are used as dielectric, hydraulic, heat transfer, power transmission, and damping fluids. They are used as additives for plastics and rubbers as process and release aids, as coatings for flow and level control, and as antifoam agents in process streams. They also have uses in acoustical applications including ultrasonic sensors and sonar buoys. Because of their light-refractive index matches with fiber-optic materials, they are used in optoelectronics. One supplier, Gelest, divides them into six general classes:

1. conventional fluids
2. thermal fluids
3. organic compatible fluids
4. fluoro fluids
5. hydrophilic and polar fluids
6. low-temperature fluids

The conventional fluids are poly(dimethylsiloxanes) with varied molecular mass, and the others are derivatives that enhance a property of importance for a particular use. These modifications often cause some degradation of some of the other properties.

Conventional polysiloxane fluids (about $16–20/kg in 16-kg lots in 1998) (54) range in viscosities from less than 5 centistokes (cSt) to 60,000 cSt. (Viscosities higher than 60,000 cSt cost more.) With reference to the viscosity of common items:

$1\ \text{cSt} = \text{water}$
$100\ \text{cSt} = \text{olive oil}$
$1000\ \text{cSt} = \text{castor oil}$
$5000\ \text{cSt} = \text{corn syrup}$
$10,000\ \text{cSt} = \text{honey}$
$100,000\ \text{cSt} = \text{sour cream}$

Conventional polysiloxane fluids up to 2,500,000 cSt are available. Flash points for the various viscosities range from below 0 °C for the lowest molecular mass to 315 °C for viscosities of 100 cSt and above. The average molecular masses for these fluids range from 770 (10 cSt) to 116,500 (60,000 cSt). The average molecular mass for the largest fluid is 423,000. The glass transition temperature of −128 °C is fairly constant, and above 30,000 average molecular mass most of the other properties (other than viscosity) stay about the same even though the entanglements must continue to increase. For example, the viscosity, density, and thermal conductivity temperature coefficients as well as the dielectric constant and dielectric strength stay virtually identical for molecular masses from less than 30,000 to 260,000. Polysiloxane fluid characteristics of select average molecular masses are shown in Table 4.4. Even though polymeric organic esters have been synthesized that have comparable pour points, the viscosity changes with temperature in the esters are from one to two orders of magnitude greater than for the polysiloxanes. Other thermal parameters that do not change appreciably above 10 cSt include specific heat (1.51 ± 0.04 J/g/°C), heat of formation (−10.08 kJ/g), and heat of combustion (25.65 kJ/g above 50 cSt).

The polysiloxane fluids are soluble in methylene chloride, chlorofluorocarbons, diethyl ether, xylene, methyl ethyl ketone, and supercritical carbon dioxide. The low-viscosity polysiloxanes are also soluble in acetone, ethanol, dioxane

TABLE 4.4 Characteristics of Conventional Silicone Fluids[a].

	$\overline{M}$					
	770	6000	28,000	63,000	116,000	260,000
Viscosity (cSt)	5	100	1000	10,000	60,000	600,000
Viscosity temp coeff	0.54	0.60	0.61	0.61	0.61	0.61
Pourpoint °C	−65°	−65°	−50°	−48°	−42°	−41°
Specific gravity	0.918	0.966	0.971	0.974	0.976	0.978
Refractive index	1.3970	1.4025	1.4034	1.4035	1.4035	1.4035
Coeff therm expand	11.2×10^{-4}	←9.3×10^{-4}→		←9.2×10^{-4}→		
Coeff therm conduc	2.8×10^{-4}	3.7×10^{-4}	←3.8×10^{-4}→			
Surface tension	19.7	20.9	21.2	21.5	21.5	21.6
Dielectric constant	2.60	← 2.75 →				
Dielectric strength, V/mil	375	← 400 →				
Flashpoint °C	135°	← 315° →				

[a]Based on data from Arkles, 1998, *op cit.*

and dihexyladipate. However, they are all insoluble in methanol, cyclohexanol, ethylene glycol, water, and many other polar solvents.

Whereas polysiloxanes are considered fairly nonreactive, hydrogen fluoride attacks the silicon-oxygen bond with volatile byproducts, free-radical reactions with the methyl groups form crosslinked materials that increase viscosity, and with peroxy compounds gelation can occur. To overcome this reactivity, thermal silicone fluids with phenyl groups replacing some of the methyl groups decreases the oxidative attacks on the silicon-methyl bonds as well as thermal attacks on both the silicon-methyl and silicon-oxygen bonds. For example, although a conventional polysiloxane may gel in less than 10 h at 250 °C in air, poly(methylphenylsiloxane) of only slightly lower viscosity has a gel time of 1500–2000 h. Although the stability at higher temperatures is markedly improved, the pour point is raised from −50 °C to −20 °C, the surface tension is raised from 21 to 28, the flash point is lowered from 315 °C to 300 °C, and the viscosity temperature coefficient is raised from 0.61 to 0.88. This and other high-phenyl-content siloxane copolymer fluids are utilized as heat-exchange fluids, dielectric coolants, impregnates for sintered metal bearings, and base oils for high-temperature fluids. For lower-temperature pour points with some high-temperature improvement, methylphenylsiloxane-dimethylsiloxane copolymers can be used. For example, one such copolymer (8–12% methylphenyl derivative) with 30,000 cSt viscosity has a pour point of −70 °C. On the other hand, its gelation time is less than 100 h. Other modifications include a tetrachlorophenylsilsesquioxane-dimethylsiloxane copolymer with a −73 °C pour point but with a 270-h gelation time. These thermal polysiloxanes are appreciably more expensive than the conventional polysiloxanes and are only used when especially needed.

Organic compatible polysiloxane fluids are synthesized using alkyl groups or aromatic substituted alkyl groups in place of some of the methyl groups of the conventional polysiloxane fluids. Poly(methyloctylsiloxane), poly(methyltetradecylsiloxane), alkylmethylsiloxane and methyl(2-phenylpropyl) siloxane copolymers, and alkylmethylsiloxane and dimethylsiloxane copolymers are four commercially available examples. Alkyl chain modification improves the compatibility with organic materials and improves lubrication properties relative to the conventional polysiloxanes, but the viscosity temperature coefficient increases and the oxidative stability decreases. Poly(octylmethylsiloxane) is useful as a lubricant for soft metals (aluminum, zinc, and copper) and for rubber and plastics (especially when these are in contact with steel or aluminum). The alkylmethyl polysiloxanes act as wetting and leveling agents in coating and ink formulations by reducing the surface tension of these nonaqueous solutions. These types of polysiloxanes have uses in the cosmetic industry for providing compatibilities as well. Again, the costs are appreciably higher than for the conventional polysiloxanes.

Fluoro substituted polysiloxanes such as poly[methyl(3,3,3-trifluoropropyl) siloxane] provide materials that are useful from −40 to 230 °C under severe environmental conditions. They are not miscible with fuels or oils (or water) and have been used in mechanical vacuum pumps where moisture, oxygen, and organic contaminants may be encountered. They are also excellent lubricants under extreme pressure conditions. The pressure and fuel nonmiscibility has led to automotive and aerospace lubrication applications. The specific gravity of these polymers ranges from 1.25 to 1.30, which makes them useful as floatation media for inertial guidance systems. Their solubility in supercritical carbon dioxide may be exploited in the future as well.

Hydrophilic polysiloxanes have some or complete compatibility with water but do not have the thermal stability noted for the conventional polysiloxanes because the organic block or side chains provide sites for degradation. Examples include polyalkylene oxide block copolymer with poly(dimethylsiloxane), where the polyalkylene block can be either ethylene oxide or mixed propylene oxide and ethylene oxide; and hydroxyalkyl substitution, where approximately two hydroxyethyleneoxypropyl groups are substituted per polysiloxane chain. They are widely used as surfactants and emulsifiers. By appropriate balance in the copolymer, the desired properties are obtained. With the ethylene oxide block copolymers, complete water miscibility requires that about 75% of the block copolymer be ethylene oxide. Such formulations can be used as anti-fog treatment for glass and optical surfaces and to facilitate wetting and the spread of developers on lithographic plates. The nonmiscible copolymers are used as lubricants in metal on plastic applications and as fiber lubricants. Other uses also exist. The hydroxylic substitution polysiloxanes are also used for anti-fog and anti-static applications.

Amphiphilic polysiloxanes that possess both hydrophilic and oleophilic sites use a combination of the organic compatibility and hydrophilic substitutions, such as a methyldodecylsiloxane and (hydroxypolyalkyleneoxypropyl)methylsiloxane copolymer. Amphiphilic polysiloxanes can form stable water-in-oil emulsions allowing formulation of a wide variety of creams and gels and can be used as surface treatments for the dispersion of polar particles in hydrocarbons.

Low-temperature polysiloxane fluids include diethylsiloxane homopolymers with triethylsiloxy terminal groups and branched methyl-T-branched poly(dimethylsiloxane). Both of these materials provide fluids at −80 °C. The diethylsiloxane polymer of 1300–2000 molecular mass (and viscosity of 200–400 cSt) has a pour point of −96 °C and the comparable-molecular-mass methyl-T-branched poly(dimethylsiloxane) (viscosity of 50–60 cSt) has a pour point of −85 °C. The standard poly(dimethylsiloxane), polyfluorosiloxanes, and petroleum oils are all frozen before −80 °C.

Other inorganic polymers that have been suggested for high-temperature fluid and lubrication duty have included polymeric metal phosphonates. Although the chromium(III) derivatives were suggested to have sufficient inertness and thermal stability for such uses, the cost of a lubricant containing such a high percentage of chromium would be prohibitive. None of the less expensive metals has phosphonates with sufficient inertness to meet these demands.

Solid film lubricants include a number of two-dimensional inorganic polymers. Molybdenum disulfide is the most notable and highly used solid lubricant next to graphite. Both are layer-lattice solids in which the bonding between the layers consists of relatively weak London dispersion forces whereas the bonding within the layers is made up of strong covalent bonds. Both MoS_2 and graphite have high melting points, high thermal stability, low evaporation rates, good radiation resistance, and effective friction-lowering ability.

Soft metals (3-D polymers) such as gallium, indium, thallium, lead, tin, gold, silver, and even copper and a number of alloys make very good lubricants for hard metals. Low shear strength, strong bonding to substrate metals, good lubricity, and high thermal conductivity make them almost ideal lubricants. Even sodium and the sodium-potassium alloy have been used as lubricants and heat-transfer materials in the absence of air.

Finally, if we consider glasses (silica, silica-soda-lime, borosilicate, and aluminosilicate) as inorganic polymers, softening glass is used as a lubricant for extruding and forming high-temperature processes with steel and nickel-base alloys, titanium and zirconium alloys up to 1000 °C, far too high for hydrocarbon or siloxane lubricants. Such glass lubrication is also used sometimes for copper extrusion.

In summary, if 2-D and 3-D inorganic polymeric species are included, the uses of inorganic materials as high-temperature and special-use lubricants are quite extensive.

4.6 INORGANIC POLYMERS AS LITHOGRAPHIC RESISTS

Lithographic resist materials are important in microelectronic device preparation, for example, computer chips (55–57). A schematic representation of a positive resist process is shown in Figure 4.11. Providing more features on the same or smaller chip area has continued the rapid advance of such devices. Just a

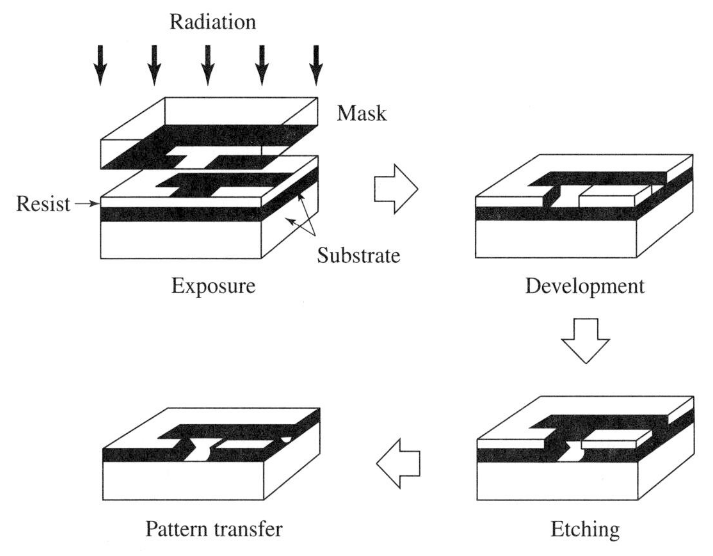

Figure 4.11 A schematic representation of a lithographic process with a positive resist. A typical example is a silicon wafer with an oxidized (silica) surface covered by the radiation-sensitive resist. The resist is covered by a mask or a mask image is projected onto the resist surface. The development of the resist may be by either a solvent or plasma. Similarly, the etching of the silica layer may be by either HF solvent or a second plasma. A metal such as copper or gold can be added after the etching step to provide conductivity on the semiconducting silicon. Silica provides insulation to separate the conducting elements (56).

few years ago the minimum size feature on such a device was of the order of 5 microns, whereas today features of less than 0.5 microns (500 nm) are common. To etch features in a silicon chip of that size or smaller requires shorter-wavelength light. The novolac-diazonaphthoquinone resist films that have been the mainstay for many years (when 436- and 365-nm Hg lines were used) have inherent low sensitivity and absorb too much light at the practical minimum thickness of about 0.7 microns at the shorter wavelengths (248 nm KrF excimer laser) needed for the further diminishing of feature size. Thus other resists must be used at the shorter wavelengths. (X-ray and electron-beam radiation sources provide even shorter wavelength irradiation than vacuum ultraviolet ones.)

The materials must have appropriate sensitivity, contrast, resolution, etching resistance, purity, and manufacturability. The resists must

1. be soluble in a solvent that allows for the deposition of a uniform defect-free thin film on the substrate;
2. be thermally stable to withstand the temperatures used with device processes (stability to $>150\,^{\circ}$C);
3. be rigid enough to provide high-fidelity transfer during the pattern transfer of the resist image to the substrate ($T_g > 90\,^{\circ}$C and adhesion to the substrate);

4. be reactive enough to facilitate pattern differentiation after irradiation; and
5. for photo exposure, have absorption characteristics that allow imaging throughout the thickness of the resist film.

Photo bleaching is the best solution to the last requirement. That is, a photo reaction that shifts the wavelength of absorption from the irradiating wavelength becomes more transparent as it reacts and allows radiation to penetrate through the film.

Historically, the first resist materials were partially inorganic (e.g., ammonium dichromate) although the polymers (egg albumin, gelatin, or gum arabic) were water soluble and of biological origin. The photosensitive dichromate ion spontaneously reduces and causes crosslinking that immobilizes the biological polymer. Assuming the chromium species reduce to chromium(III), the inertness is logical. Naturally, this is a multistep process. Dichromate lithography has also been used for over 100 years in the printing industry. Note that these resists are negative resists, the resist stays on the substrate where the light has impinged on the resist.*

More recently, synthetic organic polymers have been the mainstay of chip resists as noted above; however, polydialkylsilanes are radiation-sensitive positive resists and bleach at 313 nm (Fig. 4.12) (58). This bleaching is concurrent with polymer scission. Polysilanes have a distinct advantage because they are

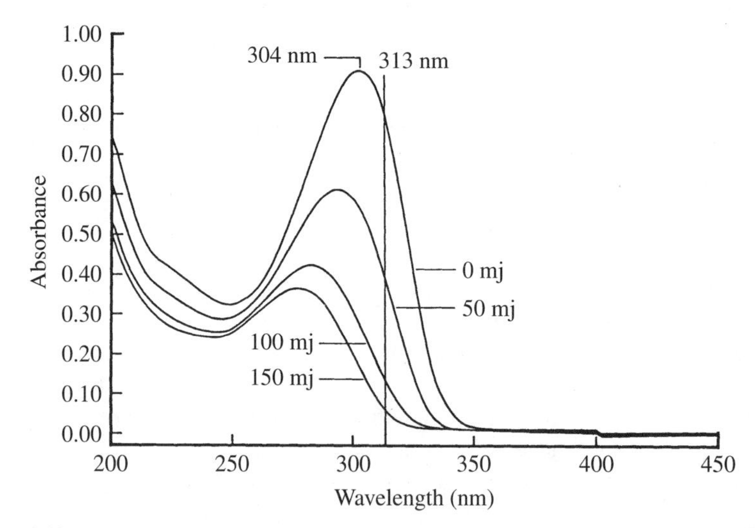

Figure 4.12 The photochemical bleaching of poly(hexylmethylsilane) at 313 nm. mj = millijoule (Reprinted with permission from Ref. 58; (c) 1987 American Chemical Society).

* The photochemical reduction and labilization of a cobalt(III) coordination compound with synthetic organic copolymers containing crosslinkable epoxide groups is a more recent example of using metal ion photochemistry to prepare negative resists (133).

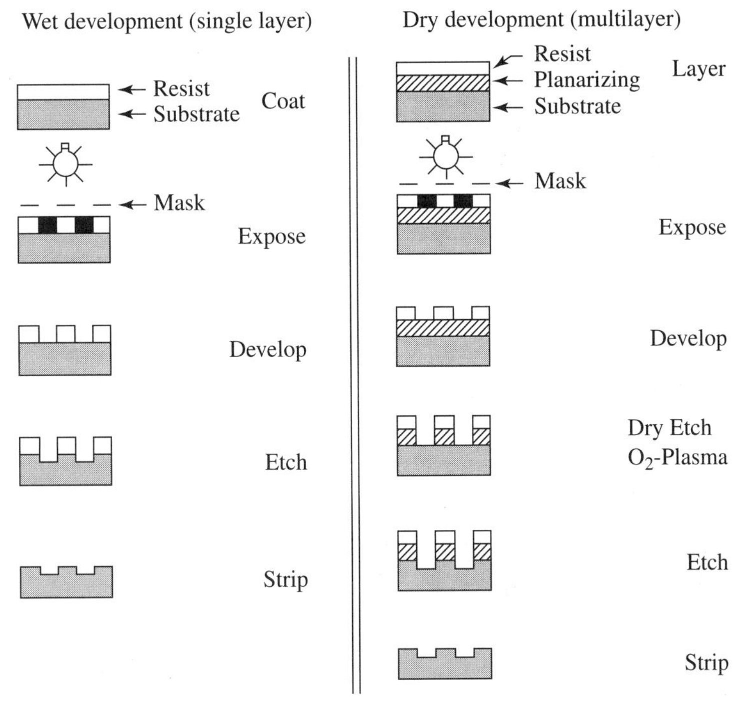

Figure 4.13 A schematic comparison between classical single layer and multilayer lithographic processes. The multilayer nature of the resist substrate (SiO_2 on Si) and the subsequent addition of a metallic conductor are not shown (Reprinted with permission from Ref. 58; © 1987 American Chemical Society).

resistant to oxygen plasma etching by forming a thin inert silica layer when the etching begins (analogous to the alumina layer that protects metallic aluminum from decomposition in ordinary metal items). This makes polysilanes particularly suitable for multilayer resist applications as shown in Figure 4.13. However, they may not be suitable at shorter wavelengths such as the 248 nm KrF excimer laser. However, they have been found suitable for high energy resists (58). This is not too surprising, because high energy radiation absorption is approximately proportional to Z^4, where Z is the atomic number of the material. The standard for electron-beam work has been poly(methylmethacrylate), in which oxygen is the heaviest atom.

An all-dry lithographic process involving silanes and siloxanes has also been suggested (Fig. 4.14) (56) in which a coating of polymethylsilane is generated on the substrate by plasma deposition. These films are very sensitive to ultraviolet light and can be patterned in an entirely dry process. Irradiation in air causes photo-oxidation of the plasma polymerized methylsilane (PPMS) and converts it to a polymeric methyl siloxane material (PPMSO) (see Fig. 4.14.) The unexposed portions of the film are easily removed by a halogen plasma, which does not attack the PPMSO regions. Annealing by an oxygen plasma provides a silica

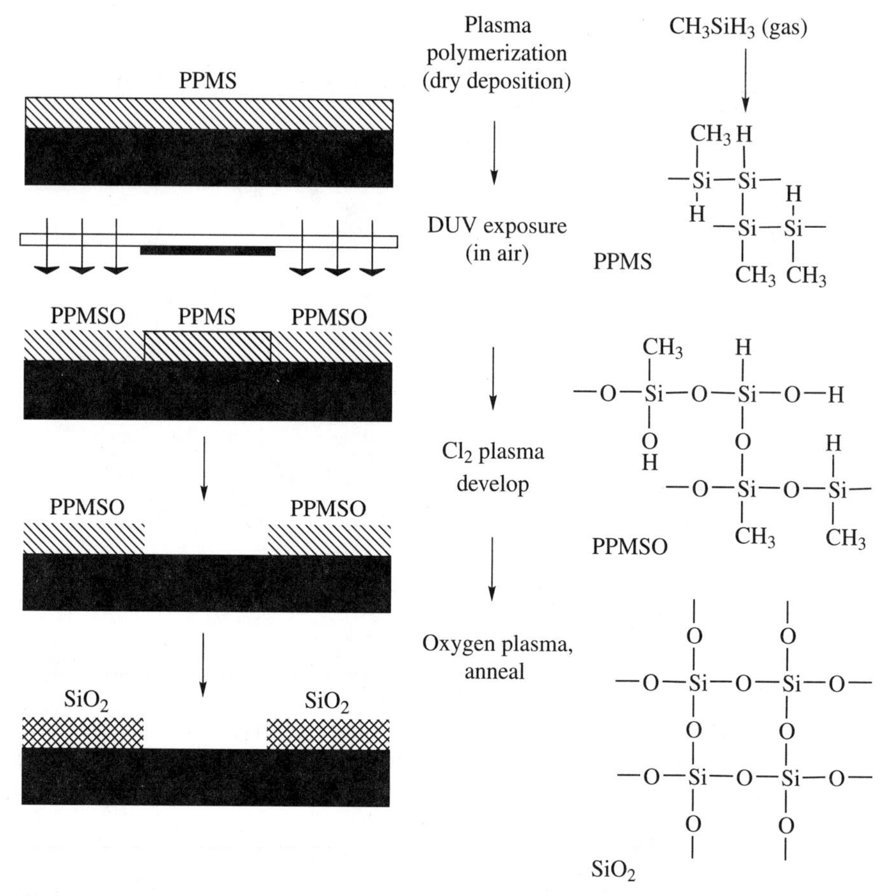

Figure 4.14 The process sequence for plasma polymerized methyl silane resists. A metallic conductor would be added between the SiO_2 insulators (58).

layer in the unexposed region. The achieved resolution is as good as the mask that defines the exposure. The films bleach during exposure so no antireflection layer is required, and the thin (0.2 micron) conformal film avoids a depth-of-focus problem that occurs with thicker films. Because no organic solvents are required, this is "green" chemistry.

Another approach to resist development is photoacid generation followed by appropriate acid-catalyzed polymerization or depolymerization of an oligomer or polymer, respectively. McEwen's photoactive sulfonium or iodonium salts are the usual proton source. The oligomer or polymer must be transparent at the wavelength of irradiation. Organic polymers have dominated this approach to resist chemistry, but when shorter wavelengths (e.g., 157-nm laser) are used for further diminution of the device features, most organic materials absorb in that region. However, polysiloxanes are transparent in this region and are in the process of being developed for this spectral region (59).

Sulfur- and metal-containing polymers also produce much higher G values than organic resist polymers (where G is the number of chemical events per 100 eV absorbed) (60–62). Because of earlier disasters with alkali metals in the vacuum electron-beam systems, a reluctance to put metal-containing polymers in such systems has negated the use of the high-G value metal-containing polymers as electron-beam resists.

4.7 INORGANIC POLYMERS AS PRECERAMICS

Inorganic polymers can be designed to provide nonoxide ceramics through pyrolysis (63–67) as shown schematically in Eq. 4.7.

$$\text{–MY–X–(MY–X)}_{n-1} + \text{heat} \rightarrow \text{nMX(s)} + n/2\text{Y}_2 \tag{4.7}$$

where M = Si and X = C or M = Al or B and X = N, etc. and Y_2 = volatile byproduct. This can also be extended to other ceramics such as Si_3N_4 and B_4C.

4.7.1 Silicon Carbide from Polycarbosilanes

Controlled pyrolysis of organic polymers is used to prepare commercial carbon materials such as glassy carbon foams, graphitic cloth, fibers, and carbon-carbon composites. In fact, this is the only reasonable processing path to some of those materials. Using the same approach for inorganic ceramics (instead of high-temperature sintering or melt processing) might provide fibers, foams, and coatings as well as normal bulk solids for a variety of ceramic materials. In fact, Yajima developed routes to silicon carbide from polysilanes via the Kumada rearrangement (Eq. 4.8) that led to the commercial silicon carbide fiber, Nicalon.

$$-\underset{\displaystyle CH_3}{\overset{\displaystyle CH_3}{\overset{|}{\underset{|}{Si}}}}-Si\equiv \xrightarrow{\geq 400°C} -\underset{\displaystyle CH_3}{\overset{\displaystyle H}{\overset{|}{\underset{|}{Si}}}}-CH_2-Si\equiv \tag{4.8}$$

The rearrangement is critical to the generation of a polycarbosilane that can be spun and further heated to generate silicon carbide fibers. The commercial procedure involves heating in one or two steps to 470 °C, thus the precursor polymer is labeled PC-470. Proof of the rearrangement is evident from infrared spectra that develops Si–C–Si bands at 1020 and 1355 cm^{-1} and a Si–H band at 2100 cm^{-1} and from the ultraviolet spectra that show the disappearance of the $\sigma - \sigma^*$ absorption of the polysilane. However, the typical elemental analysis of PC-470 is $SiC_{1.77}O_{0.03}H_{3.7}$ rather than the anticipated SiC_2H_6 (or $SiC_2H_{6.10}$ based on the short chains with hydride ends). The ^{29}Si NMR experiments indicate two different Si units, one corresponding to Si bonded to three C atoms and one H atom (as anticipated for the polycarbosilane) and one corresponding to Si bonded to four C atoms. This latter type suggests extensive crosslinking and hydride loss.

The pyrolysis of PC-470 gives typical ceramic yields of about 60%. The theoretical ceramic yield for pure polycarbosilane to pure silicon carbide would be 69%. For the carbon- and hydrogen-deficient PC-470 preceramic polymer that is used, the yield should be even higher. The pyrolysis chemistry of PC-470 has been studied in detail and shows the anticipated changes: Si–H bonds disappear, then $Si–CH_3$ bonds and $Si–CH_2–Si$ groups as crosslinking and loss of methane and hydrogen occur. The ^{29}Si–NMR shows that almost all silicon atoms are now surrounded by four carbon atoms, and by 800 °C the material is essentially SiC with 2 CH groups per Si plus some excess carbon. At 1000 °C the X-ray diffraction pattern shows some β-SiC with crystallites about 2 nm in size by tunneling electron microscopy (TEM) plus some amorphous SiC-like particles plus some aromatic carbon layers piled two to three layers high. By 1200 °C the amorphous SiC phase disappears and the aromatic carbon layers form oriented rims of graphitic carbon around the SiC crystals. Even at the end of the pyrolysis (1500 °C), although the crystal size has increased to about 5–13 nm, the excess carbon prevents further improvement.

For commercially satisfactory fibers, heating in air cures the PC-470. This provides Si–O–Si crosslinks from 2 Si–H groups as shown in Eq. 4.9.

$$2{\equiv}Si\text{–}H + O_2 \rightarrow {\equiv}Si\text{–}O\text{–}Si{\equiv} + H_2O \qquad (4.9)$$

The result is a cured preceramic fiber that gives 80–85% ceramic yields (about 20% higher than the uncured PC-470). The conversion of the cured polymer to the ceramic occurs between 500 and 800 °C, and the oxygen is retained. Thus, Nicalon is not pure SiC. It contains 10–15% oxygen and consists of a continuous Si–C–O major phase with about half of the Si plus small SiC crystallites of less than 3 nm and free graphitic carbon particles of less than 1 nm in size. Although much of the oxygen can be removed as CO above 1200 °C and larger β-SiC crystallites ($\sim$7 nm) occur at 1400 °C. However, these changes have deleterious effects on the fiber properties.

The Yajima or Nicalon polymer has good spinnability, has a high ceramic yield, and has high mechanical properties for a ceramic fiber. Nicalon fibers of 10- to 20-micron diameter have tensile strengths of $>$2.3 GPa and elastic moduli of $\sim$200 GPa. (At 1200 °C the tensile strength drops to 0.56 GPa and the elastic moduli drop 50% by 1500 °C.) The disadvantages include the multistep synthetic procedure required for this fiber synthesis, the undefined polymer structure, the oxidative curing step, and the presence of both free C and the oxygen from the curing step. Even so, it does show that ceramic fibers can be prepared from inorganic polymers.

Improvements on the Nicalon silicon carbide seem obvious and many have been tried. However, the most successful was Interrante and co-workers' use of a preceramic polymer that has just one C per Si. The obvious polymers are the polycarbosilane, poly(silaethylene) or $(–CH_2–SiH_2–)_n$, and polymethylsilane $(SiCH_3H)_n$. The first polymer, when pure, provides a high ceramic yield of 85–90% and forms pure SiC below 1000 °C (68, 69). In its less pure form in which the $(–CH_2–SiCl_2–)_n$ precursor is prepared by a Grignard reaction with Cl_3SiCH_2Cl, the hydride polymer prepared by $LiAlH_4$ reduction has $SiH_3CH_2–$, $–SiH_2CH_2–$, $=SiHCH_2–$, and $\equiv SiCH_2–$ units in the approximate ratios of 2:8:20:11. This hyperbranched polymer with a carbosilane backbone is then crosslinked with either vinyl or allyl groups. At the 2.5–10% levels, these crosslinks do not give any substantial excess carbon on pyrolysis at 1000 °C with a 70–80% overall ceramic yield. Because of the high yield, the stability in air, and its fluidity, the product has found applications as a matrix source for silicon-carbide-fiber-reinforced/SiC matrix composites as well as for joining silicon carbide ceramics. The allyl crosslinked preceramic polymer is being manufactured by Starfire Systems in Watervliet, NY. The pure version, synthesized by ring-opening polymerization of the dimeric (Cl_2SiCH_2) with a K_2PtCl_6 catalyst followed by the same $LiAlH_4$ reduction, is too expensive for commercialization, even with its higher ceramic yield.

The other 1:1 possibility, polymethylsilane $(SiCH_3H)_n$, can provide 70–75% ceramic yields of fairly pure SiC at 900 °C, although it contains some of the $(C_5H_5)_2Ti(CH_3)_2$ or analogous Zr catalyst used in the polymer synthesis. A slight variation using a $[(CH_3SiH)_{0.8}(CH_3Si)_{0.2}]_n$ crosslinked copolymer and 0.6% $(C_5H_5)_2ZrH_2$ catalyst in hexane under reflux for 2 h caused even more crosslinking (the fraction of CH_3SiH units decreased from 0.8 to about 0.6). Pyrolysis under argon (1500 °C, 5 h) of the orange solid residue after solvent removal produced a black ceramic in 72% ceramic yield with 98% SiC, 1.6% ZrC, and 0.4% Si with good properties.

Other examples of SiC precursors can be given, such as the use of the thermolysis of $[Si(i\text{-butyl})_2CH_2]_n$ to approximate the $(SiH_2CH_2)_n$ composition. About 75% of the i-butyl groups are eliminated as i-butene, producing a copolymer of about half (SiH_2CH_2) and about half $[Si(i\text{-butyl})HCH_2]$ units at 420 °C, and further heating eliminates the remaining i-butyl groups and hydrogen to yield a nearly stoichiometric silicon carbide at about 900 °C. Other modifications include the deliberate addition of another tetravalent metal, titanium, to disrupt crystallization for uses requiring amorphous ceramic type material. Note that simple linear polymers generally fail to provide a substantial ceramic yield unless a precuring crosslinking step is involved.

4.7.2 Silicon Nitride Preceramic Polymers

Synthesizing usable silicon nitride with preceramic polymers has also had its problems. One natural problem is the lack of a good preceramic polymer with a 3:4 Si: N stoichiometry. Instead, polysilazanes with a 1:1 Si: N stoichiometry plus an ammonia atmosphere during the heating phase have provided Si_3N_4 as shown schematically in Eq. 4.10.

$$H_2SiCl_2 \cdot 2\text{pyridine} \xrightarrow{NH_3} \text{“}\{H_2SiNH\}_n\text{”} \xrightarrow{NH_3,\ \Delta} Si_3N_4 \qquad (4.10)$$

Dichlorosilane without the pyridine adduct has been known to detonate under some conditions, the intermediate polymer is undoubtedly more complicated than indicated by the formula given, and ammonia or hydrazine can be used to adjust the Si: N ratio during pyrolysis. Tonen Corporation has produced silicon nitride fibers of about 10-micron diameter that have high (2.5 GPa) tensile strength and a high (250 GPa) tensile modulus using this method. A drop in the mechanical properties occurs $\geq 1300\,^{\circ}C$ as crystallization of the fibers occurs analogous to the Nicalon SiC observations.

A wide variety of carbon-containing silicon/nitrogen preceramic polymers have also taken advantage of the ammonia atmosphere. For example, CH_3SiHCl_2 plus ammonia gives cyclic $(CH_3SiHNH)_n$ oligomers that form crosslinked polysilazane polymers when treated with potassium hydride (and hydrogen loss). These polymers can be represented as $[(CH_3SiHNH)_a(CH_3SiN)_b]_n$. Pyrolysis of these polymers in an inert gas stream yields a black silicon carbonitride. Actually, evidence of crystalline SiC and Si_3N_4 and C is obtained on heating to $1500\,^{\circ}C$. However, in an ammonia gas stream, white Si_3N_4 with less than 0.5% C is obtained with evidence of reaction starting at $400\,^{\circ}C$. The carbon "kick out" or "burn out" reaction:

$$C + NH_3 \xrightarrow[\sim 1000\,^{\circ}C]{} HCN + H_2 \tag{4.11}$$

The decomposition of ammonia to nitrogen and hydrogen is competitive; thus, excess ammonia is necessary. The nitride can be used as a binder for ceramic powders, in coating formulations, and as fibers by dry-spinning.

4.7.3 Other Preceramic Polymers

Boron nitride fibers, films, and composites are usually prepared by classical methods, although considerable effort has been expended to provide modified properties using preceramic polymers. Crosslinking borazine $(\text{–BH–NH–})_3$, inorganic benzene, with NH bridges takes advantage of the volatility of nitrogen compounds to provide stoichiometric BN. Boranes are functionalized with unsaturated organic groups to provide a site for crosslinking. These boranes react with functionalized amines or ammonia to provide preceramic polymers that can be pyrolyzed under ammonia to prepare BN materials. The reason for the interest in alternate routes to BN fibers, for example, is that the ammonia treatment and hot stretching at $2200\,^{\circ}C$ required by the classical method is quite expensive. Synthetic pathways for boron-containing preceramic polymers are shown in Figure 4.15 (70). Polymers (II), (IV), (VI), and (VII) provide BN when pyrolyzed. Polymers (II) and (IV) can be pyrolyzed in an inert atmosphere, but (VI) and (VII) require pyrolysis at $1000\,^{\circ}C$ under ammonia to obtain BN. Polymer (VIII) provides a Si–B–C ceramic plus some SiO_2 on pyrolysis, and polymer (IX) provides $Si_3B_3N_7$.

Analogous reactions for the synthesis of aluminum-containing preceramic polymers are shown in Figure 4.16 (70). Precursors labeled (III) and (IV) are

i. Polyvinylborazine:

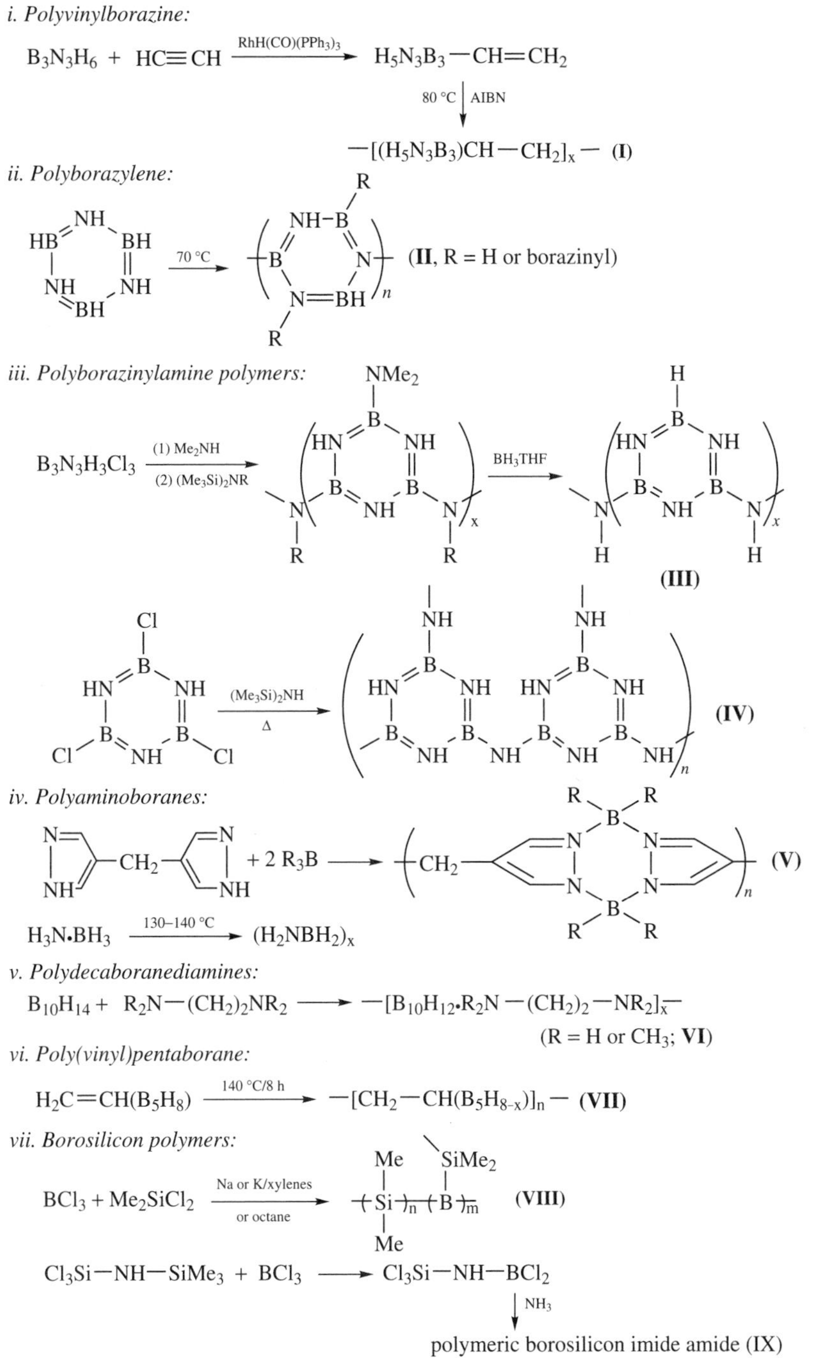

$$B_3N_3H_6 + HC\equiv CH \xrightarrow{RhH(CO)(PPh_3)_3} H_5N_3B_3{-}CH{=}CH_2 \xrightarrow[\text{AIBN}]{80\,°C} {-}[(H_5N_3B_3)CH{-}CH_2]_x{-} \quad \textbf{(I)}$$

ii. Polyborazylene:

$$B_3N_3H_6 \xrightarrow{70\,°C} \textbf{(II, R = H or borazinyl)}$$

iii. Polyborazinylamine polymers:

$$B_3N_3H_3Cl_3 \xrightarrow[(2)\ (Me_3Si)_2NR]{(1)\ Me_2NH} \xrightarrow{BH_3THF} \textbf{(III)}$$

$$\xrightarrow[\Delta]{(Me_3Si)_2NH} \textbf{(IV)}$$

iv. Polyaminoboranes:

$$+\ 2\ R_3B \longrightarrow \textbf{(V)}$$

$$H_3N{\cdot}BH_3 \xrightarrow{130\text{–}140\,°C} (H_2NBH_2)_x$$

v. Polydecaboranediamines:

$$B_{10}H_{14} + R_2N{-}(CH_2)_2NR_2 \longrightarrow {-}[B_{10}H_{12}{\cdot}R_2N{-}(CH_2)_2{-}NR_2]_x{-}$$

(R = H or CH_3; **VI**)

vi. Poly(vinyl)pentaborane:

$$H_2C{=}CH(B_5H_8) \xrightarrow{140\,°C/8\,h} {-}[CH_2{-}CH(B_5H_{8-x})]_n{-} \quad \textbf{(VII)}$$

vii. Borosilicon polymers:

$$BCl_3 + Me_2SiCl_2 \xrightarrow[\text{or octane}]{\text{Na or K/xylenes}} {+}Si(Me)_2{+}_n{+}B(SiMe_2){+}_m \quad \textbf{(VIII)}$$

$$Cl_3Si{-}NH{-}SiMe_3 + BCl_3 \longrightarrow Cl_3Si{-}NH{-}BCl_2 \xrightarrow{NH_3} \text{polymeric borosilicon imide amide (IX)}$$

Figure 4.15 Synthetic procedures for boron-containing preceramic polymers (Reprinted with permission from Ref. 70; © CRC Press, Boca Raton, Florida).

i. Reactions between organoaluminum and NH_3 or amines:*

$$R_3Al + NH_3 \longrightarrow [R_3Al\ NH_3] \xrightarrow{\Delta} \xrightarrow{\Delta} (RAlNH)_x \text{ [e.g. } EtAlNH)_n(Et_2AlNH_2)_m(AlEt_3)_x \text{ (I)]}$$

$$R_3Al + N_2N(CH_2)_2NH_2 \longrightarrow R_3AlNH_2(CH_2)_2NH_2AlR_3 \xrightarrow{\Delta} \xrightarrow{\Delta} [RAlNH(CH_2)_2NHAlR]_x$$

(e.g. $[Et_2AlNH(CH_2)_2NHEt_2]_x$ (II), when R = Et and ring structures, i.e.,

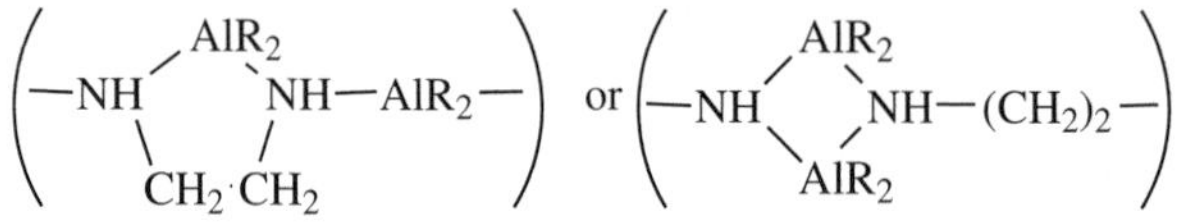

$\left(-NH \ldots AlR_2 \ldots NH-AlR_2- ;\ CH_2 \cdot CH_2 \right)$ or $\left(-NH \ldots (AlR_2)_2 \ldots NH-(CH_2)_2- \right)$ were also postulated.

ii. Polyalazane (polyaminoalane, polyiminoalane):

$$R_2AlH + R'CN \xrightarrow{\Delta} R'CN{=}NAlR_2]_2 \xrightarrow{\Delta} \text{Polyalazane } (R' = CH_3, Ph, etc.)$$

(e.g. $\{[EtNAlC_4H_9]_x[EtNAl(R'')]_y[EtNAl]_z\}_n$ (III))

$$Al + 3\ NH_2R \longrightarrow Al(NHR)_3 + 3/2\ H_2 \text{ (electrochemical reaction with Al)}$$

$$Al(NHR)_3 \xrightarrow{\Delta} -(RN-AlNHR)_x- \xrightarrow{\Delta} \text{Polymeric Precursor (IV, when R = n-propyl)}$$

$$LIAlH_4 + RNH_3Cl \longrightarrow 1/n\ (AlNHR)_n + 3\ H_2 + LICl\ (n = 2 \text{ to at least } 35)$$

iii. Polyaluminosilazane

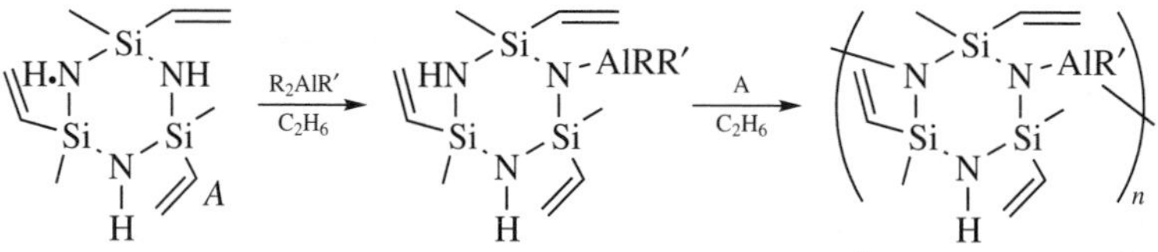

(Cyclic silazane A $\xrightarrow[C_2H_6]{R_2AlR'}$ N-AlRR' substituted silazane $\xrightarrow[C_2H_6]{A}$ polymer $(\ldots N\text{-}AlR' \ldots)_n$)

(R = Et, R′ = Et or EtO; V)

Figure 4.16 Synthetic procedures for aluminum-containing preceramic polymers. The asterisk on heading *i*: amines including aromatic di-, tri-, and tetraamines have been used (70). (Reprinted with permission from Ref. 70; © CRC Press, Boca Raton, Florida).

aluminum nitride precursors and provide a good product when pyrolyzed under NH_3. Precursors (I), (II), and (V) give multielement products or composites.

Composite ceramics, such as Si–B–N, C–Si–N, and C–Si–B–N ceramics, have been investigated by several groups. The reader is referred to reviews and on-line searching for more information on these and other combinations. Good chemistry, good ceramics, and serendipity undoubtedly will all continue to be important in the future of this intriguing field.

4.8 INORGANIC POLYMER CONDUCTIVITY

4.8.1 Main Group Inorganic Polymers

Polythiazyl [or poly(sulfur nitride)] (6, 71) is a bronze-colored solid with a metallic luster that conducts electrons quite efficiently. As anticipated for a linear polymer that is an almost planar zig-zag structure, its conductivity is quite anisotropic: $\sigma_{\parallel} = 3700$ S cm^{-1} and $\sigma_{\parallel}/\sigma_{\perp} \approx 50$ at room temperature. The $\sigma_{\parallel}$ value increases about 1000-fold when the $(SN)_x$ polymer is cooled to 4.2 K, and then at about 0.3 K the polymer becomes superconducting. At high pressures the superconductivity transition temperature can be raised to about 3 K. The quality of the solid determines the actual conductivity values and the transition temperature for superconductivity. The room temperature $\sigma_{\parallel}$ value of a poorly prepared sample might only be 1000 S cm^{-1}, and the superconductivity might not start until 0.26 K. {The SI unit for conductivity is siemens per meter (siemens = ohm^{-1}) or S m^{-1}, but S cm^{-1} or ohm^{-1} cm^{-1} are more commonly used.}

Partial bromination of $(SN)_x$ with bromine vapor yields blue-black single crystals of $(SNBr_{0.4})_x$ that have a room temperature conductivity of about 38,000 S cm^{-1} parallel to the fiber axis, which is almost an order of magnitude greater than that of the parent $(SN)_x$ polymer, but only about 8 S cm^{-1} perpendicular to the fiber axis. Further removal of bromine at 80 °C *in vacuo* yields a copper-colored $(SNBr_{0.25})_x$ species. The extremely high $\sigma_{\parallel}/\sigma_{\perp}$ ratio for the brominated polymer is thought to be due to a lower degree of order in chain packing perpendicular to the fiber axis.

A wide variety of uses have been suggested for polythiazyl $(SN)_x$, including light-emitting diodes, solar cells, and transitors (6, 72). A polythiazyl-GaAs solar cell with 6.2% efficiency has been produced, and blue light-emitting diodes have been made on ZnS with either gold or polythiazyl. The polythiazyl-ZnS combination has a 100-fold increase in quantum efficiency relative to the gold-ZnS diode. A power output of 440 watt-hours/kg at 200-mA current has been obtained for a lithium cell with a polythiazyl electrode. Patents have been issued for the use of polythiazyl as a battery cathode, coatings on an image recording sheet, an explosive initiator, an electroconductive resin, multilayered wiring patterns, and light-emitting diodes. A number of other electrode uses have also been suggested. Of course, uses not related to its conductivity have been suggested, including using it as an etching mask and as sublimable polishing grains for semiconductor devices.

Almost all other main-group inorganic polymers are either insulators or semiconductors with measurable band gaps unless doped either to remove electrons from the filled valence band or to add electrons to the empty upper conduction band. That is, for metal like conduction the top electron energy level of a material (the Fermi level) must be only a partially filled level or band. A schematic energy level diagram for the electron occupation of metals, semimetals, semiconductors, and insulators is shown in Figure 4.17 (73). Although metals with their loosely held valence electrons have conductivities of the order of 10^4 or 10^5 S cm^{-1}, molecular materials have lower conductivities ranging from insulators ($\sigma < 10^{-9}$ S cm^{-1}) to semiconductors and up to metals ($\sigma > 10^3$ S cm^{-1}). Although the metals and semimetals have partially filled bands at the Fermi level, semiconductors and insulators have filled bands and an energy gap to a level that is empty.

The temperature dependence of the two classes varies as well. Metals show an increase in conductivity as the temperature decreases, whereas a semiconductor

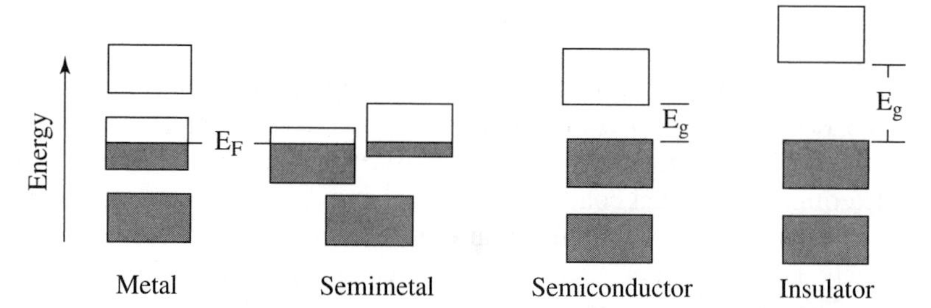

Figure 4.17 The schematic illustration of the electron occupation of the allowed energy bands for a metal, semimetal, semiconductor, and insulator. The shaded bands are filled levels up to the Fermi level (E_F). E_g is the energy gap between the valence and conduction bands (73).

shows an increase in conductivity as the temperature increases because more electrons will have the thermal energy to bridge the gap. On the other hand, metal conductivity is limited by lattice vibrations (phonons) that decrease as the temperature decreases, thus allowing better electron mobility that translates to increased conductivity.

By doping a semiconductor, the band gap can be narrowed and the conductivity increased. For example, polysilanes and polystannanes are insulators, but with doping, the conductivity can be increased up to 0.5 S/cm (6) and 0.3 S/cm (74), respectively.

Alternatively, side group substitution, for example, on an aromatic ring of a poly(arylmethylsilane), can change the energy band gap to a lesser extent (down as much as 0.14 eV from the unsubstituted phenyl polymer band gap of about 3.4 eV) (75). The ability to modify the band gap is important for the many possible uses envisioned for polysilanes, such as hole and photoconduction, electroluminescence, microlithography (noted above), and thermochromism to name a few of the possibilities.

The silicon, germanium, and tin oxo phthalocyanine polymers $[–M(Pc)–O–]_n$ (Fig. 1.14c) and the aluminum and gallium fluoro phthalocyanine polymers $[–M(Pc)–F–]_n$ (also Fig. 1.14c) are excellent examples of the extreme changes that doping can make in an inorganic polymer (76–79). An increase in conductivity of at least five orders of magnitude occurs with oxidative iodine doping. In fact, conductivities as high as almost 1 S/cm have been noted for the smaller cations, and stacks of the ligand alone produce even higher conductivities (700 S/cm at ambient temperature) that increase as the temperature is lowered (to 3500 S/cm at 1.5 K). Such an increase is consistent with metallic-type electron conduction.

4.8.2 Metal-Containing Polymers

This leads us into the extensive work of Hanack and his co-workers on oligomeric and polymeric metal phthalocyanines and related macrocycles (80–83). Through the use of bridging ligands such as tetrazine (or 1,2,4,5-tetrazabenzene) and

TABLE 4.5 Inherent Conductivities of Iron and Ruthenium Oligomers/Polymers[a].

Oligomer/Polymer	Conductivity(S cm^{-1})
$[PcFe(pyz)]_n$	1×10^{-6}
$[PcFe(tz)]_n$	2×10^{-2}
$[NcFe(pyz)]_n$	5×10^{-5}
$[NcFe(tz)]_n$	3×10^{-1}
$[PcRu(pyz)]_n$	1×10^{-7}
$[PcRu(tz)]_n$	1×10^{-2}
$[NcRu(pyz)]_n$	7×10^{-3}
$[NcRu(tz)]_n$	4×10^{-2}

[a] Pc = phthalocyaninato; Nc = 2, 3-naphthalocyaninato; pyz = pyrazine; tz = tetrazine. Powder conductivity measurements at 298 K using 4-probe method. The low solubility of the parent complexes suggests that only oligomers are formed, but no proof of molecular mass was given by Hanack and Polley, *op cit*.

ligands such as 2,3-naphthalocyaninato (with an extra aromatic ring on each ligand quadrant relative to phthalocyaninato), **inherent conductivities** up to 0.3 S/cm are possible for iron(II) polymers *without doping*.

The inherent conductivities of several polymers (or oligomers) of such macrocycles with iron(II) and ruthenium(II) are shown in Table 4.5. Because the length of these oligomers or polymers is not known, a detailed comparison is unwise. Even so, the increases tetrazine relative to pyrazine and 2,3-naphthalocyaninato relative to phthalocyaninato are general, suggesting a very low energy gap for the tetrazine bridging ligand and good interchain overlap for the naphthalene groups. Bridging cyano ligands also provide reasonable low energy gaps. For example, one solubilized 2,3-naphthalocyaninatocobalt(II) shish kebob polymer with bridging cyano ligands has an inherent conductivity of almost 10^{-1} S/cm (84).

One of the basic problems in using linear polymers as conductors involves the limited size of the polymeric molecule compared to the distance over which the electron must move for the conductivity to be observed. For example, in metal coordination polymers in which the bridging ligand is conjugated and should possess no barrier to electron mobility, the conductivity of the doped polymers is not significantly different from those of polymers in which a saturated methylene bridge occurs in the bridging ligand (cf. Table 4.6) (85). In fact, the conductivity of the zirconium polymer with the conjugated benzidine derivative (value in bold face in Table 4.6) is slightly higher than the zirconium polymers with less conjugated structures (either methylene or sulfone bridged); however, when the polymers are doped with iodine to increase the conductivity, the nonconjugated methylene bridged polymer has the higher conductivity (value in bold face in Table 4.6). For the cerium(IV) polymers, no trends are evident. Thus the bulk conductivity is *not* a function of the conjugation of the individual polymer molecules. Therefore, interchain resistance determines the bulk conductivity.

TABLE 4.6 Conductivity of Cerium and Zirconium Coordination Polymer Films[a].

Polymer	Conductivity ($S\ cm^{-1}$)	Conductivity (I_2 doped) ($S\ cm^{-1}$)
$[Zr(tsdb)]_n$	$\mathbf{5 \times 10^{-7}}$	5×10^{-4}
$[Zr(tsts)]_n$	4×10^{-7}	3×10^{-3}
$[Zr(tstm)]_n$	2×10^{-7}	$\mathbf{5 \times 10^{-3}}$
$[Ce(tsdb)]_n$	4×10^{-7}	1×10^{-3}
$[Ce(tsts)]_n$	1×10^{-7}	2×10^{-3}
$[Ce(tstm)]_n$	4×10^{-7}	1×10^{-3}

[a] H_4tsdb = N, N′, N″, N‴-tetrasalicylidene-3,3′-diaminobenzidine; H_4tsts = N, N′, N″, N‴-tetrasalicylidene-3,3′, 4,4′-tetraaminodiphenylsulfone; H_4tstm = N, N′, N″, N‴-tetrasalicylidene-3,3′,4,4′-tetraaminodiphenylmethane; cf. Figure 2.12 for ligand structures and synthesis of cerium derivatives. Conductivity measurements at 298 K. Iodine doping in closed container.

Similar observations have been made by other investigators of planar d^8 phosphine coordination polymers (86) and phthalocyanine and related macrocycle metal coordination polymers (82). On the other hand, 1,1′-bis(8-cyclopentadienyl-1-naphthyl)ferrocene polymers that have good conjugation increase from 10^{-12} S cm^{-1} to 7×10^{-3} S cm^{-1} when oxidatively doped with iodine to a 37% level (87). Other workers found that nonconjugated ferrocene units pendant to the backbone (i.e., polymeric vinylferrocene) increase in conductivity to only about 5×10^{-5} S cm^{-1} from about 10^{-10} S cm^{-1} (88).

Poly[tetrathiooxalatometallate(II)] low-molecular-mass metal-coordination oligomers behave as metallic conductors (89–91). Poly[tetrathiooxalatometallate(II)] ($\overline{DP} = 6$)$^{\otimes}$ has a conductivity of 20 S/cm *without doping* (89). Its transport properties suggest that it is a true metallic electron conductor. Unfortunately, only oligomers can be obtained before the oligomer precipitates as an insoluble material. The copper analogue with an average degree of polymerization of almost 15* has a conductivity of 3 S/cm and appropriate transport properties for an electron conductor. Several tetrathiolate derivatives of iron(II), cobalt(II), nickel(II), copper(II), palladium(II), and platinum(II) have been synthesized, but almost all of them are black insoluble oligomers. The derivatives include tetrathiosquarate, tetrathiofulvalene tetrathiolate, benzene-1,2,4,5-tetrathiolate, and naphthalene tetrathiolate in addition to tetrathiooxalate. Solubility can be increased by adding organic spacers, but the conductivity becomes very low (10^{-5}–10^{-6} S/cm) (92). Improved syntheses of the original tetrathiolates should provide much better conducting polymers.

$^{\otimes}$ The $\{[(C_2S_4)Ni]_3(C_2S_4)\}^{2-}$ anion by analysis requires 6 polymerization steps; c.f., Fig. 2.2.
* The authors used the $\overline{n}$ value of 7.8 as $\overline{DP}$. For step-growth polymerization $\overline{DP} = 2\overline{n} - 1$ or 14.6; cf. Fig. 2.2 and Section 2.1.

The future should bring new approaches to the preparation of conducting inorganic polymers. One approach would be to use conducting counter ions, analogous to the use of tetracyanoethene (TCNE) and tetracyano-*p*-quinodimethane (TCNQ) as bridges in oligomeric organometallic systems (93). Also, Manners suggests that some of the ring-opening polymerization organometallics can be designed to provide anisotropic semiconductors (94, 95).

4.9 NONLINEAR OPTICS METAL-CONTAINING POLYMERS

Metal-containing polymers may have an edge on third-order nonlinear optics (NLO) materials in the future. Some of the rigid-rod polymers appear to exceed the currently used third-order materials (96).

The basic concepts of NLO have been described by Marder in some detail (96). The equations generated in explaining the nonlinear effects at the molecular level are often simplified (97) to:

$$(\mu - \mu_o) = \alpha \cdot \mathrm{E} + \beta \cdot \mathrm{EE} + \gamma \cdot \mathrm{EEE} \tag{4.12}$$

where

μ = the total dipole moment

μ_o = the permanent dipole moment

E = the applied external electric field

α = linear polarizability

β = second-order polarizability (also called first-order hyperpolarizability)

γ = third-order polarizability (also called second-order hyperpolarizability)

When E is obtained from an applied optical field, the terms β and γ represent the nonlinear response of the molecule.

At the bulk level the analogous expansion is

$$\mathrm{P} = \chi^{(1)} \cdot \mathrm{E} + \chi^{(2)} \cdot \mathrm{EE} + \chi^{(3)} \cdot \mathrm{EEE} \tag{4.13}$$

where

P = the bulk polarization

$\chi^{(n)}$ = the *n*th-order susceptibilities

E = applied energy as before

Ordinary polarization is well understood as related to “loose” electrons, as in π-electrons and electrons in large atoms and ions.

The second order (or β term) is related to the asymmetry of the electronic field of the molecule and only becomes important with high-energy laser radiation. Extended aromatic molecules with a donor group at one end and an acceptor group at the other end (e.g., an amine and a nitro group, respectively) are very effective. To be effective in the bulk, however, the molecules must align themselves to obtain the desired additive effect. Also, it is helpful if the molecules are transparent at the laser wavelength. The second-order effect can be used to obtain frequency doubling through second harmonic frequency mixing, light-modulation through a linear electro-optical effect, and intensity enhancement through optical rectification. It has been suggested (98) that the one-dimensional coordination polymer $KTiO(PO_4)$ (also known as KTP) has a high NLO coefficient, a high optical damage threshold, a low threshold power, and a low phase-matching sensitivity that should make it better than the two materials currently used, KH_2PO_4 and $LiNbO_3$.$KTiO(PO_4)$ consists of a zigzag Ti=O–Ti=O–chain with unequal bond lengths (<1.75 vs. >2.10 Å) to give the noncentrosymmetric structure required for second-order NLO. Phosphate groups complete a crosslinked distorted octahedral structure about each titanium(IV) ion in the $(\text{–Ti=O})_n$ chain.

Other one-dimensional metal coordination polymers have been investigated as well. For example, $[(\text{salen})M^{III}\text{-}\mu\text{-}O_3SCH_2NHC_5H_5N]_n$, where M = Cr, Mn, Fe, Co, shows small amounts of second-order NLO properties, but the adjacent salen rings in each chain are almost perpendicular to each other, which causes a diminished efficiency.

The third-order (γ) term is responsible for the generation of new frequencies using third harmonics. The nature of the best types of molecules is still not totally settled; however, linear or rigid-rod metal-containing polymers show considerable promise (96). Other solid-state polymeric coordination compounds are also quite promising (98, 99) as well as composites consisting of inorganic oxides and conjugated polymers (97) and selected phthalocyaninato metal species (83). This is an area in which inorganic polymers may become increasingly important.

4.10 LUMINESCENT INORGANIC POLYMERS

Although a number of polymeric inorganic solids exhibit luminescence, we will restrict our discussion to synthetic linear polymers.

4.10.1 Ruthenium Polymers for Solar Energy Conversion

Ruthenium(II) polymer research for solar energy conversion has progressed less than the research and applications of monomer ruthenium(II) coordination compounds on semiconductors (100) and research on dendrimer ruthenium(II) coordination species (101, 102). The problem of having over 2.1 eV per photons fed into the ruthenium coordination system when a photon is absorbed means that sufficient energy is available for bond breaking within the coordination species as well as for "splitting" water. (Water splitting theoretically needs only 1.23 eV.) The large amount of energy available tends to cause the slow decomposition ruthenium species. Conversion of one diimine to 8-quinolinolato lowers the transition to 1.85 eV (103), and encapsulation of ruthenium(II) within a hexaimine cryptate can slow the decomposition to a negligible level

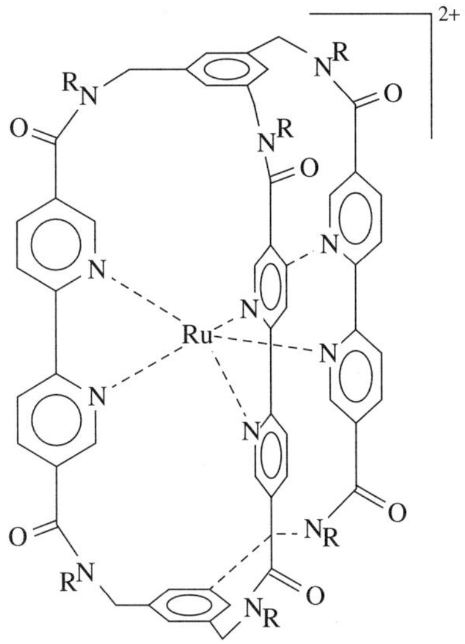

Figure 4.18 Encapsulated ruthenium coordination compound to avoid slow photo-decomposition.

(Fig. 4.18). However, none of the polymer systems suggested to date seems to have used more than tridentate ligands. Examples of research progress over the past few years include a number of electropolymerization studies by Meyer. For example, poly[4-(2-aminoethyl)styrene], prepared by living anionic polymerization with about 18 repeat units has been derivatized by amide coupling to [Ru^{II}(4-vinyl-4′-methyl-2,2′-bipyridine)$_2$(4-methyl-2,2′-bipyridine-4′-carboxylic acid)]$^{2+}$. The product PS-[(CH_2CH_2NHCO-bpy-4-methyl)(4-vinyl-4′-methyl-2,2′-bipyridine)$_2Ru^{II}]_{18}^{36+}$ undergoes reductive electropolymerization (at the vinyl groups — Type III type) and gives open porous films that appear to retain the electron-transfer and excited-state properties of the original ruthenium species slightly modified by the film (104). A number of related Type III species have also been investigated by Meyer and others as well (105–107). Two examples and an energy-minimized structure for one of the anchored ruthenium(II) species are shown in Figure 4.19.

Also, the potential applications of chiral ruthenium(II) polymers to induce chiral photosynthesis and to provide optically active luminescence have brought forth new chiral ruthenium coordination polymers. As noted in Chapter 3, stereoselective condensation reactions between Δ- and Λ-bis(1,10-phenanthroline-5,6-dione)bipyridineruthenium(II) ions and Δ- and Λ-bis(1,10-phenanthroline-5,6-diamine)bipyridineruthenium(II) ions produces homochiral Δ- and Λ-coordination polyelectrolytes and a meso Δ,Λ-coordination polyelectrolyte (Fig. 4.20) (108). Tor and Glaser have also prepared chiral ruthenium(II) polyelectrolytes (109), and a review article on this synthetic concept for ruthenium(II) oligomers from dimers to decamer dendrimers has appeared (110).

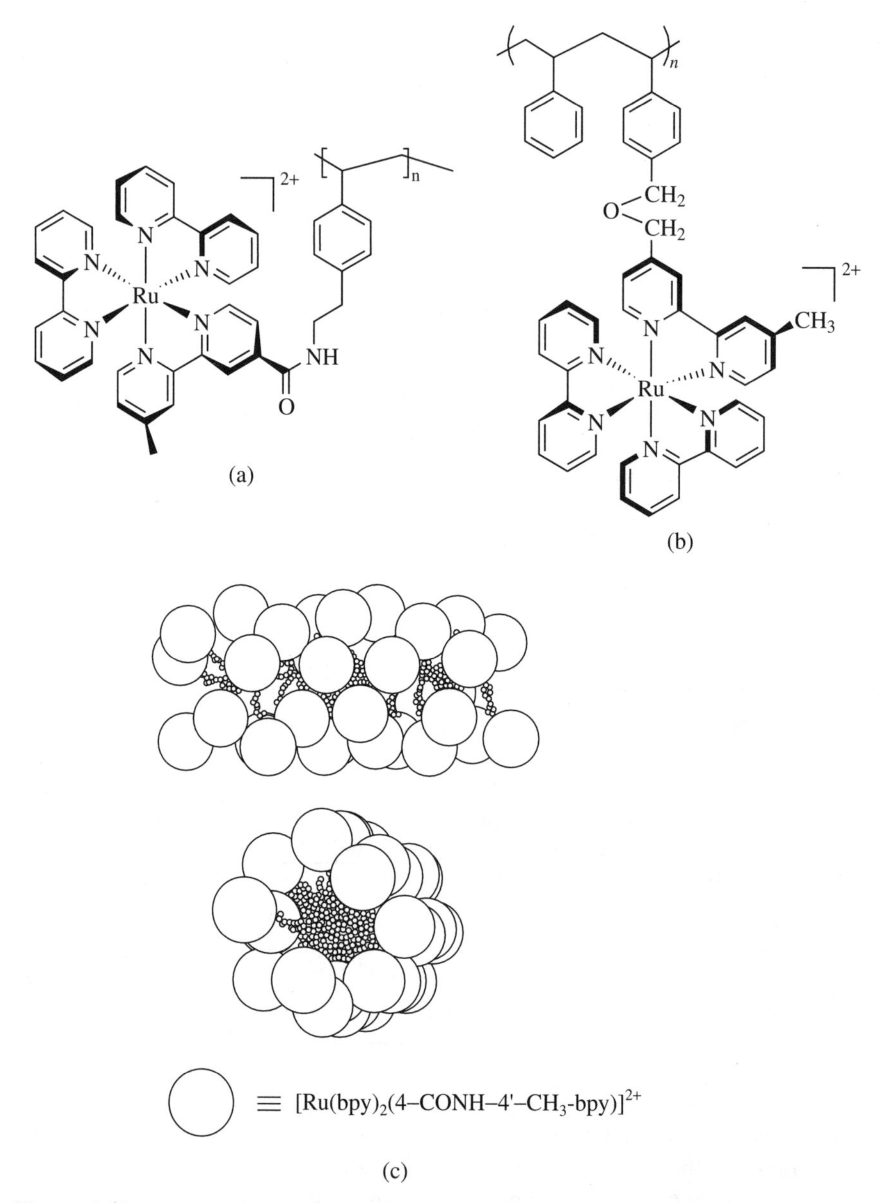

Figure 4.19 Anchored ruthenium(II) polymers: (a) amide linkage to bis(2,2′-bipyridine)(4-carbonyl-4′-methyl-2,2′-bipyridine)ruthenium(II) (reprinted with permission from Ref. 106; © 1998 American Chemical Society); (b) similar species with ether linkage (reprinted with permission from Ref. 105; © 1999 American Chemical Society); and (c) energy-minimized end-on and side views of polystyrene amide linked to a ruthenium polymeric cation, probably a $[(PS\text{-}4\text{-derivatized-}4'\text{-methyl-}2,2'\text{-bipyridine})(2,2'\text{-bipyridine})_2Ru^{II}]_{30}^{60+}$ type ion, based on modified MM2 parameters and CAChe software (reprinted with permission from Ref. 106; © 1998 American Chemical Society). [(c) is not the $Ru^{II}]_{18}^{36+}$ type ion claimed in the paper — more than 18 Ru centers are shown.]

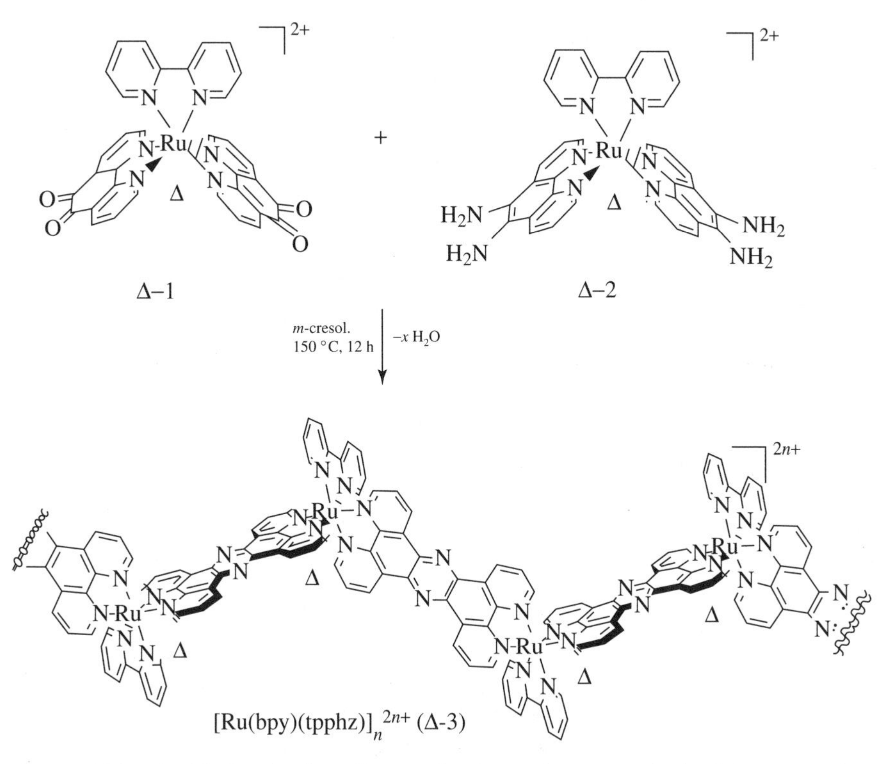

Figure 4.20 A chiral ruthenium(II) coordination polymer synthesized in *m*-cresol at 150 °C for 12 h under nitrogen (Ref. 108; reproduced by permission of The Royal Society of Chemistry).

4.10.2 Other Luminescent Metal Polymers

A brief mention of other luminescent metal polymers was made in Chapter 3 (Section 3.4.6). Although early in the 1960s a push was made to make lasers with europium(III) monomeric complexes, the tetrakis(2,4-penatanedionato)europium species used was too labile to be practical. More recently, anchored macrocyclic catalysts have been suggested ($t_{1/2} \geq 10^6$ s), and even the tetradentate Schiff-base species appear to be quite inert with $t_{1/2} \geq 10^4$ s (111).

4.10.3 Silicon Luminescent Materials

Novel blue-light-emitting molecules containing 3,3,3′,3′-tetramethyl-3,3′-disila-indeno[2,1-a]indene units were synthesized by the intramolecular addition of disilanes to acetylenes. Whereas the chromophore is on the side chain, polysiloxane polymers with this unit as a side chain have been synthesized and characterized (112). The structure is shown in Figure 4.21.

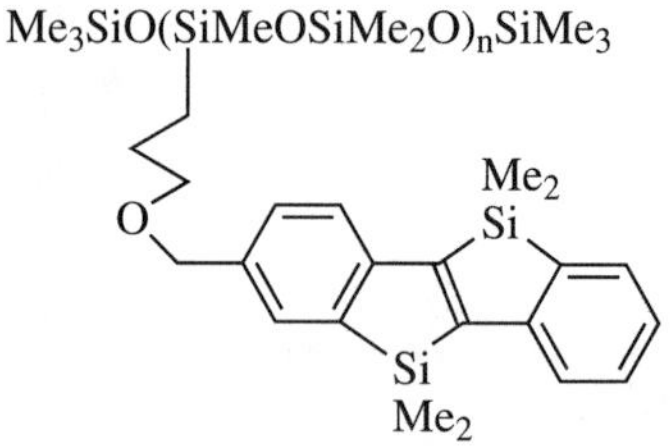

$Me_3SiO(SiMeOSiMe_2O)_nSiMe_3$

O, Me_2Si, SiMe_2

Figure 4.21 A luminescent silicon polymer.

4.11 MAGNETIC METAL-COORDINATION POLYMERS

All metal-coordination and organometallic polymers that have unpaired electrons are termed paramagnetic. Of particular interest here are the cooperative magnetic effects that can be obtained in metal-coordination and organometallic polymers; that is, ferromagnetic, antiferromagnetic, and ferrimagnetic. A summary of magnetic ordering is shown schematically in Figure 4.22. Ferromagnetic ordering thermally changes to the random orientation of paramagnetism at the Curie temperature, T_C.

The temperature at which antiferromagnetic and ferrimagnetic polymers change to random paramagnetism is called the Neel temperature. Thus all ordered magnetic systems convert to random orientations if the temperature is high enough. Naturally, some species decompose before reaching that temperature.

Because ferromagnetic and antiferromagnetic effects both depend on order, rigid-rod metal-coordination polymers provide excellent vehicles for large anisotropic magnetic properties. The anisotropic behavior of several rigid-rod polymers is shown in Figure 4.23.

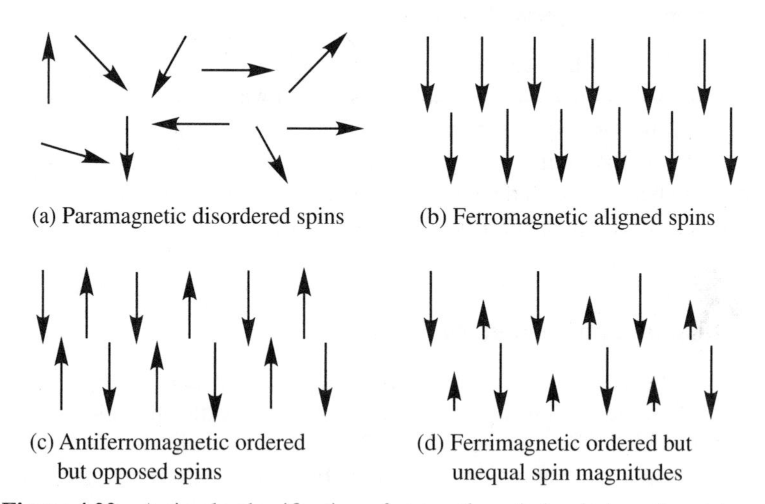

Figure 4.22 A simple classification of magnetic ordering in two dimensions.

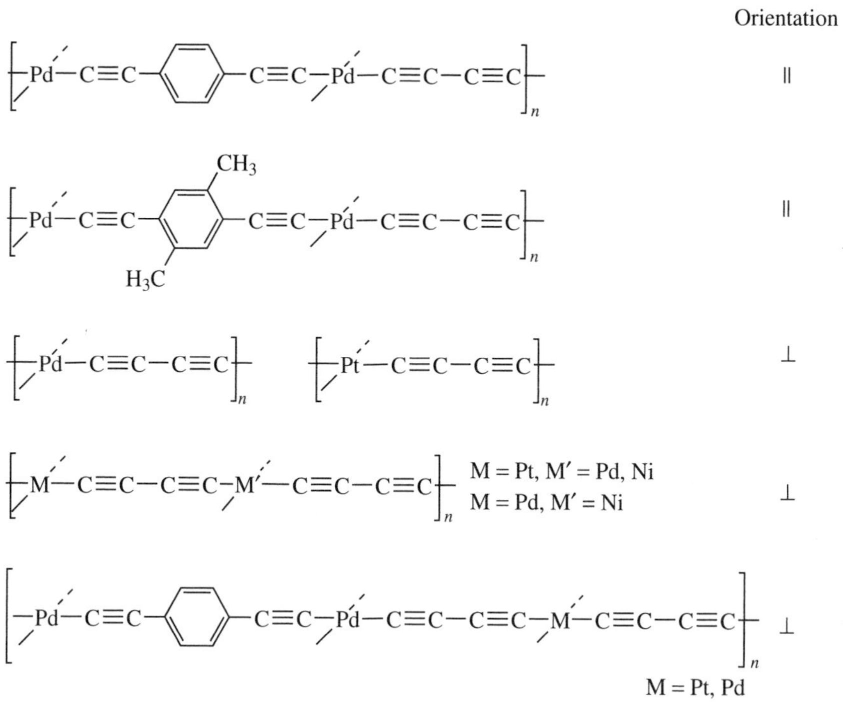

Orientation

Pd—C≡C—C6H4—C≡C—Pd—C≡C—C≡C, n — ||

Pd—C≡C—(CH3)2C6H2—C≡C—Pd—C≡C—C≡C, n — ||

Pd—C≡C—C≡C, n; Pt—C≡C—C≡C, n — ⊥

M—C≡C—C≡C—M′—C≡C—C≡C, n; M = Pt, M′ = Pd, Ni; M = Pd, M′ = Ni — ⊥

Pd—C≡C—C6H4—C≡C—Pd—C≡C—C≡C—M—C≡C—C≡C, n; M = Pt, Pd — ⊥

Figure 4.23 Anisotropic magnetism in rigid-rod polymers. The orientation of the chains relative to a strong magnetic field is shown in the right column.

An area of exploration that may find practicality in magnetic device applications is the spin-change that some octahedral iron(II) centers undergo as a function of temperature. Because iron(II) is a d^6 ion, it can exist in a diamagnetic t_{2g} ground state when surrounded by strong-ligand-field ligands. However, if the iron(II) is just barely spin-paired at room temperature, heating can convert it to the spin-free paramagnetic state with four unpaired electrons per iron(II) center. In polymeric molecules, such as the one shown in Figure 4.24, a concerted process of spin-change occurs as the spin-change also causes a size change (113). This provides considerable hysteresis, as shown in the figure. The hysteresis loop is valuable for computer memory applications, where laser heat is used to flip the state. The width of the loop is twice the coersive field, the term that is commonly used when discussing such devices. A much larger value (and one that spans room temperature) would be needed to provide useful computer memory devices. However, the need for uniform magnetic particles of smaller and smaller size is currently pushing the limit for solids (114) and may eventually lead to molecular based magnetic systems. Metal coordination polymers provide a logical basis for this application.

Another approach to new magnetic materials that involves organometallic polymers is the thermolysis of the poly(ferrocenylsilanes) obtained from the

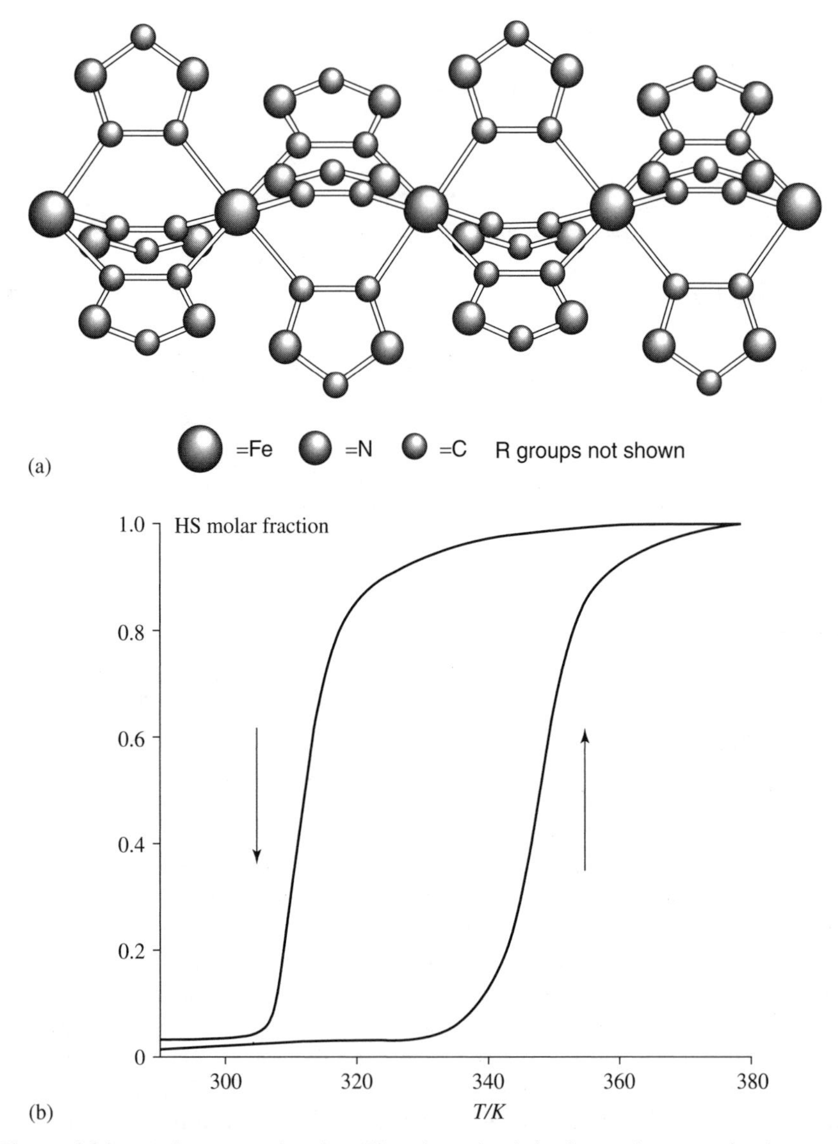

Figure 4.24 (a) Structure of an iron(II) polymeric chain that undergoes spin-change as the temperature is changed; and (b) the temperature dependence of the high-spin (HS) molar fraction of the polymer as a nitrate salt (113). Reproduced from *Chemistry in Britain* by permission of the Royal Society of Chemistry.

R_2Si bridged ferrocenes by ring-opening synthesis (Section 2.3.2). Manners and his co-workers claim to be able to make shaped ceramics that can be tuned between superparamagnetic and ferromagnetic states by controlling the pyrolysis conditions with iron nanoclusters homogeneously dispersed in a carbosilane-graphic-silicon nitride matrix (115). Similarly, they found that pyrolyzing the dimethyl [1]silaferroceneophane inside channels of mesoporous silica (MCM-41) produced superparamagnetic iron nanoparticles of 50–64 Å coated with a thin 4- to 6-Åoxide layer (116). Earlier they had noted that hydrocarbon bridged ferrocene polymers form ferromagnetic iron carbide ceramics at 600 °C and when oxidized with tetracyanoethylene (TCNE) significant antiferromagnetic interactions could be observed. These may be just the beginning of a wide range of iron-containing ceramics with variable magnetic products.

4.12 INORGANIC POLYMERS AS CATALYSTS

We looked at the structures of two anchored metal catalysts early in Chapter 1 (Fig. 1.2). However, with the exception of a few cases in which polysiloxanes and polyphosphazenes are used as the polymer chains with ferrocenyl or dye side chains, anchored or immobilized catalysts fall in the realm of organic polymers. Nevertheless, the immobilization of catalysts on polymer supports is a growth area. The idea is to combine the advantages of homogeneous and heterogeneous catalysis. Standard homogeneous catalysts often have a problem with catalyst recovery. In fact, sometimes the catalyst is just left in the polymer, even when it contains a platinum metal, because it is too difficult to remove the catalyst. Homogeneous and anchored catalysts are used for many types of reactions because they are more active than heterogeneous catalysts. Types of reactions performed well by such catalysts include hydrogenation, oxidation, polymerization, and activation of small molecules (117). Readers interested in anchored catalysts on organic polymer substrates can find a number of review articles and at least two books written about anchored catalysts and translated from earlier Russian editions on synthesis (118) and applications (117).

One recent example of a polyphosphazene with ferrocenyl side groups being used as a glucose detector merits special attention. Electrochemical studies have shown that these functionalized polymers function as effective electron-transfer mediators for accurate determination of glucose using carbon paste electrodes containing glucose oxidase. The use of the enzyme precludes interferences (22, 52).

Earlier use of poly[bis(aryloxy)phosphazene] for immobilizing enzymes is a classic example of a potential commercial use for polyphosphazenes. The advantages noted above are true here, too. Immobilizing the enzyme should make it easier to separate the products from the enzyme. Also, if done in a porous medium like alumina, the reactants will stay in contact with the enzyme longer. In one case a thin film of the polyphosphazene was deposited on the surface of highly porous alumina. The phenoxy groups of the polyphosphazene were nitrated with 90% nitric acid and reduced with sodium dithionite or hydrosulfite ($Na_2S_2O_4$), and the enzyme was then coupled with glutaric dialdehyde $CHOCH_2CH_2CH_2CHO$, which condenses with both the amine of the polymer and an amine on the enzyme (132).The enzymes trypsin and glucose 6-phosphate dehydrogenase were immobilized in this fashion. The authors project use in biochemical and fermentation industries and medical diagnostic equipment (see the preceding paragraph!) and eventually for the construction of an artificial liver. In a related earlier study, reduction of a nitrophenoxy group to an amine used hydrogen with a platinum dioxide catalyst. The amine was diazotized with nitrous acid and hydrochloric acid and then dopamine, *dl*-norepinephrine, or *dl*-epinephrine was coupled to the polyphosphazene. The activity was retained according to animal tests (119).

Polysilanes have also been noted previously as catalytic species for radical polymerizations (Chapter 2). They have been used as polymerization catalysts for making high-molecular weight-vinyl polymers (120). We have also noted the photocatalysis of the ruthenium(II) species in Section 4.10.1 and the projected use of other ions for the photo activation. Undoubtedly, more catalytic uses of great magnitude will come to fruition in the future.

4.13 MISCELLANEOUS USES

We have not exhausted the multitude of uses to which inorganic polymers are being put or the projected future uses for them. For example, siloxane liquid crystals can be synthesized, with the most important applications being retardation plates, cholesteric reflectors and filters, storage materials for optical information, agents for preventing unauthorized copying of documents, and pigments for iridescent and polarizing coatings (121). Ferroelectric liquid crystal polymers including silicon in the main chain of the polymers also have been reported (122). A number of examples of metal coordination polymers as liquid crystals also exist, including copper(II) (123) and erbium polymers (124).

Silica aerogels can serve as nanoglues (125) and as oxygen sensors when incorporating fluorescent 2,7-diazapyrenium moieties (126). Oxygen can be sensed near the speed of open-air diffusion with this aerogel system.

Langmuir-Blodgett films of a copper Schiff-base coordination polymer can be used for hydrocarbon vapor sensing (127). Two dimensional arrays of pyridine substituted porphyrins linked with palladium dichloride bridges (128), copper "wreaths" with 12 copper(I) ions held in interlocking organic strands (128), and dendrimers with up to 54 metal ions (129–131) (and going higher each time a new report appears) all make the future of inorganic polymers look rosy indeed.

REFERENCES

1. Stevens, M. P. in Polymer Chemistry: *An Introduction*; 3rd ed., Oxford University Press: Oxford, 1999, pp 425–446.
2. Sheats, J. E., Carraher, C. E., Jr., Pittman, C. U., Jr., Zeldin, M., Culbertson, B. M. in *Metal-Containing Polymeric Materials*; Pittman, C. U., Jr., Carraher, C. E., Jr., Zeldin, M., Sheats, J. E. and Culbertson, B. M., Ed., Plenum Press: NY, 1996, pp 3–37.
3. Roy, A. K. in *Kirk-Othmer Encyclopedia of Chemical Technology*; Kroschwitz, J. I. and Howe-Grant, M., Ed., John Wiley & Sons: New York, 1995; Vol. 14, pp 504–23.

4. Mark, J. E., Allcock, H. R., West, R. *Inorganic Polymers*; Prentice Hall: Englewood Cliffs, NJ, 1992.
5. Allcock, H. R., Lampe, F. W. in *Contemporary Polymer Chemistry*; Prentice-Hall: Engelwood Cliffs, NJ, 1990, pp 196–231 and references therein.
6. Carraher, C. E., Pittman, C. U., Jr. "Inorganic Polymers," preprint of review for Ullmann's Encyclopedia of Industrial Chemistry revision, 2000.
7. Clarson, S. J., Mark, J. E. in *Polymeric Materials Encyclopedia*; Salamone, J. C., Ed., CRC Press: Boca Raton, 1996; Vol. 10, pp 7663–77.
8. Anonymous *Reactive Silicones: Forging New Polymer Links* pamphlet; 2.0 ed., Gelest, Inc.: Tullytown, PA, 1999.
9. Auner, N., Fearon, G., Weis, J. in *Organosilicon Chemistry III: From Molecules to Materials*; Wiley-VCH: Weinheim, 1998, pp 471ff Part II Silicon Based Materials.
10. Rich, J., Cella, J., Lewis, L., Stein, J., Singh, N., Rubinsztajn, S., Wengrovius, J. in *Kirk-Othmer Encyclopedia of Chemical Technology*; 4th ed., Kroschwitz, J. I. and Howe-Grant, M., Ed., John Wiley & Sons: New York, 1997; Vol. 22, pp 82–142.
11. Clarson, S. J., Mark, J. E. in *Polymeric Materials Encyclopedia*; Salamone, J. C., Ed., CRC Press: Boca Raton, 1996; Vol. 10, pp 7663–77; *cf.* also Liles, D. T., pp 7694–99; Grant, R. D., pp 7699–7705; Tomanek, pp 7706–11.
12. West, R. in *Encyclopedia of Inorganic Chemistry*; King, R. B., Ed., John Wiley & Sons: New York, 1994; Vol. 6, pp 3389–3404.
13. Mark, J. E. in *Silicon-Based Polymer Science: A Comprehensive Resource*; Zeigler, J. M. and Fearon, F. W. G., Ed., American Chemical Society: Washington, DC, 1990; Vol. 224 Advances in Chemistry Series, pp 47–68.
14. Lawson, D. F. in *Handbook of Polymer Science and Technology*; Cheremisinoff, N. P., Ed., Marcel Dekker, Inc.: New York, 1989; Vol. 2, pp 203–42.
15. Arkles, B. *CHEMTECH* 1983, **13**, 542.
16. Allcock, H. R. *Polym Preprints* 2000, **41**, 553.
17. Matyjaszewski, K., White, M. L. in *Polymeric Materials Encyclopedia*; Salamone, J. C., Ed., CRC Press: Boca Raton, FL, 1996; Vol. 9, pp 6556–63.
18. Wisian-Neilson, P. in *Encyclopedia of Inorganic Chemistry*; King, R. B., Ed., John Wiley & Sons: Chichester, 1994; Vol. 6, pp 3371–89.
19. Books, J. T. in *Kirk-Othmer Encyclopedia of Chemical Technology*; Kroschwitz, J. I. and Howe-Grant, M., Ed., John Wiley & Sons: New York, 1993; Vol. 8, pp 1022–30.
20. Mark, J. E. Allcock, H. R., West, R. in *Inorganic Polymers*; Mark, J. E., Allcock, H. R. and West, R., Ed., Prentice Hall: Englewood Cliffs, NJ, 1992, pp 61–140.
21. Allen, C. W., Hneihen, A. S. *Phosphor. Sulfur, Silicon* 1999, **146**, 213.
22. Allen, C. W., Myer, C. N. "Phosphazene polymers containing the redox active N-methyl-N-ferrocenylamino substituent," Northeast Regional Meeting, American Chemical Society, Storrs, CT, June 2000, paper 212.
23. Allen, C. W. *J. Fire Sci.* 1993, **11**, 320.
24. Pape, P. G. in *Polymeric Materials Encyclopedia*; Salamone, J. C., Ed., CRC Press: Boca Raton, 1996; Vol. 10, pp 7636–39.
25. Palmer, R. A., Klosowski, J. M. in *Kirk-Othmer Encyclopedia of Chemical Technology*; Kroschwitz, J. I. and Howe-Grant, M., Ed., John Wiley & Sons: NY, 1997; Vol. 21, pp 654–66.

26. Mittal, K. L. *Silanes and Other Coupling Agents*; Mittal, K. L., Ed., VSP: Utrecht, 1992.
27. Plueddemann, E. P. *Silane Coupling Agents*; 2nd ed., Plenum Press: New York, 1991.
28. Arkles, B. *CHEMTECH* 1977, **7**, 766.
29. McCarthy, T. J., *Chimia* 1990, **44**, 316 and references cited therein.
30. Wang, B., Archer, R. D. *Polym. Mater. Sci. Engr.* 1988, **59**, 120; 1988, **60**, 710.
31. Archer, R. D., Wang, B. *Polym. Mater. Sci. Engr.* 1989, **61**, 101.
32. Archer, R. D., Wang, B. *Chem. Mater.* 1993, **5**, 317.
33. Archer, R. D., Tong, W., Tan, U.-K. *Poly. Mater. Sci. Engr.* 1996, **75**, 383.
34. Byrd, H., Hollowary, C. E., Pogue, J. *Polym. Prepr.* 1999, **40(1)**, 167.
35. Katz, H. E., Schilling, M. L., Chidsey, C. E. D., Putvinski, T. M., Hutton, R. S. *Chem. Mater.* 1991, **3**, 699.
36. Tesk, J. A., Antonucci, J. M., Eichmiller, F. C., Kelly, J. R., Rupp, N. W., Waterstrat, R. W., Fraker, A. C., Chow, L. C., George, L. A., Stansbury, J. W., Parry, E. E. in *Kirk-Othmer Encyclopedia of Chemical Technology*; Kroschwitz, J. I. and Howe-Grant, M., Ed., John Wiley & Sons: NY, 1993; Vol. 7, pp 996–1022.
37. Teraoka, F., Takahashi, J. *Dent. Mater.* 2000, **16**, 145 and references cited therein.
38. Haug, S. P., Andres, C. J., Moore, B. K. *J. Prosthetic Dentistry* 1999, **81**, 431.
39. Matsumura, H. in *Polymer Materials Encyclopedia*; Salamone, J. C., Ed., CRC Press: Boca Raton, 1996, pp 1834, 1836 lists 16 silanes and 1 siloxane oligomer used as dental adhesives.
40. Zhang, C., Laine, R. M. *Polym. Prepr.* 1997, **38(2)**, 120 and references therein.
41. Antonucci, J. M., Fowler, B. O., Sansbury, J. W. *Polym. Prepr.* 1997, **38(2)**, 118.
42. Paul, P. P., Timmons, S. F., Machowski, W. J. *Polym. Prepr.* 1997, **38(2)**, 124.
43. Wei, Y., Jin, D. *Polym. Prepr.* 1997, **38(2)**, 122.
44. Ellingsen, J. E., Rolla, G. *Scand. J. Dent. Res.* 1994, **102**, 26.
45. Lindén, L.-Å., Rabek, J. F. *J. Appl. Polym. Sci.* 1993, **50**, 1331.
46. Prosser, J. J., Powis, D. R., Wilson, A. D. *J. Dent. Res.* 1986, **65**, 146.
47. Kunzler, J., Ozark, R. *J. Appl. Polym. Sci.* 1997, **65**, 1081.
48. Lai, Y.-C., Wilson, A. C., Zantos, S. G. in *Kirk-Othmer Encyclopedia of Chemical Technology*; Kroschwitz, J. I. and Howe-Grant, M., Ed., John Wiley & Sons: New York, 1993; Vol. 7, pp 200–218.
49. Alvord, L., Court, J., Davis, T., Morgan, C. F., Schindhelm, K., Vogt, J., Winterton, L. *Optometry Vision Sci.* 1998, **75**, 30.
50. Gill, I., Ballesteros, A. *J. Am. Chem. Soc.* 1998, **120**, 8587.
51. Hendry, S. P., Cardosi, M. F., Turner, A. P. F., Neuse, E. W. *Anal. Chim. Acta* 1993, **281**, 453.
52. Myer, C. N., Allen, C. W. *Polym. Prepr.* 2000, **41(1)**, 558.
53. Siegmann, D. W., Brenner, D., Colvin, A., Polner, B. S., Strother, R. E., Carraher, C. E., Jr. in *Inorganic and Metal-containing Polymeric Materials*; Sheats, J. E., Carraher, C. E., Jr., Pittman, C. U., Jr., Zeldin, M. and Currell, B., Ed., Plenum Press: New York, 1990, pp 335–361 and references therein.

54. Arkles, B. *Silicon, Germanium, Tin and Lead Compounds: Metal Alkoxides, Diketonates, and Carboxylates*; Gelest, Inc.: Tullytown, PA, 1998.
55. Hinsberg, W. D., Wallraff, G. W., Allen, R. D. in *Kirk Othmer Concise Encyclopedia of Chemical Technology*; Kroschwitz, J. I. and Howe-Grant, M., Ed., John Wiley & Sons, Inc.: New York, 1999, pp 1208–1211.
56. Reichmanis, E., Neenan, T. X. in *Chemistry of Advanced Materials: An Overview*; Interrante, L. V. and Hampden-Smith, J. J., Ed., Wiley-VCH, Inc.: New York, 1998, pp 99–141.
57. Reichmanis, E., Novembre, A. E., Tarascon, R. G., Shugard, A., Thompson, L. F. in *Silicon-based polymer science: a comprehensive resource*; Zeigler, J. M. and Fearon, F. W. G., Ed., American Chemical Society: Washington, DC, 1990; Vol. 224 Advances in Chemistry Series, pp 265–281.
58. Miller, R. D., Rabolt, J. F., Sooriyakumaran, R., Fleming, W., Fickes, G. N., Farmer, B. L., Kuzmany, H. in *Inorganic and Organometallic Polymers: Macromolecules containing Silicon, Phosphorus, and Other Inorganic Elements*; Zeldin, M., Wynne, K. J. and Allcock, H. R., Ed., American Chemical Society: Washington, DC, 1987; No. 360 in ACS Symposium Series, pp 43–60.
59. Willson, C. G. "Polymers for microlithography: The race goes on," American Chemical Society National Meeting, New Orleans, March 2000, MACR 44.
60. Archer, R. D., Tramontano, V. J., Ochaya, V. O., West, P. V., Cumming, W. G. in *Inorganic and Metal-Containing Polymeric Materials*; Carraher, C. E., Jr., Pittman, C. U., Jr., Sheats, J. E., Zeldin, M. and Currell, B., Ed., Plenum Publishing Co.: New York, 1991, pp 161–71.
61. Archer, R. D., Hardiman, C. J., Lee, A. Y. in *Photochemistry and Photophysics of Coordination Compounds*; Yersin, H. and Vogler, A., Ed., Springer-Verlag: Berlin, 1987, pp 285–290.
62. Archer, R. D., Hardiman, C. J., Grybos, R., Chien, J. C. W., *Transition and inner transition metal chelate polymers for high energy lithographic resists*, U. S. Patent no. 4,693,957, Sept. 15, 1987.
63. Livage, J., Sanchez, C., Babonneau, F. in *Chemistry of Advanced Materials*; Interrante, L. V. and Hampden-Smith, M. J., Ed., Wiley-VCH, Inc.: New York, 1998, pp 429–448.
64. Seyferth, D. in *Silicon-based polymer science: a comprehensive resource*; Zeigler, J. M. and Fearon, F. W. G., Ed., American Chemical Society: Washington, DC, 1990; Vol. 224 Advances in Chemistry Series, pp 565–591.
65. Atwell, W. H. in *Silicon-based polymer science: a comprehensive resource*; Zeigler, J. M. and Fearon, F. W. G., Ed., American Chemical Society: Washington, DC, 1990; Vol. 224 Advances in Chemistry Series, pp 593–606.
66. Kanner, B., King, R. E. I. in *Silicon-based polymer science: a comprehensive resource*; Zeigler, J. M. and Fearon, F. W. G., Ed., American Chemical Society: Washington, DC, 1990; Vol. 224 Advances in Chemistry Series.
67. Shen, Q., Interrante, L. V., "Polycarbosilanes," in *Silicon-containing Polymers: The Science and Technology of Their Synthesis and Applications*; Ando, W., Chojnowski, J., and Jones, R. Ed., Kluwer Publishing Co.: NY, 2000, in press.
68. Whitmarsh, C. K., Interrante, L. V., *Silicon carbide from polycarbosilane*, U. S. Patent no. 5,153,295, 1992.

69. Liu, Q., Wu, H. J., Lewis, R., Maciel, G. E., Interrante, L. V. *Chem. Mater.* 1999, **11**, 2038.
70. Mariam, Y. H., Feng, K. in *Polymeric Materials Encyclopedia*; Salamone, J. C., Ed., CRC Press: Boca Raton, FL, 1996; Vol. 9, pp 7215–28.
71. Greenwood, N. N., Earnshaw, A. in *Chemistry of the Elements*; Pergamon Press: Oxford, 1984, pp 860–861.
72. Banister, A. J., Gorrell, I. B. *Adv. Mater.* 1998, **10**, 1415 based on Carraher and Pittman, 2000, *op cit.*
73. Cassoux, P., Miller, J. S. in *Chemistry of Advanced Materials: An Overview*; Interrante, L. V. and Hampden-Smith, J. J., Ed., Wiley-VCH, Inc.: New York, 1998, pp 19–72.
74. Imori, T., Lu, V., Cai, H., Tilley, T. D. *J. Am. Chem. Soc.* 1995, **117**, 9931.
75. Cleij, T. J., King, J. K., Jenneskens, L. W. *Macromolecules* 2000, **33**, 89–96.
76. Pietro, W. J., Marks, T. J., Ratner, M. A. *J. Am. Chem. Soc.* 1985, **107**, 5387.
77. Marks, T. J. *Angew. Chem. Int. Ed. Engl.* 1990, **29**, 857.
78. Schramn, C. J., Scaringe, R. P., Stojakovic, D. R., Hoffmann, K. M., Ibers, J. A., Marks, T. J. *J. Am. Chem. Soc.* 1980, **102**, 6702.
79. Kuznesof, P. M., Nohr, R. S., Wynne, K. J., Kenney, M. E. *J. Macromol. Sci.-Chem.* 1981, **A16**, 299.
80. Hanack, M. in *Metal-Containing Polymeric Materials*; Pittman, C. U. J., Carraher, C. E., Jr., Zeldin, M., Sheats, J. E. and Culbertson, B. M., Ed., Plenum Press: New York, 1996, pp 331–336.
81. Hanack, M. *Synthetic Metals* 1995, **71**, 2275; also *New Matls: Conj. Dbl. Bond Syst.* 1995, **191**, 13.
82. Hanack, M., Pohmer, J. *Polym. Mater. Sci. Eng.* 1994, **71**, 391.
83. Schultz, H., Lehmann, H., Rein, M., Hanack, M. in *Mertal Complexes with Tetrapyrrole Ligands II*; Buchler, J. W., Ed., Springer-Verlag: Berlin, 1990; Vol. Structure and Bonding 74, pp 41–146.
84. Hanack, M., Polley, R., Knecht, S., Schlick, U. *Inorg. Chem.* 1995, **34**, 3621.
85. Archer, R. D., Chen, H., Cronin, J. A., Palmer, S. M. in *Metal-Containing Polymeric Materials*; Pittman, C. U., Jr., Carraher, C. E., Jr., Zeldin, M., Sheats, J. E. and Culbertson, B. M., Ed., Plenum Press: New York, 1996, pp 81–91.
86. Wang, P.-W., Fox, M. A. *Inorg. Chem.* 1994, **33**, 2938.
87. Rosenblum, M., Nugent, H. M., Jang, K. S., Labes, M. M., Cahalane, W., Klemarczyk, P., Reiff, W. M. *Macromolecules* 1995, **28**, 6330.
88. Neef, C. J., Glatzhofer, D. T., Nicholas, K. M. *J. Polym. Sci. A-Polym Chem.* 1997, **35**, 3365.
89. Reynolds, J. R., Chien, J. C. W., Lillya, C. P. *Macromolecules* 1987, **20**, 1184.
90. Hoyer, E., Synthesis of metal tetrathiooxalates as conducting polymers, private communication, 1983.
91. Lund, J., Hoyer, E., Hazell, R. G. *Acta Chem. Scand.* 1982, **B36B**, 207.
92. Wang, F., Reynolds, J. R. *Macromolecules* 1990, **23**, 3219.
93. Diaz, C., Arancibia, A. *Polyhedron* 2000, **19**, 137.
94. Manners, I. *Pure Appl. Chem.* 1999, **71**, 1471.
95. Manners, I. *Ang. Chem. Intl. Ed. Engl.* 1996, **35**, 1602.

96. Marder, S. R. in *Inorganic Materials*; Bruce, D. W. and O'Hare, D., Ed., John Wiley & Sons: Chichester, 1992; Vol. 1, pp 116–164.
97. Prasad, P. N., Bright, F. V., Narang, U., Wang, R., Dunbar, R. A., Jordan, J. D., Gvishi, R. in *Hybrid Organic-Inorganic Composites*; Mark, J. E., Lee, C. Y.-C. and Bianconi, P. A., Ed., American Chemical Society: Washington, DC, 1995; Vol. ACS Symposium Series 585, pp 317–330.
98. Chen, C.-T., Suslick, K. S. *Coord. Chem. Rev.* 1993, **128**, 293.
99. Lin, W., Evans, O. R., Wang, Z., Ma, L. "Coordination polymer approach to second-order nonlinear optical materials," American Chemical Society National Meeting, Anaheim, CA, March 1999, INOR 50.
100. Kalyanasundaram, K., Grtzel, M. *Coord. Chem. Rev.* 1998, **177**, 347.
101. Venturi, M., Serroni, S., Juris, A., Campagna, S., Balzani, V. in *Dendrimers*; Vögtle, F., Ed., Springer: Berlin, 1998; *Structure and Bonding*, Vol. 197, pp 193 and references cited therein.
102. Issberner, J., Vogtle, F., DeCola, L., Balzani, V. *Chem.-Eur. J.* 1997, **3** , 706.
103. Warren, J. T., Johnston, D. H., Turro, C. *Inorg. Chem. Commun.* 1999, **2** , 354 based on Web of Science abstract.
104. Leasure, R. M., Kajita, T., Meyer, T. J. *Inorg. Chem.* 1996, **35**, 5962.
105. Worl, L. A., Jones, W. E., Strouse, G. F., Younathan, J. N., Danielson, E., Maxwell, K. A., Sykora, M., Meyer, T. J. *Inorg. Chem.* 1999, **38**, 2705 and references 6–14.
106. Friesen, D. A., Kajita, T., Danielson, E., Meyer, T. J. *Inorg. Chem.* 1998, **37**, 2756 and references 1–20 therein.
107. Carraher, C. E., Jr., Murphy, A. T. *Polym. Mater. Sci. Engr.* 1997, **76**, 409.
108. Chen, J., MacDonnell, F. M. *J. Chem. Soc. Chem. Commun.* 1999, 2529.
109. Tor, Y., Glazer, E. C. *Polym. Prepr.* 1999, **40(1)**, 513.
110. MacDonnell, F. M., Kim, M.-J., Bodige, S. *Coord. Chem. Rev.* 1999, **185**, 535.
111. Hatwell, K. R., Ph. D. Dissertation, University of Massachusetts, Amherst, 1999.
112. Ma, Z. X., Ijadi-Maghsoodi, S., Barton, T. J. *Polym. Prepr.* 1997, **38(2)**, 249.
113. Kahn, O. *Chem. Britain* 1999, February, 24.
114. Jacoby, M. *Chem. Engr. News* 2000, **78(24)**, 37.
115. MacLachlan, M. J., Ginzburg, M., Coombs, N., Coyle, T. W., Raju, N. P., Greedan, J. E., Ozin, G. A., Manners, I. *Science* 2000, **287**, 1460.
116. MacLachlan, M. J., Ginzburg, M., Coombs, N., Raju, N. P., Greedan, J. E., Ozin, G. A., Manners, I. *J. Am. Chem. Soc.* 2000, **122**, 3878.
117. Pomogailo, A. D. *Catalysis by Polymer-Immobilized Metal Complexes*; Gordon and Breach: The Netherlands, 1998.
118. Pomogailo, A. D., Savost'yanov, V. S. *Synthesis and Polymerization of Metal-Containing Monomers*; CRC Press: Boca Raton, 1994.
119. Allcock, H. R., Hymer, W. C., Austin, P. E. *Macromolecules* 1983, **16**, 1401.
120. Pittman, C. U., Jr., Carraher, C. E., Jr., Reynolds, J. R. in *Encyclopedia of Polymer Science and Engineering*; Kroschwitz, J. I., Mark, H. F., Bikales, N. M., Overberger, C. G. and Menges, G., Ed., John Wiley and Sons: NY, 1987; Vol. 10, pp 541–594.

121. Kreuzer, F.-H., Haberle, N., Liegeber, H., Maurer, R., Stohrer, J., Weis, J. in *Organosilicon Chemistry III. From Molecules to Materials*; Auner, N. and Weis, J., Ed., Wiley-VCH: Weinheim, 1998, pp 567–585.
122. Cooray, N. F., Fujimoto, H., Kakimoto, M., Yoshio, I. *Macromolecules* 1997, **30**, 3169.
123. Serrano, J. L., Oriol, L. *Polym. Prepr.* 1996, **37(1)**, 56.
124. Haase, W., Soto Bustamante, E. A., Grossmann, S., Werner, R., Galyametdinov, Y. G. *Polym. Prepr.* 1996, **37(1)**, 64.
125. Morris, C. A., Anderson, M. L., Stroud, R. M., Merzbacher, C. I., Rolison, D. R. *Science* 1999, **284**, 622.
126. Leventis, N., Elder, I. A., Rolison, D. R., Anderson, M. L., Merzbacher, C. I. *Chem. Mater.* 1999, **11**, 2837.
127. Wilde, J. N., Wigman, A. J., Nagel, J., Oertel, U., Beeby, A., Tanner, B., Petty, M. C. *Acta Polymer.* 1998, **49**, 294.
128. Dagani, R. *Chem. Engr. News* 1998, **76(23)**, 35.
129. Frey, H., Lach, C., Lorenz, K. *Adv. Mater.* 1998, **10**, 279.
130. Gorman, C. *Adv. Mater.* 1998, **10**, 295.
131. Rehahn, M. *Acta Polymer.* 1998, **49**, 201.
132. Allcock, H. R., Kwon, S. *Macromolecules* 1986, **19**, 1502.
133. Kutal, C., Willson, C. G. *J. Electrochem. Soc.* 1987, **134**, 2280.

EXERCISES

4.1. The sections in this chapter give just an introduction to the various uses of inorganic and organometallic polymers. Choose one of the sections and do an in-depth search for more details on the topic. Several methods can be used to update references to a topic. First, taking the major references provided for the section using Science Citation Index or the online Web of Science web site, find references that have recently cited the references given in this chapter for the topic of concern. A second suggestion is to use the subject index of either CAS OnLine or Web of Science. If you obtain too many references, use a combination of terms separated by "AND" to refine your search. Conversely, if you get very few references, try other terms that might be used for indexing—for example, searching "siloxane OR silicone" will produce more references than either term alone.

4.2. Acetoxy (the anion of acetic or ethanoic acid) is only one silane derivative that can undergo moisture curing. The text lists enoxy, oxime, alkoxy, and amine groups. Write the structural formulas for each of these groups.

4.3. Write the reactions for the following siloxane cure reactions (a-c. are based on Fig. 4.3):

a. a vinyl derivative with a hydride derivative,

b. a hydride derivative with a silanol derivative,

c. an amine derivative with an epoxy derivative, and

d. the borate thermal cure noted for poly(dimethylsiloxane) in Section 2.1.3.

4.4. Titanium dioxide has replaced toxic white lead pigments in paints in the United States and many other countries. Suggest nontoxic alternates to:

the bright cadmium yellow pigment, and

the red-orange red lead pigment.

EPILOGUE

A wide variety of synthetic methods have been developed to help solve the low-solubility problems associated with transition-metal coordination polymers. The characterization of these soluble inorganic polymers is straightforward, and enough methods are available to ensure good molecular mass determinations for such polymers. Even so, relatively few soluble metal coordination polymers have been synthesized and evaluated.

The prospect of long-chain polymers through ring-opening polymerization has been realized for metallocenes, and new synthetic methods for main group polymers are proliferating. Even so, enormous possibilities still exist for better synthetic methodology for a vast array of inorganic polymers. Thus inorganic polymer synthesis is a fundamental research area for the new millenium.

Numerous uses for inorganic polymers have been developed. For many of these uses, the inorganic polymers have advantages over their organic counterparts, but often relatively high cost has precluded the use of the inorganic polymers, except where the cost differential is less than the advantage perceived for the polymer. As simpler synthetic methods and larger scale use occur, the differential will diminish and more uses will flourish. For some uses where no organic counterpart is satisfactory, the inorganic polymers have been accepted. Inorganic polymers will undoubtedly be featured in future high-tech nanoscale materials, where cost may be less of a factor.

Thus the ultimate fate of inorganic polymers rests on future practitioners in the field and the uses for which the polymers are found to be superior over their organic counterparts. If recent developments are a guide to the future, the future for inorganic polymers is very bright indeed.

INDEX